T0335533

Mathematical Models in Science

Mathematical Models in Science

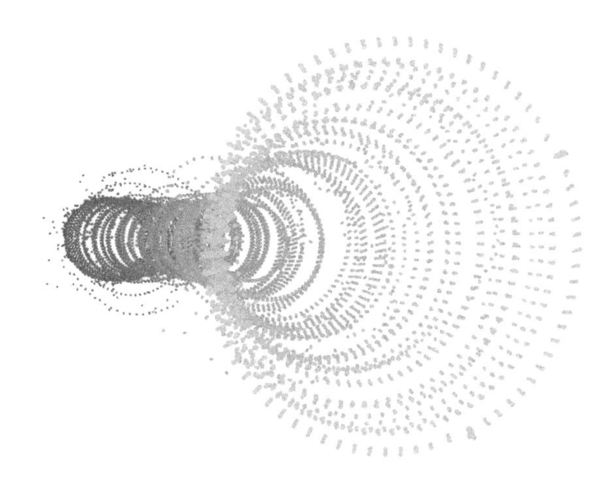

Olav Arnfinn Laudal
University of Oslo, Norway

NEW JERSEY • LONDON • SINGAPORE • BEIJING • SHANGHAI • HONG KONG • TAIPEI • CHENNAI • TOKYO

Published by

World Scientific Publishing Europe Ltd.
57 Shelton Street, Covent Garden, London WC2H 9HE
Head office: 5 Toh Tuck Link, Singapore 596224
USA office: 27 Warren Street, Suite 401-402, Hackensack, NJ 07601

Library of Congress Cataloging-in-Publication Data
Names: Laudal, Olav Arnfinn, author. | Laudal, Olav Arnfinn. Geometry of time-spaces.
Title: Mathematical models in science / Olav Arnfinn Laudal, University of Oslo, Norway.
Description: New Jersey : World Scientific, [2021] | Includes bibliographical references and index.
Identifiers: LCCN 2021015319 | ISBN 9781800610279 (hardcover) |
ISBN 9781800610286 (ebook for institutions) | ISBN 9781800610293 (ebook for individuals)
Subjects: LCSH: Mathematical physics. | Geometry, Algebraic. |
General relativity (Physics) | Quantum theory.
Classification: LCC QC20.7.A37 L39 2021 | DDC 530.1563/5--dc23
LC record available at https://lccn.loc.gov/2021015319

British Library Cataloguing-in-Publication Data
A catalogue record for this book is available from the British Library.

For any available supplementary material, please visit
https://www.worldscientific.com/worldscibooks/10.1142/Q0302#t=suppl

Desk Editors: Aanand Jayaraman/Michael Beale/Shi Ying Koe

Typeset by Stallion Press
Email: enquiries@stallionpress.com

Printed in Singapore

To
Inger

Acknowledgment

This work could not have been done, at least not so quickly, under the strain of the COVID-19 pandemic, without the very professional help of Professor Arvid Siqveland. He helped the author and the Editors by being the intermediate between us, translating the old style Latex version into the WS-book system. In doing this work, he also corrected several mistakes and many typos. As a coauthor of the basic reference Eriksen *et al.* (2017), he also served as a much needed referee.

Contents

Chapter 1

Introduction

Mindful of the well-known quotation,

> Vos calculs sont corrects, mais votre physique est abominable: (Albert Einstein 1927 à George Lemaître),

I shall start by declaring my cards.

1.1 Philosophy

In this note, I shall try to explain my belief in the usefulness of mathematical models in science. I assume that this will seem as an utter waste of time to most mathematicians and, I guess also, to most contemporary physicists. And very few will notice the difference if I replace mathematical models with "purely mathematical models." I don't blame them. I shall have to work a little to explain the difference that I see.

The present struggle in theoretical physics, to find a reasonable common ground for general relativity theory (GRT) and quantum theory (QT), has spurred a kind of opposition, making philosophers of science (PS) talk about abuse of mathematics in physics, see Earman (1989), for a good exposition. Most of this opposition may be considered as pretty far-fetched, or simply due to a rather poor background in modern mathematics. But there are certainly voices that should be listened to. Some of these question the formalism, or the unnecessary abstraction of the models in use. Some are even more specifically attacking the kind of Geometry used in models, or they are suspicious of those "that find more inspiration in "self-consistent" mathematical abstractions, than in nature" (Meschini, 2008).

However, a story explaining the workings of Nature on a real phenomenon, should, since Galilei, be written up in a mathematical language, depending more and more on (different forms of) Geometry. I assume that almost everyone having thoughts about this, would agree.

This story could reasonably be called a mathematical model of our understanding of the Nature we live in, so where is the problem?

Part of the problem lies, obviously, in our different understanding of the concept "Nature," the workings of which we want to describe. And maybe also, in our very different relations to the concept of "Understanding." Suppose the story is written in a perfectly coherent mathematical language; is that enough for the one that understands the mathematics, to understand the subject matter? Obviously not, there is a problem of translation, from the mathematical language to some unknown protolanguage, still haunting us. A point in space, what is that? Except a very small stone that once caught my eyes?

So, how do we understand the workings of Nature? Well, we have the mathematical models, like GRT and QT, that have been amazingly useful during the last centuries. We may think we understand the physical reality underlying both, but we cannot hide the fact that they refuse to be fused into one theory, based on one model, explaining all of physics. And this lack of unity seems to bother most of us.

Let us look at gravitation, and the GRT. Since Einstein's field equation is dependent upon the stress–mass–charge tensor, see Section 10.2, and the mathematical model for this tensor is written in the language of differential geometry, Einstein's Hole problem (see Earman, 1989), becomes a philosophical problem. Is it the physics, expressed in the equation, that is the problem, or is it the Differential Geometry? Einstein and all his co-workers used differential geometry, instead of, say Algebraic Geometry. Now, differential geometry has (too) many functions, admitting the possibility of forming a mathematical hole in the "tensor," algebraic geometry does not. Is this telling us that, maybe, Newton and Leibniz's differential calculus is not the best mathematical tool for understanding the problem?

QT is also, in most textbooks, and certainly for most physicists, a theory of differential operators acting on very complicated complete (normed) vector spaces, usually just referred to as "the Hilbert space." However, in an obvious sense, it is fundamentally an algebraic theory. And the natural geometric picture, might well be algebraic geometry, although then necessarily a non-commutative version of the classical algebraic geometry.

In GRT and QT, there are also problems having nothing to do with the mathematical tools used. Are the notion of a Material Object, sitting in the Space of GRT, and the State Function, sitting somewhere in Hilbert Space, in QT, representing something in the real world? Are the protolanguage translations reasonably clear? Do they exist?

Many more problematic questions can be asked. Clearly, we talk about all of this with the help of our Memory, and we consider ourselves as Conscientious living subjects. Memory is, in some circles, considered as a result of natural processes in the theory of gravitational waves (see Sormani, 2017), and consciousness is being debated in mathematical biology (see Penrose and Hameroff, 2011).

The purpose of this book is to look at philosophically difficult situations like those mentioned above, and offer a definition of a Purely Mathematical Model, based upon a choice of mathematical tools, via which one might, in some situations, come closer to an understanding of Nature.

I propose that this understanding of Nature (whatever that is), will have to be related to such a model, coupled with a way of checking the result of its application to some reasonably "accepted" natural phenomena. Again a formidable task, leading to the notion of "experiment," including the very non-trivial notion of a "measuring device." We shall have to leave this till later, much later, which obviously will put strains on most physicists for which this coupling of theory and experiment is the core of (their) science.

According to the famous work, *The Rise of Scientific Philosophy*, of Reichenbach (1961), the models I shall present might be called rationalist, in line with Pythagoras, Plato, Descartes, and Kant, and the rest of the authors of the "Speculative Philosophy" that he dismisses. I shall protest, I am never pronouncing the word "truth," nor "certainty," and I shall never want to. The rest of this book will therefore be concerned with mathematical statements, although mixed with some "interpretations," that look fun, and who knows, may be interesting.

But first, let us start with the notion of a "general mathematical model" of "natural phenomena," and see what this leads to.

1.2 Mathematical Models

In the book, Eriksen *et al.* (2017), I wrote, based upon the introduction of Laudal (2011), the following:

"If we want to study a natural phenomenon, called π, we must in the present scientific situation, describe π, in some mathematical terms, say as a mathematical object X, depending upon some parameters in such a way that the changing aspects of π would correspond to altered parameter-values for X. This object would be a *model for* π if, moreover, X with any choice of parameter-values would correspond to some, possibly occurring, aspect of π."

The phenomenon π in question is then, obviously, assumed to be in some sense measurable in terms of numbers. There must be different "aspects" of π that can be described, determined, and classified, by using sequences of "numbers," called parameter values.

Obviously, there are lots of phenomena we would like to "understand" and share with others, that are not of this "quality." A symphony, or a feeling of anxiety, may be communicated by either some written notes, or by a painting of Munch, but it is not (yet) possible to model these phenomena the way we have assumed in the quotation. Thus, we do not plan to make mathematical models covering phenomena like "Symphonies" or "Anxieties," certainly not yet!

However, if I am staying in an ordinary room of a usual flat in a Western city, and ask myself about the phenomenon, Space, I might refer to René Descartes, and present as my model, the Euclidean–Descartes model $X = E^3$, with specified coordinates (x_1, x_2, x_3), and metric $\sum_{i=1}^{3} dx_i^2$ included.

Clearly, I have now seen that the real world, in which I would not hesitate to include both the Symphony and the feeling of Anxiety, is not easily modeled. Moreover, I have found one phenomenon for which I feel I have a very good mathematical model: Space.

That good feeling turns out to be of short duration. The real problem with the notion of space is that we conceived it empty. And we assumed that we might furnish it as we wanted, without problems. For obvious reasons, we would very much like to fill it with objects, like molecules, moons, and stars, and it should be a medium for transmission of light and force, thus for all kinds of communication between objects. In the old days we had the notion of Ether, filling space, being a medium for forces, and taking care of the communication between the known elements. Today we have just a Differential Manifold for GRT, and a Hilbert Space for QT. I propose that we first try to find a space, provided with a common Ether, of some sort, for these theories, and then start asking the obvious questions of existence and reality.

Two mathematical objects, $X(1)$ and $X(2)$, corresponding to the same aspect of π would be called equivalent. Assuming that all models are algebraic objects of some sort, the set $\mathcal{P}$ of the objects π would correspond to (possibly a quotient of) the *moduli space* $\mathfrak{M}$ of the mathematical model X. The study of the natural phenomena π and their changing aspects, would then be equivalent to the study of the geometry of the moduli space $\mathfrak{M}$. In particular, the notion of *time* would, in agreement with Aristotle and St. Augustin (see Augustin, 1861; Laudal, 2005), correspond to some *metric* defined for this space.

Obviously, this moduli space would not have any reason to look like "our familiar three-dimensional space," modeled above, but we shall see that it may be the starting point for the construction of a kind of Ether, both Space and Room for Furniture and Communication.

But first, the above definition of a *model* is clearly circular, as all general definitions of this kind will have to be; in this case since the *aspect* of π would have to be defined in terms of something, and here simply in terms of the parameters of the model, the mathematical object X. Nevertheless, the wording above turns out to be helpful for comparing our point of view with the different mathematical models in science in use today. See Chapter 14, End Words, for a discussion of some of these, their notions of reality, non-locality, and later contributions to a kind of Theory of Everything.

So, this could be a first sketch of the basic ingredients of a Mathematical Model in Science. It consists of two parts, first the fixing of a mathematical model of the phenomenon we want to study, then the working out of the moduli space of these models, and its eventual metric structure. We shall show that this new space, in a situation of particular interest in physics, has much more structure than our

Euclidean three-dimensional space, and may show us the way to the Ether, the Cosmos of our dreams.

1.3 Geometry of the Space of Models

It turns out that to obtain a reasonably complete, theoretical framework for studying the phenomenon π, or the model X, together with its *dynamics*, we need a replacement for the differential calculus of the Newton model. We should instead work with algebraic geometry, and introduce the notion of *dynamical structure*, defined for the moduli space $\mathfrak{M}$, assumed to be a union of some, possibly non-commutative, affine schemes $\mathrm{Spec}(A)$, see Chapter 3. This is done via the construction, see Section 2.1, of a universal non-commutative *Phase Space*-functor, $\mathrm{Ph}(-) : \mathbf{Alg}_k \to \mathbf{Alg}_k$, where $\mathbf{Alg}_k$ is the category of associative k-algebras A, k any field. There is a universal injection, $i : A \to \mathrm{Ph}(A)$, together with a universal i-derivation, $d : A \to \mathrm{Ph}(A)$, such that for any representation $\rho_0 : A \to \mathrm{End}_k(V)$, where V is a k-vector space, any ρ_0-derivation $\xi : A \to \mathrm{End}_k(V)$ decomposes into the composition of the universal derivation d and a unique homomorphism, or representation, $\rho_1 : \mathrm{Ph}(A) \to \mathrm{End}_k(V)$. Obviously, we may iterate the Ph-construction and obtain the diagram

$$
\begin{array}{ccccccccc}
A & \xrightarrow{i_0^0} & \mathrm{Ph}(A) & \xrightarrow{i_p^1} & \mathrm{Ph}^2(A) & \xrightarrow{i_p^2} & \cdots & \longrightarrow & \mathrm{Ph}^\infty(A) \\
\downarrow{\scriptstyle \rho_0} & \swarrow{\scriptstyle \rho_1} & & & & & & & \\
\mathrm{End}_k(V) & & & & & & & &
\end{array}
$$

where for each integer n, the symbol i_p^n, for $p = 0, 1, \ldots, n$ signifies a universal family of A-morphisms between $\mathrm{Ph}^n(A)$ and $\mathrm{Ph}^{n+1}(A) := \mathrm{Ph}(\mathrm{Ph}^n(A))$ that will be constructed in Chapter 2, and $\mathrm{Ph}^\infty(A)$ is the direct (inductive) limit of the resulting diagram.

Now, the set of representations $\mathrm{Rep}(A)$ of A has a local structure. Any $\rho_0 : A \to \mathrm{End}_k(V)$ has a tangent space in $\mathrm{Rep}(A)$ given by $\mathrm{Ext}_A^1(V, V) = \mathrm{Der}_k(A, \mathrm{End}_k(V))/\,\mathrm{Triv}$. Therefore, any tangent to ρ_0 is given by an extension ρ_1 of ρ_0. We may continue, finding a sequence $\{\rho_n\}, 0 \leq n$, of extensions of ρ_0 to $\mathrm{Ph}^n(A)$.

This amounts to specifying a tangent or a *momentum*, ξ_0, in $\mathrm{Rep}(A)$, an acceleration vector, *momentum* ξ_1, and any number of higher-order *momenta* ξ_n. Thus, specifying a $\mathrm{Ph}^\infty(A)$-representation extending ρ_0 implies specifying a *formal curve* through ρ_0, the base-point of the *miniversal deformation space* of the corresponding A-module V.

The $\mathrm{Ph}(-)$-functor extends to the category of schemes, and its infinite iteration

$$\mathrm{Ph}^\infty(-) := \varinjlim_{n\geq 0}\{\mathrm{Ph}^n(-), i^n\},$$

is outfitted with a universal *Dirac derivation*, $\delta \in \mathrm{Der}_k(\mathrm{Ph}^\infty(-), \mathrm{Ph}^\infty(-))$, see Section 2.2.

Any extension of the representation ρ_0 to a representation $\rho_\infty : \mathrm{Ph}^\infty(A) \to \mathrm{End}_k(V)$ should therefore be considered as the Taylor series of an extension, or deformation, $\rho(t)$ of $\rho_0 \in \mathrm{Rep}(A)$ parametrized with *time* t.

A dynamical structure defined on an associative k-algebra $A \in \mathbf{Alg}_k$ is now a δ-stable ideal $\sigma \subset \mathrm{Ph}^\infty(A)$, and its quotient $A(\sigma) := \mathrm{Ph}^\infty(A)/(\sigma)$, with its induced Dirac derivation. The structure we are interested in is the *space*

$$\mathbf{U} := \mathrm{Ph}^\infty(\mathfrak{M})/\sigma,$$

corresponding to an open affine covering by subspaces of $\mathfrak{M}$, of the type, $U(\sigma) := \mathrm{Simp}(A(\sigma))$ (see Laudal, 2005, 2008, see also Laudal, 2011).

If A is commutative, say generated by the elements $t_i \in A$, such that $[t_i, t_j] = 0$, then in $\mathrm{Ph}(A)$, we would have $[dt_i, t_j] = [dt_i, t_j] + [t_i, dt_j] = 0$, so that $g^{i,j} := [dt_i, t_j] = [dt_j, t_i]$ is a family of symmetric elements of $\mathrm{Ph}(A)$.

In this commutative case, we shall see that for any metric g, defined in $\mathrm{Spec}(A)$, with dual $g^{i,j} \in A$, the dynamical structure σ_g generated by the subset $(g^{i,j} - [dt_i, t_j]) \subset \mathrm{Ph}(A)$ will furnish models for a "relativistic quantum theory," where time is encoded in the metric g.

But now we observe that there may be an action of a Lie algebra $\mathfrak{g}_0$ on $\mathbf{U}$, such that the dynamics of $\mathcal{P}$ really corresponds to that of the quotient $\mathbf{U}/\mathfrak{g}_0$. To any *open affine* subset $U(\sigma)$ of $\mathbf{U}$, there would be associated a, not necessarily commutative, affine k-algebra

$A(\sigma)$, and an action of the Lie algebra $\mathfrak{g}_0$, on A such that the non-commutative quotient $\mathrm{Simp}(A(\sigma))/\mathfrak{g}_0$ identified with the system of simple representations of an algebra $A(\sigma)$ with a $\mathfrak{g}_0$-connection, see Section 2.1, would contain all the available information about the structure of $\mathbf{U}$.

An element of $A(\sigma)$ would be called an *observable*. Wishing to measure the *value* of an observable $a \in A(\sigma)$ leads to the study of the eigenvectors and their eigenvalues of $\rho(a)$, for all the $\mathfrak{g}_0$-invariant representations of $A(\sigma)$, which as we have claimed is the same as the representations of $A(\sigma)$ with $\mathfrak{g}_0$-connections.

Again, we observe that for those representations $\rho : A(\sigma) \to \mathrm{End}_k(V)$, that we find interesting, there may be given symmetries, inducing identities for the values of the observables of the phenomena studied. These symmetries may be represented by an action of an A-Lie algebra, $\mathfrak{g}$, on ρ, i.e. an A-Lie homomorphism $\mathfrak{g} \to \mathrm{End}_A(V)$, and the Space we are interested in, should be given by the category of representations of $A(\sigma)$ with $\mathfrak{g}_0$-connections, where the "states" of interest are those marked with the eigenvalues of the Cartan sub-algebra $\mathfrak{h}$ of $\mathfrak{g}$.

We propose that this system is the "Ether" we would like to have, containing the objects, the forces, and serving as a medium for all communications related to the study of our phenomena.

1.4 Cosmology

With this philosophy in mind, and stimulated by the results in deformation theory obtained in Laudal (1979, 1986), we embarked, in a series of papers, see Laudal (2000, 2002, 2003), on the study of moduli spaces of representations (modules) of associative algebras, in general, and on their quotients, modulo Lie algebra actions. Here is where *invariant theory* and *non-commutative algebraic geometry* enter the play. The Dirac derivation translates into a vector field on these moduli spaces, and gives us the equations of motions that we need.

In Laudal (2005, 2008), see also Laudal (2011), we introduced a *Toy Model*, used to illustrate the general theory, and to connect to present day physics. It was shown to generalize both general relativity and quantum field theory. In particular, the definition of time fits well with the notion of time in both QT and in GRT, where it made the space of velocities compact.

In this book, this Toy Model has become the main figure, furnishing a (nice, but maybe not too realistic) mathematical model for a Big Bang scenario for the universe. The title, *Cosmos and its Furniture* of the paper Laudal (2013), refers to this model, to its *geometry*, defined by a metric, its dynamical structure defined by a Dirac derivation, and to its *material content*, called its *furniture*.

The basic idea is that, if one chooses to take the Big Bang idea seriously, one would, probably, have to accept the presence of a *singularity* at the *start of the universe*. But then one might guess upon a mathematical model for this singularity, and use deformation theory, and the machinery described above, to unravel the universe that we see. This is what we do, first in Chapter 5, to link to the ideas of Laudal (2005), then in Chapter 7. The idea is that the notion of a three-dimensional world seems to be basic to our species, and certainly not trivial. We start with the basic singularity in dimension 3. It is composed of a single point, and a three-dimensional tangent space, with affine algebra given by

$$U = k[x_1, x_2, x_3]/(x_1, x_2, x_3)^2.$$

The base space of the miniversal deformation of this algebra, considered as an associative algebra, turns out to contain the above Toy Model, and a lot of structure, with strange and maybe interesting interpretations in physics. In particular, it seems that it contains a mathematically reasonable set-up for a version of GRT, together with a QFT, complete with a probabilistic set-up, based on the notion of time, satisfying the Wightman axioms, see Section 11.3, and thus furnishing a basis for the Standard Model. There are also obvious relations to the age-long debate in Philosophy of Science, on the absolute, versus relational, theories of space and time (see, e.g. Earman (1989), and see again Chapter 14, End Words).

1.5 Organization of the Work: Leitfaden

This book is organized as follows:

Chapter 2, "Dynamics," is an introduction to the general method proposed in Laudal (2011). There are three main ingredients, moduli theory, dynamical structures, and gauge theory. We include a short reminder of these ingredients and of the most important technical tools for handling them, such as the basic notion of the non-commutative *phase space functor* Ph, for associative k-algebras, including some purely mathematical consequences like the co-simplicial structure of the infinitely iterated Ph^*, the resulting definition of a universal *Dirac Derivation*, and the corresponding relations to de Rham theory together with a short excursion into the Jacobian Conjecture.

Chapter 3, "Non-Commutative Algebraic Geometry." Here, we recall the basics of non-commutative deformation theory for families of representations $\mathfrak{V}$ of a k-algebra, the ring $O(\mathfrak{V})$ of *observables* of this family, and its use in non-commutative algebraic geometry.

Chapter 4, "The Dirac Derivation and Dynamical Structures." This is where we introduce the basic tools for the application of algebraic geometry in Physics. The main ingredient is the generic dynamical structure defined by a metric, and the introduction of the notions of global and local gauge groups related to invariant theory in our version of non-commutative algebraic geometry. The Ph-construction allows the study of a second-order momentum of representations, and relating third-order changes of objects to changes of the metric, giving ideas of gravitational waves and some ideas of Penrose. Together, these give us the link between GRT and QFT, which is the purpose of this text.

Chapter 5, "Time–Space and Space–Times," contains a reworking of the Toy Model, introducing Black Holes and their importance for the notion of mass and its origin. Here, we find a reasonable explanation for the PT equivalencies in GRT/QFT. In particular, we introduce the notion of *furniture*, and the related Schrödinger equation, generalizing the Heat and the Navier–Stokes equations.

Chapter 6, "Entropy," contains a short introduction to an algebraic version of Entropy, that will be of some importance for the choice of the main cosmological model. In particular, the notion of an Arrow of Time, its origin, and the relation of this origin to the notion of information of objects in the Universe. Related to this, we add a computation of the moduli-space of finite-dimensional representations of the infinite phase space of a polynomial algebra.

Chapter 7, "Cosmology, Cosmos, and Cosmological Time," is then concerned with the deformation theory of the four-dimensional associative algebra U, geometrically a fat point in 3-space, and with the structure of its miniversal base space, our model for the Big Bang. The fact that this space contains the Toy Model provides new and unsuspected structures for this model. The main result is the existence, produced by the deformation theory of associative algebras, of a canonical gauge Lie algebra bundle, containing the gauge algebras of the Standard Model and a strange *super symmetry* relating the spaces of, our versions of, *bosonic* and *fermionic* fields.

Chapter 8, "The Universe as a Versal Base Space," in which we propose that the Universe is not made out of nothing in the Big Bang, but is well rooted in our mental notion of a three-dimensional world.

Chapter 9, "Worked Out Formulas," contains worked out formulas that we need for the final chapters summing up: In particular, here is our version of gluons, quarks, and charge, treated together with the missing equivalence C, in the CPT story in Chapter 5.

Chapter 10, "Summing Up the Model." We propose a mathematical fusion of General Relativity Theory, Quantum Field Theory, and Yang–Mills. Then we comment on Black Energy and Mass.

Chapter 11, "Particles, Fields, and Probabilities." Here, we propose a model for Elementary Particles as furnished by the Toy Model, where the probabilities introduced to handle the uncertainties of the QFT are seen to stem from the metric and the extensions of its quadratic form, to all objects of the theory.

Chapter 12, "Interactions," treats the notion of interaction and decay in the language of non-commutative deformation of families of modules. In particular, we obtain a possible purely mathematical model for the Weak Interaction and for Entanglement, with possibly interesting new properties.

Chapter 13, "Comparing the Toy Model with the Standard Model," takes a bird's eye view of the present state of the Standard Model compared with the Toy Model, and presents a list of well-known structural problems of SM, touched upon in this book.

Chapter 14, "End Words," in which we discuss the relationship between our set-up and the many models of non-commutative geometries proposed for the purpose of making physics a unified whole. We start with the very well-known non-commutative geometry (NCG), based upon von Neumann and Gelfand's operator theory, and developed by Alain Connes and his School. We then comment on several proposals for a Quantum Geometry, motivated by the needs for necessary "Planck scale corrections" of both quantum field theory and general relativity. This leads to some non-scientific thoughts about the present status of quantum theory, in Weinberg's tapping, and about the recent, more speculative, literature in cosmology.

Finally, this book has been circulating, in different versions, as PDF or PP's from conferences in Europe over many years. Part of the story is included in the book Laudal (2011), and in related papers, Laudal (1986, 2000, 2002, 2003, 2004, 2005, 2008, 2013, 2014), Laudal and Pfister (1988), other parts have been published in the book Eriksen *et al.* (2017). However, the present version is certainly not the last.

Chapter 2

Dynamics

2.1 The Phase Space Functor

Given a k-algebra A, denote by $A/k - \mathbf{alg}$ the category where the objects are homomorphisms of k-algebras $\kappa : A \to R$, and the morphisms, $\psi : \kappa \to \kappa'$, are commutative diagrams

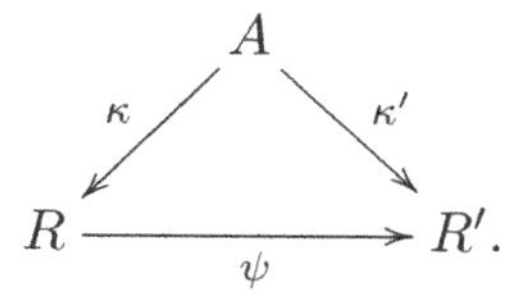

Consider the functor

$$\mathrm{Der}_k(A, -) : A/k - \mathbf{alg} \longrightarrow \mathbf{Sets}$$

defined by $\mathrm{Der}_k(A, \kappa) := \mathrm{Der}_k(A, R)$. It is representable by a k-algebra morphism, $\iota : A \longrightarrow \mathrm{Ph}(A)$, with a *universal family* given by a universal derivation, $d : A \longrightarrow \mathrm{Ph}(A)$. It is easy to construct $\mathrm{Ph}(A)$. In fact, let $\pi : F \to A$ be a subjective homomorphism of algebras, with $F = k\langle t_1, t_2, \ldots, t_r\rangle$ freely generated by the t_i's, and put $I = \ker \pi$. Let

$$\mathrm{Ph}(A) = k\langle t_1, t_2, \ldots, t_r, dt_1, dt_2, \ldots, dt_r\rangle/(I, dI),$$

where dt_i is a formal variable, for $i = 1, \ldots, r$. Clearly there is a homomorphism $i_0' : F \to \mathrm{Ph}(A)$ and a derivation $d' : F \to \mathrm{Ph}(A)$, defined by putting $d'(t_i) = \text{cl}(dt_i)$, the equivalence class of dt_i. Since i_0' and d' both kill the ideal I, they define a homomorphism

$$i_0 : A \to \text{Ph}(A)$$

and a derivation

$$d : A \to \text{Ph}(A).$$

To see that i_0 and d have the universal property, let $\kappa : A \to R$ be an object of the category $A/k - \mathbf{alg}$. Any derivation $\xi : A \to R$ defines a derivation $\xi' : F \to R$, mapping t_i to $\xi'(t_i)$. Let $\rho_{\xi'} : k\langle t_1, t_2, \ldots, t_r, dt_1, dt_2, \ldots, dt_r\rangle \to R$ be the homomorphism defined by

$$\rho_{\xi'}(t_i) = \kappa(\pi(t_i)), \quad \rho_{\xi'}(dt_i) = \xi(\pi(t_i)).$$

Then $\rho_{\xi'}$ sends I and dI to zero, so defines a homomorphism $\rho_\xi : \text{Ph}(A) \to R$, such that the composition with $d : A \to \text{Ph}(A)$ is ξ. The uniqueness is a consequence of the fact that the images of i_0 and d generate $\text{Ph}(A)$ as k-algebra.

2.1.1 *First properties*

Clearly $\text{Ph}(-)$ is a covariant functor on $k - \mathbf{alg}$, and we have the identities

$$d_* : \text{Der}_k(A, A) = \text{Mor}_A(\text{Ph}(A), A)$$

and

$$d^* : \text{Der}_k(A, \text{Ph}(A)) = \text{End}_A(\text{Ph}(A)),$$

the last one associating d to the identity endomorphism of $\text{Ph}(A)$. In particular, we see that i_0 has a co-section, $\sigma_0 : \text{Ph}(A) \to A$, corresponding to the trivial (zero) derivation of A.

Now let V be a right A-module with structure morphism

$$\rho(V) =: \rho : A \to \text{End}_k(V).$$

We obtain a universal derivation

$$u(V) =: u : A \longrightarrow \text{Hom}_k(V, V \otimes_A \text{Ph}(A))$$

defined by $u(a)(v) = v \otimes d(a)$.

Let U and V be right A-modules, and consider the long exact sequences of Hochschild cohomology,

$$\begin{aligned}0 &\to \mathrm{Hom}_A(U,V) \to \mathrm{Hom}_k(U,V)\\ &\xrightarrow{\iota} \mathrm{Der}_k(A, \mathrm{Hom}_k(U,V)) \xrightarrow{\kappa} \mathrm{Ext}^1_A(U,V) \to 0,\end{aligned} \tag{2.1}$$

$$\begin{aligned}0 &\to \mathrm{Hom}_A(V, V \otimes_A \mathrm{Ph}(A)) \to \mathrm{Hom}_k(V, V \otimes_A \mathrm{Ph}(A))\\ &\xrightarrow{\iota} \mathrm{Der}_k(A, \mathrm{Hom}_k(V, V \otimes_A \mathrm{Ph}(A))) \xrightarrow{\kappa} \mathrm{Ext}^1_A(V, V \otimes_A \mathrm{Ph}(A)) \to 0,\end{aligned}$$

where the map ι is defined, for any $\Phi \in \mathrm{Hom}_k(U,V)$ and $a \in A$, by

$$\iota(\Phi)(a) = [\Phi, a] = \rho(U)\Phi - \Phi\rho(V),$$

and κ is the obvious map,

$$\begin{aligned}\kappa : \mathrm{Der}_k(A, \mathrm{Hom}_k(U,V)) &\to \mathrm{Der}_k(A, \mathrm{Hom}_k(U,V))/\,\mathrm{Triv}\\ &= \mathrm{Ext}^1_A(U,V),\end{aligned}$$

and where $\mathrm{Triv} := \mathrm{im}(i) \subset \mathrm{Der}_k(A, \mathrm{Hom}_k(U,V))$ is the subset of those derivations δ for which there exist a $\Phi \in \mathrm{Hom}_k(U,V)$, such that for all $a \in A$, $\delta(a) = [\Phi, a]$.

We obtain the non-commutative Kodaira–Spencer class

$$c(V) := \kappa(u(V)) \in \mathrm{Ext}^1_A(V, V \otimes_A \mathrm{Ph}(A))$$

inducing the Kodaira–Spencer morphism

$$\mathrm{ks} := \mathrm{ks}(\rho) : \Theta_A := \mathrm{Der}_k(A,A) \longrightarrow \mathrm{Ext}^1_A(V,V)$$

via the identity d_*. If $c(V) = 0$, then the exact sequence above proves that there exist a $\nabla \in \mathrm{Hom}_k(V, V \otimes_A \mathrm{Ph}(A))$ such that $u = \iota(\nabla)$. This is just another way of proving that $c(V)$ is the obstruction for the existence of a connection

$$\nabla : \mathrm{Der}_k(A,A) \longrightarrow \mathrm{Hom}_k(V,V).$$

The same exact sequences furnish a proof for the following.

Lemma 2.1. *Let $\rho : A \to \mathrm{End}_k(V)$ be an A-module and let $\delta \in \mathrm{Der}_k(A, \mathrm{End}_k(V))$ map to 0 in $\mathrm{Ext}^1_A(V,V)$, i.e. assume $\kappa(\delta) = 0$, then there exists an element, $Q_\delta \in \mathrm{End}_k(V)$, such that for all $a \in A$,*

$$\delta(a) = [Q_\delta, \rho(a)].$$

If V is a simple A-module, $ad(Q_\delta)$ is unique.

Definition 2.1. For any representation $\rho : A \to \mathrm{End}_k(V)$, put

$$\mathfrak{g}_V := \mathfrak{g}_\rho = \{\gamma \in \mathrm{Der}_k(A) |\, \mathrm{ks}(\gamma) = \kappa(\delta\rho) = 0\}.$$

Note that $\mathfrak{g}_V$ is a Lie sub-algebra of $\mathrm{Der}_k(A)$.

Obviously this means that there is always a connection,

$$\nabla : \mathfrak{g}_V \to \mathrm{End}_k(V).$$

These simple results will be important in relation to invariant theory, gauge groups in physics, and quantizations, covered in the next sections.

Any $\mathrm{Ph}(A)$-module W, given by its structure map

$$\rho(W) =: \rho_1 : \mathrm{Ph}(A) \longrightarrow \mathrm{End}_k(W),$$

corresponds bijectively to an induced A-module structure $\rho_0 : A \to \mathrm{End}_k(W)$, together with a derivation $\delta_\rho \in \mathrm{Der}_k(A, \mathrm{End}_k(W))$, defining an element $[\delta_\rho] \in \mathrm{Ext}^1_A(W, W)$. Fixing this last element we find that the set of $\mathrm{Ph}(A)$-module structures on the A-module W is in one-to-one correspondence with $\mathrm{End}_k(W)/\mathrm{End}_A(W)$. Conversely, starting with an A-module V and an element $\delta \in \mathrm{Der}_k(A, \mathrm{End}_k(V))$, we obtain a $\mathrm{Ph}(A)$-module V_δ. It is then easy to see that the kernel of the natural map

$$\mathrm{Ext}^1_{\mathrm{Ph}(A)}(V_\delta, V_\delta) \to \mathrm{Ext}^1_A(V, V)$$

induced by the linear map

$$\mathrm{Der}_k(\mathrm{Ph}(A), \mathrm{End}_k(V_\delta)) \to \mathrm{Der}_k(A, \mathrm{End}_k(V))$$

is the quotient

$$\mathrm{Der}_A(\mathrm{Ph}(A), \mathrm{End}_k(V_\delta))/\mathrm{End}_k(V)$$

and the image is a subspace $[\delta_\rho]^\perp \subseteq \mathrm{Ext}^1_A(V, V))$, which is rather easy to compute, see examples below.

Example 2.1. (i) Let $A = k[t]$, then obviously $\mathrm{Ph}(A) = k\langle t, dt\rangle$, and d is given by $d(t) = dt$, such that for $f \in k[t]$ we find $d(f) = J_t(f)$ with the notations of Laudal (2003), i.e. the non-commutative derivation of f with respect to t. One should also compare this with the non-commutative Taylor formula of loc.cit. If $V \simeq k^2$ is an A-module defined by the matrix $X \in \mathrm{M}_2(k)$, and $\delta \in \mathrm{Der}_k(A, \mathrm{End}_k(V))$ is defined in terms of the matrix $Y \in \mathrm{M}_2(k)$, then the $\mathrm{Ph}(A)$-module V_δ is the $k\langle t, dt\rangle$-module defined by the action of the two matrices $X, Y \in \mathrm{M}_2(k)$, and we find

$$e_V^1 := \dim_k \mathrm{Ext}_A^1(V,V) = \dim_k \mathrm{End}_A(V) = \dim_k\{Z \in \mathrm{M}_2(k) \mid [X,Z] = 0\},$$

$$e_{V_\delta}^1 := \dim_k \mathrm{Ext}_{\mathrm{Ph}(A)}^1(V_\delta, V_\delta) = 8 - 4 + \dim\{Z \in \mathrm{M}_2(k) \mid [X,Z] = [Y,Z] = 0\}.$$

We have the following inequalities:

$$2 \le e_V^1 \le 4 \le e_{V_\delta}^1 \le 8.$$

(ii) Let $A = k[t_1, t_2]$, then we find

$$\mathrm{Ph}(A) = k\langle t_1, t_2, dt_1, dt_2\rangle / ([t_1, t_2], [dt_1, t_2] + [t_1, dt_2]).$$

In particular, we have a surjective homomorphism,

$$\mathrm{Ph}(A) \to k\langle t_1, t_2, dt_1, dt_2\rangle / ([t_1, t_2], [dt_1, dt_2], [t_i, dt_i] - 1),$$

the right-hand algebra being the Weyl algebra. This homomorphism exists in all dimensions. We also have a surjective homomorphism

$$\mathrm{Ph}(A) \to k[t_1, t_2, \xi_1, \xi_2],$$

i.e. onto the affine algebra of the classical phase space.

2.1.2 *The deformation functor of representations*

Recall that

$$\mathrm{Ext}_A^1(V,V) = \mathrm{Der}_k(A, \mathrm{End}_k(V)) / \mathrm{Triv},$$

where Triv is the subspace of derivations, defined for $\phi \in \mathrm{End}_k(V)$ by sending $a \in A$ to $a\phi - \phi a$. This is the tangent space of the miniversal deformation space of V as an A-module, and we see that the non-commutative space $\mathrm{Ph}(A)$ also parametrizes the set of *generalized momenta*, i.e. the set of pairs of an A-module V, and a tangent vector of the formal moduli of V, at that point.

We claim that $\mathrm{Ph}(A)$ is relatively easy to compute. In particular, if $A = k[x_1, \ldots, x_n]$ is the polynomial algebra, we have,

$$\mathrm{Ph}(A) = k\langle x_1, \ldots, x_n, dx_1, \ldots, dx_n\rangle / ([x_i, x_j], [x_i, dx_j] + [dx_i, x_j]).$$

Note that any rank 1 representation of $\mathrm{Ph}(A)$ is represented by a pair (q, p) of a closed point q of $\mathrm{Spec}(k[\underline{x}])$ and a tangent p at that point. We shall need the following formulas.

Theorem 2.1. *Given two such points, (q_i, p_i), $i = 1, 2$, we find*

$$\dim_k \mathrm{Ext}^1_{\mathrm{Ph}(A)}(k(q_1, p_1), k(q_2, p_2)) = 2n, \ \textit{for} \ (q_1, p_1) = (q_2, p_2),$$

$$\dim_k \mathrm{Ext}^1_{\mathrm{Ph}(A)}(k(q_1, p_1), k(q_2, p_2)) = n, \ \textit{for} \ q_1 = q_2, \ p_1 \neq p_2,$$

$$\dim_k \mathrm{Ext}^1_{\mathrm{Ph}(A)}(k(q_1, p_1), k(q_2, p_2)) = 1, \ \textit{for} \ q_1 \neq q_2.$$

Moreover, there is a generator of

$$\mathrm{Ext}^1_{\mathrm{Ph}(A)}(k(q_1, p_1), k(q_2, p_2)) = \mathrm{Der}_k(\mathrm{Ph}(A), \mathrm{Hom}_k(k(q_1, p_1)), k(q_2, p_2)))/\,\mathrm{Triv},$$

uniquely characterized by the tangent line defined by the vector $\overline{q_1 q_2}$.

Proof. Assume for convenience that $n = 3$. Put $x_j(q_i, p_i) := q_{i,j}$, $dx_j(q_i, p_i) := p_{i,j}$, $\alpha_j = q_{1,j} - q_{2,j}$, $\beta_j = p_{1,j} - p_{2,j}$.

See that for any element $\alpha \in Hom_k(k(q_1, p_1), k(q_2, p_2))$, we have

$$x_j\alpha = q_{1,j}\alpha, \ \alpha x_j = q_{2,j}\alpha, \ dx_j\alpha = p_{1,j}\alpha, \ \alpha dx_j = p_{2,j}\alpha,$$

with the obvious identification. Any derivation

$$\delta \in \mathrm{Der}_k(\mathrm{Ph}(A), \mathrm{Hom}_k(k(q_1, p_1), k(q_2, p_2)))$$

must satisfy the relations

$$\delta([x_i, x_j]) = [\delta(x_i), x_j] + [x_i, \delta(x_j)] = 0,$$

$$\delta([dx_i, x_j] + [x_i, dx_j]) = [\delta(dx_i), x_j] + [dx_i, \delta(x_j)] + [\delta(x_i), dx_j] + [x_i, \delta(dx_j)] = 0.$$

Using the above left–right action-rules, the result follows from the long exact sequence (2.1) computing $\mathrm{Ext}^1_{\mathrm{Ph}\,A}$. The two families of relations above give us two systems of linear equations.

The first, in the variables $\delta(x_1), \delta(x_2), \delta(x_3)$, with matrix

$$\begin{pmatrix} -\alpha_2 & \alpha_1 & 0 \\ -\alpha_3 & 0 & \alpha_1 \\ 0 & -\alpha_3 & \alpha_2 \end{pmatrix}.$$

The second, in the variables $\delta(x_1), \delta(x_2), \delta(x_3), \delta(dx_1), \delta(dx_2), \delta(dx_3)$, with matrix

$$\begin{pmatrix} -\beta_2 & \beta_1 & 0 & -\alpha_2 & \alpha_1 & 0 \\ -\beta_3 & 0 & \beta_1 & -\alpha_3 & 0 & \alpha_1 \\ 0 & -\beta_3 & \beta_2 & 0 & -\alpha_3 & \alpha_2 \end{pmatrix}.$$

In particular, we see that the *trivial* derivation given by

$$\delta(x_i) = \alpha_i, \ \delta(dx_j) = \beta_j,$$

satisfies the relations, and the generator of $\mathrm{Ext}^1_{\mathrm{Ph}\,A}(k(q_1, p_1), k(q_2, p_2))$ is represented by

$$\delta(x_i) = 0, \ \delta(dx_j) = \alpha_j.$$

This is, in an obvious sense, the "tangent vector" $-\overline{q_1, q_2}$. □

It is easy to extend this result from dimension 3 to any dimension n. Note that this result shows that the "Space" of all rank 1 representations of $\mathrm{Ph}(A)$ is the classical phase space of $\mathrm{Spec}\,A$, of dimension $2n$, but endowed with an extra structure. Between two different points, corresponding to either one point in $\mathrm{Spec}(A)$ and two different tangents, or to two different points in $\mathrm{Spec}(A)$, there are respectively a subspace of dimension n, and of dimension 1 of "ext-tangents." This will be important in the search for the "Ether" mentioned in Chapter 1.

2.1.3 *Blow-ups and desingularizations*

The A-algebra $\mathrm{Ph}(A)$ is graded by defining, for $a \in A$,

$$\deg(a) = 0, \deg(d(a)) = 1.$$

By definition, any $\mathrm{Ph}(A)$-representation $\rho : \mathrm{Ph}(A) \to \mathrm{End}_k(V)$, corresponds to a representation $\rho_0 : A \to \mathrm{End}_k(V)$ together with a derivation of A into $\text{End}_k(V)$, which again induces a tangent direction in the moduli space of representations of A at the point ρ. In complete generality, we have a map that we shall call *The General Blowing Up Map*,

$$\mathfrak{bu} : \text{Simph}_A(\text{Ph}(A)) \to \text{Simp}(A),$$

where $\text{Simph}_A(\text{Ph}(A))$ is the set of simple graded $\text{Ph}(A)$-modules and the mapping is onto the 0th component.

The corresponding morphism in the commutative case is

$$\mathfrak{bu} : \text{Proj}_A(\text{Ph}(A)) \to \text{Spec}(A),$$

the *Blowing Up Map*. By the universal property of $\text{Ph}(A)$ it is clear that the fibre of $\mathfrak{bu}$ at a (k-point) $x \in \text{Spec}(A)$ is $\text{Proj}(T(x)) \simeq \mathbb{P}^{n-1}$, where $T(x)$ is the tangent space of $\text{Spec}(A)$ at the point x, supposed to be of embedding dimension n.

Therefore, any vector field ξ on $\text{Spec}(A)$, i.e. any derivation $\xi \in \text{Der}_k(A, A)$, defines a canonical section of $\mathfrak{bu}$,

$$\sigma(\xi) : D(\xi) \to \text{Proj}_A(\text{Ph}(A)),$$

defined in the open sub-scheme $D(\xi)$, where ξ is non-trivial. The blow-up of $\text{Spec}(A)$ defined by ξ is now the closure $\text{Spec}(A, \xi)$ of the image of $\sigma(\xi)$. Thus,

$$D(\xi) \subset \text{Spec}(A, \xi) \subset \text{Proj}_A(\text{Ph}(A)).$$

Blowing up the origin in the affine n-space would then correspond to the blowing up of $\text{Spec}(k[x_1, \ldots, x_n])$ defined by the derivation $\xi = \sum x_i \delta_{x_i}$.

In the commutative, general case, if $A = k[x_1, \ldots, x_n]/(r_1, \ldots, r_s)$, consider the Jacobian matrix

$$J = \left(\frac{\partial r_i}{\partial x_j}\right),$$

and let J_α be a maximal sub-determinant. $J_\alpha \neq 0$ in an open subset U of $\text{Spec}(A) \setminus \text{Sing}(A)$ is supposed to be non-empty. Compute the solutions of the linear system of equations

$$\sum_{j=1}^{n} \delta_{x_j}(r_i)dx_j = 0, \quad i = 1, \dots, s.$$

We find solutions of the form

$$dx_l = \sum_{i}^{d} c_i^l / J_\alpha dx_i, \quad l = d+1, \dots, n,$$

for some d, and the derivations of A of the form

$$\xi_i := J_\alpha \delta_{x_i} - \sum_{l=d+1}^{n} c_i^l \delta_{x_l}$$

are all non-trivial in U. The corresponding blow-ups of A look like

$$A(\xi_i) = \mathrm{Ph}(A) / \left(\left\langle J_\alpha dx_l = \sum_{i}^{d} c_i^l dx_i, l = d+1, \dots, n \right\rangle \right).$$

This gives us a possible easier road to desingularization, since the Ph-operation is canonical and may be iterated.

All this can easily be generalized to perform much more complex blow-ups of non-commutative affine algebraic schemes defined by the associative k-algebra A.

Denote, generally, $\mathfrak{m}_A := \ker(\pi)$ where $\pi : \mathrm{Ph}(A) \to A$ corresponds to the trivial derivation of A. Obviously, $\mathfrak{m}_A$ is the ideal of $\mathrm{Ph}(A)$ generated by the elements $\{d(a), a \in A\}$.

Consider now a bilateral ideal $\mathfrak{a} \subset A$, and a bilateral sub-ideal $\mathfrak{b} \subset \mathfrak{m}_A$. The bilateral ideal $(\mathfrak{a}, \mathfrak{b}) = \mathfrak{a}(\mathfrak{m}_A) + (\mathfrak{b}) \subset \mathrm{Ph}(A)$ defines a sub-scheme

$$\mathfrak{bl}(\mathfrak{a}, \mathfrak{b}) \subset \mathrm{Simph}_A(\mathrm{Ph}(A))$$

which is our general *blow-up* of a non-commutative algebraic scheme.

The blowing up of a closed sub-scheme $Y = \mathrm{Spec}(B) \subset X := \mathrm{Spec}(A)$ is gotten by considering the ideal $\mathfrak{a} = \ker(\pi)$, for $\pi : A \to B$, and the bilateral ideal $\mathfrak{b} = \ker\{\mathrm{Ph}(\pi) : \mathfrak{m}_A \to \mathfrak{m}_B\}$. The fiber over Y of the *blow down* map, is the projectivization of the normal bundle $N_Y(X)$.

Now, the blowing down of a sub-scheme $Y = \mathrm{Spec}(B) \subset X := \mathrm{Spec}(A)$ is gotten by considering, together with the points of $X - C$, the simple representations $\rho_x : A \to k(x)$, also the representation $\rho_0 : A \to B$, in the way we construct the general non-commutative algebraic geometry, see Chapter 3.

2.1.4 *Chern classes*

It is, I think, well known that in the commutative case, the Kodaira–Spencer class gives rise to the Chern characters. In this general case, we shall prepare for a result mimicking the Chern–Simons classes. Let us assume given a representation, $\rho_0 : A \to \mathrm{End}_k(V)$, and a momentum at ρ_0, i.e. an extension, $\rho_1 : \mathrm{Ph}(A) \to \mathrm{End}_k(V)$ of ρ_0. Consider now the class in $\mathrm{ch}^n(\rho_1) \in \mathrm{HH}^n(A, \mathrm{End}_k(V))$ defined by the following Hochschild co-chain, the k-linear map

$$\mathrm{ch}^n : A^{\otimes n} \to \mathrm{End}_k(V),$$

defined by

$$\mathrm{ch}^n(a_1 \otimes a_2 \ldots, \otimes a_n) = \rho_1(da_1 da_2 \ldots da_n) \in \mathrm{End}_k(V).$$

It is easy to see that ch^n is a co-cycle since

$$\delta(\mathrm{ch}^n)((a_1 \otimes a_2 \ldots \otimes a_{n+1})) = a_1 \rho_1((da_2 da_3 \ldots da_{n+1})) \quad (2.2)$$

$$+ \sum_1^n (-1)^i \rho_1((da_1 \ldots d(a_i a_{i+1}) \ldots da_{n+1}))$$

$$+ (-1)^{n+1} \rho_1((da_1 \ldots da_n) a_{n+1} = 0. \quad (2.3)$$

One may define the nth *Generalized Chern–Simons Class* of ρ_1 as the class

$$\mathrm{csh}^n(\rho_1) := 1/n!\ \mathrm{ch}^n(\rho_1) \in \mathrm{HH}^n(A, \mathrm{End}_k(V)).$$

2.2 The Iterated Phase Space Functor Ph* and the Dirac Derivation

The phase space construction may be iterated. Given the k-algebra A we may form the sequence, $\{\mathrm{Ph}^n(A)\}_{0 \le n}$, defined inductively by

$$\mathrm{Ph}^0(A) = A,\ \mathrm{Ph}^1(A) = \mathrm{Ph}(A), \ldots, \mathrm{Ph}^{n+1}(A) := \mathrm{Ph}(\mathrm{Ph}^n(A)).$$

Let $i_0^n : \mathrm{Ph}^n(A) \to \mathrm{Ph}^{n+1}(A)$ be the canonical embedding, and let $d_n : \mathrm{Ph}^n(A) \to \mathrm{Ph}^{n+1}(A)$ be the corresponding derivation. Since the composition of i_0^n and the derivation d_{n+1} is a derivation $\mathrm{Ph}^n(A) \to \mathrm{Ph}^{n+2}(A)$, corresponding to the homomorphism

$$\mathrm{Ph}^n(A) \to^{i_0^n} \mathrm{Ph}^{n+1}(A) \to^{i_0^{n+1}} \mathrm{Ph}^{n+2}(A),$$

there exists by universality a homomorphism $i_1^{n+1} : \mathrm{Ph}^{n+1}(A) \to \mathrm{Ph}^{n+2}(A)$ such that

$$i_0^n \circ i_1^{n+1} = i_0^n \circ i_0^{n+1},$$

and such that

$$d_n \circ i_1^{n+1} = i_0^n \circ d_{n+1}.$$

Note that we here compose functions and functors from left to right. Clearly, we may continue this process constructing new homomorphisms,

$$\{i_j^n : \mathrm{Ph}^n(A) \to \mathrm{Ph}^{n+1}(A)\}_{0 \leq j \leq n},$$

such that

$$i_p^n \circ i_0^{n+1} = i_0^n \circ i_{p+1}^{n+1}$$

with the property

$$d_n \circ i_{j+1}^{n+1} = i_j^n \circ d_{n+1}.$$

We find, see Laudal (2011), the following identities:

$$\begin{aligned} i_p^n i_q^{n+1} &= i_{q-1}^n i_p^{n+1}, \quad p < q \\ i_p^n i_p^{n+1} &= i_p^n i_{p+1}^{n+1} \\ i_p^n i_q^{n+1} &= i_q^n i_{p+1}^{n+1}, \quad q < p. \end{aligned}$$

To see this, notice that d_n by definition is an i_0^n-derivation, and start with the definition of i_1^1,

$$i_0^0 \circ d_1 = d_0 \circ i_1^1, i_0^0 \circ i_0^1 = i_0^0 \circ i_1^1.$$

Then since d_2 is an i_0^2-derivation, $i_1^1 \circ d_2$ is an $i_1^1 \circ i_0^2$-derivation, so by universality, there exists an i_2^2 such that

$$i_1^1 \circ d_2 = d_1 \circ i_2^2, i_1^1 \circ i_0^2 = i_0^1 \circ i_2^2.$$

Now, take the last equality, and compose with d_3, then we find

$$i_1^1 \circ i_0^2 \circ d_3 = i_1^1 \circ d_2 \circ i_1^3 = d_1 \circ i_2^2 \circ i_1^3,$$
$$i_0^1 \circ i_2^2 \circ d_3 = i_0^1 \circ d_2 \circ i_3^3 = d_1 \circ i_1^2 \circ i_3^3.$$

Moreover, we find that

$$i_0^1 \circ i_2^2 \circ i_1^3 = i_1^1 \circ i_0^2 \circ i_1^3 = i_1^1 \circ i_0^2 \circ i_0^3,$$
$$i_0^1 \circ i_1^2 \circ i_3^3 = i_0^1 \circ i_0^2 \circ i_3^3 = i_0^1 \circ i_2^2 \circ i_0^3 = i_1^1 \circ i_0^2 \circ i_0^3.$$

Since $\mathrm{im}(i_0^1)$ and $\mathrm{im}(d_1)$ generate $\mathrm{Ph}^2(A)$, this proves that we have

$$i_2^2 \circ i_1^3 = i_1^2 \circ i_3^3.$$

Together with the formula above, using induction we get

$$i_0^0 \circ i_0^1 = i_0^0 \circ i_1^1.$$

Thus, the $\mathrm{Ph}^*(A)$ is a semi-co-simplicial k-algebra with a co-section h_0 onto A. And it is easy to see that h_0 together with the corresponding co-sections $h_p : \mathrm{Ph}^{p+1}(A) \to \mathrm{Ph}^p(A)$, for $\mathrm{Ph}^p(A)$ replacing A, form a trivializing homotopy for $\mathrm{Ph}^*(A)$. Thus, we have

$$H^n(\mathrm{Ph}^*(A)) = 0, \forall n \geq 0,$$

i.e. Ph^{*+1} is a co-simplicial resolution of A. Therefore, for any object,

$$\kappa : A \to R \in A/k - \mathbf{alg}$$

the co-simplicial algebra above induces simplicial sets

$$\mathrm{Mor}_k(\mathrm{Ph}^*(A), R),\ \ \mathrm{Mor}_A(\mathrm{Ph}^*(A), R),$$

and one should be interested in the homotopy. See also that this generalizes to a canonical functor,

$$\mathrm{Spec} : (k - \mathbf{alg}^{\Delta})^{op} \longrightarrow \mathbf{SPr}(k),$$

where $(k - \mathbf{alg})^{\Delta}$ is the category of co-simplicial k-algebras, and $\mathbf{SPr}(k)$ is the category of simplicial pre-sheaves on the category of k-schemes enriched by any Grothendieck topology. As usual, the embedding of the category of k-algebras in the category of co-simplicial algebras is defined simply by giving any k-algebra a constant co-simplicial structure. The fact that $\text{Ph}^*(A)$ is a resolution of A is therefore simply saying that

$$\text{Spec}(\text{Ph}^*(A)) \to \text{Spec}(A)$$

is a week equivalence in $\mathbf{SPr}(k)$.

This might be a starting point for a theory of homotopy for k-schemes. We may, for any k-algebra R, consider the simplicial set

$$\text{Mor}_k(\text{Ph}^*(A), R),$$

and also the set of extensions

$$\begin{aligned}\text{Mor}_A(\text{Ph}^*(A), R) = \{\rho_k : \text{Ph}^k(A) \to R \mid i_0^{k-1} \circ \rho_k \\ = \rho_{k-1}, k \geq 1\},\end{aligned}$$

and the corresponding simplicial k-vector-space

$$\begin{aligned}\text{Der}_k(\text{Ph}^*(A), R) = \{\xi_0 \circ \xi_1 \circ \ldots \circ \xi_r | \xi_i \in der_k(\text{Ph}^i(A), R), \rho_{i+1}(d_i) \\ = \xi_{i+1}, i \geq 0\}.\end{aligned}$$

For $R = A$, and $\rho_0 = id$, we have,

$$\begin{aligned}\text{Mor}_A(\text{Ph}^n(A), A) = \{\xi_0 \circ \xi_1 \circ \ldots \circ \xi_r | \xi_i \in \text{Der}_k(\text{Ph}^i(A), A), \rho_{i+1}(d_i) \\ = \xi_{i+1}, i \geq 0\}.\end{aligned}$$

Since Ph is a functor, and Ph^{*+1} is a co-simplicial resolution of A, we may apply this to any scheme X given in terms of an affine covering $\mathbf{U}$ and obtain an algebraic homology (or cohomology) with converging spectral sequences,

$$\begin{aligned}E^1_{pq} &= H_p({H_{\mathbf{U}}}^{-q}(\text{Der}_k(\text{Ph}^*(A), A))), \\ E_{q,p} &= H^{-q}_{\mathbf{U}}(H_p(\text{Der}_k(\text{Ph}^*(A), A))).\end{aligned}$$

If we, in $\text{Mor}_A(\text{Ph}^n(A), A)$, identify $\xi \sim \alpha\xi, \alpha \in k^*$, we obtain a rational cohomology with converging spectral sequences,

$$\begin{aligned}E_1^{pq} &= H^p(H^q_{\mathbf{U}}(\text{Mor}_A(\text{Ph}^n(A), A), \mathbf{Q})), \\ E_2^{q,p} &= H^q_{\mathbf{U}}(H^p(\text{Mor}_A(\text{Ph}^n(A), A), \mathbf{Q})).\end{aligned}$$

Remark 2.1. The above suggests that we are closing in on Stacks and Motives. Any reasonable cohomology theory defined on the category of k-schemes is now seen to be defined on the image category of the functor Spec, so probably extendable to $\mathbf{SPr}(k)$, therefore, coming with a homotopy theory attached. Moreover, suppose we, instead of the example $R = A$ above, considered the category (actually an ordered set) of morphisms, $\mathfrak{a}(A) := \{A \to A/\mathfrak{p}_i\}$, for some family of irreducible bilateral ideals, corresponding in the commutative case to families of sub-schemes of $X = \mathrm{Spec}(A)$. By Theorem 4.2.4 of Laudal (1979) there is for any finite subcategory $\mathfrak{V} \subset \mathfrak{a}(A)$, a formal moduli $H(\mathfrak{V})$, for the deformation functor $\mathsf{Def}_{\mathfrak{V}}$, of the category of morphisms $\mathfrak{V}$, with the algebra A trivially deformed, provided the corresponding cohomology groups of the deformation theory are countably generated.

Moreover, we may globalize this to hold for any scheme, X, and in particular to any projective scheme over k, for which we know that the cohomology groups of the deformation theory will be of finite dimension, implying that the formal moduli $H(\mathfrak{V})$ will be finitely generated formal k-algebras. Formally, the theory will be of the same nature for schemes as for algebras, and so to minimize the place and problems with hanging on to dull dual descriptions, we shall just describe the affine case. Of course, in this case the formal moduli $H(\mathfrak{V})$ will not, in general, be finitely generated formal k-algebras, but this should not be of much trouble for most mathematicians.

Obviously, if $\mathfrak{W} \subset \mathfrak{V}$, there is a natural morphism,

$$\pi(\mathfrak{V}, \mathfrak{W}) : H(\mathfrak{V}) \to H(\mathfrak{W}),$$

usually just called π, or name omitted. Given a commutative diagram,

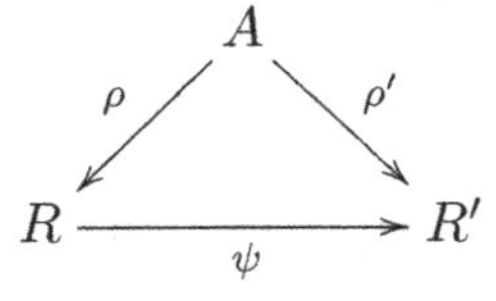

we have a diagram of canonical morphisms,

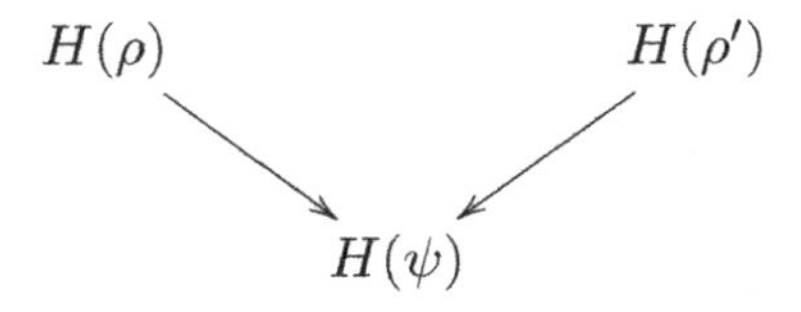

where we have put $H(\psi) := H(\psi : \rho \to \rho')$. and For any diagram like

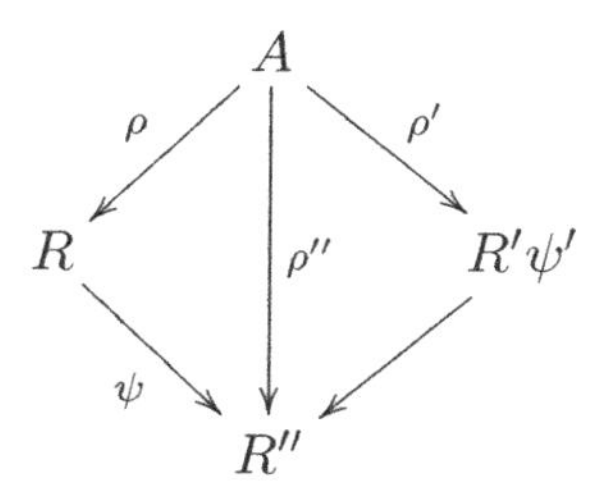

we find a diagram of canonical morphisms,

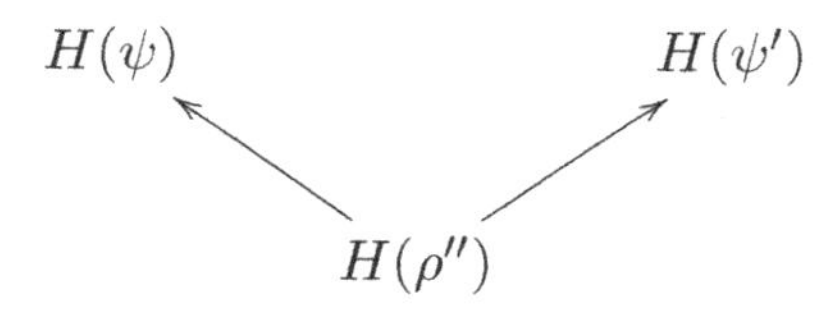

with the same abbreviation as above.

A *prime cycle* in the *motive of* A should be any object $(\rho, H(\rho)), \rho \in \mathbf{Irr}(A)$. The set of which we shall call $h(A)$. A *cycle* should then be a linear combination over some abelian group of such prime cycles.

We should like to define the *intersection product* of cycles as a bilinear product of cycles. If ρ and ρ' are prime cycles, then we define the intersection as the sum

$$\rho \cap \rho' = \sum \alpha(\rho, \rho')\rho'', \;\; \alpha(\rho, \rho') := |H(\rho, \psi) \otimes_{H(\rho'')} H(\rho', \psi')|.$$

We would like to compare it to the Serre intersection formula,

$$\rho \cap \rho' = \sum_{0}^{\infty} (-1)^i \operatorname{Tor}_i^A(R, R'),$$

in the commutative case. But this demands a certain work which will be postponed.

Anyway, the notion of motive, over the rationals, should be given by the set of finite *cycles*,

$$M(A) = \operatorname{FinMap}(h(A), \mathbf{Q}),$$

divided out with some equivalence relation, which we shall come back to.

Consider now the co-simplicial algebra,

$$A \overset{i_0^0}{\to} \mathrm{Ph}(A) \overset{i_p^1}{\to} \mathrm{Ph}^2(A) \overset{i_p^2}{\to} \mathrm{Ph}^3(A) \overset{i_p^3}{\to} \cdots$$

where, for each integer n, the symbol i_p^n, for $p = 0, 1, \ldots, n$ signifies the family of A-morphisms between $\mathrm{Ph}^n(A)$ and $\mathrm{Ph}^{n+1}(A)$ defined above. The system of k-algebras and homomorphisms of k-algebras $\{\mathrm{Ph}^n(A), i_j^n\}_{n,0\leq j\leq n}$ has an inductive (direct) limit, defined by

$$\mathrm{Ph}^\infty(A) = \varinjlim_{n\geq 0}\{\mathrm{Ph}^n(A), i_j^n\}$$

together with homomorphisms

$$i_n : \mathrm{Ph}^n(A) \longrightarrow \mathrm{Ph}^\infty(A)$$

satisfying

$$i_j^n \circ i_{n+1} = i_n, \quad j = 0, 1, \ldots, n.$$

Moreover, the family of derivations, $\{d_n\}_{0\leq n}$, define a unique Dirac derivation

$$\delta : \mathrm{Ph}^\infty(A) \longrightarrow \mathrm{Ph}^\infty(A)$$

such that $i_n \circ \delta = d_n \circ i_{n+1}$. Put

$$\mathrm{Ph}^{(n)}(A) := \mathrm{im}\ i_n \subseteq \mathrm{Ph}^\infty(A).$$

The k-algebra $\mathrm{Ph}^\infty(A)$ has a descending filtration of two-sided ideals, $\{\mathbf{F}_n\}_{0\leq n}$ given inductively by

$$\mathbf{F}_1 = \mathrm{Ph}^\infty(A) \cdot \mathrm{im}(\delta) \cdot \mathrm{Ph}^\infty(A)$$

and

$$\delta \mathbf{F}_n \subseteq \mathbf{F}_{n+1}, \quad \mathbf{F}_{n_1}\mathbf{F}_{n_2}\cdots\mathbf{F}_{n_r} \subseteq \mathbf{F}_n, \quad n_1 + \cdots + n_r = n$$

such that the derivation δ induces derivations $\delta_n : \mathbf{F}_n \longrightarrow \mathbf{F}_{n+1}$. Using the canonical homomorphism $i_n : \mathrm{Ph}^n(A) \longrightarrow \mathrm{Ph}^\infty(A)$, we pull the filtration $\{\mathbf{F}_p\}_{0\leq p}$ back to $\mathrm{Ph}^n(A)$, obtaining a filtration of each $\mathrm{Ph}^n(A)$ with

$$\mathbf{F}_1^n = \mathrm{Ph}^n(A) \cdot \mathrm{im}(\delta) \cdot \mathrm{Ph}^n(A),$$

and inductively,

$$\delta\mathbf{F}_p^n \subseteq \mathbf{F}_{p+1}^{n+1}, \quad \mathbf{F}_{p_1}^n\mathbf{F}_{p_2}^n \cdots \mathbf{F}_{p_r}^n \subseteq \mathbf{F}_p^n, \quad p_1 + \cdots + p_r = p.$$

Definition 2.2. Let

$$\mathbf{D}(A) := \varprojlim_{n\geq 1} \mathrm{Ph}^\infty(A)/\mathbf{F}_n,$$

the completion of $\mathrm{Ph}^\infty(A)$ in the topology given by the filtration $\{\mathbf{F}_n\}_{0\leq n}$. The k-algebra $\mathrm{Ph}^\infty(A)$ will be referred to as the k-algebra of higher differentials, and $\mathbf{D}(A)$ will be called the k-algebra of formalized higher differentials. Put

$$\mathbf{D}_n := \mathbf{D}_n(A) := \mathrm{Ph}^\infty(A)/\mathbf{F}_{n+1}.$$

Clearly, δ defines a derivation on $\mathbf{D}(A)$, and an isomorphism of k-algebras,

$$\epsilon := \exp(\delta) : \mathbf{D}(A) \to \mathbf{D}(A)$$

and in particular, an algebra homomorphism,

$$\tilde{\eta} := \exp(\delta) : A \to \mathbf{D}(A)$$

inducing the algebra homomorphisms

$$\tilde{\eta}_n : A \to \mathbf{D}_n(A)$$

which, by killing, in the right-hand algebra, the image of the maximal ideal $\mathfrak{m}(\underline{t})$ of A corresponding to a point $\underline{t} \in \mathrm{Simp}_1(A)$, induces a homomorphism of k-algebras

$$\tilde{\eta}_n(\underline{t}) : A \to \mathbf{D}_n(A)(\underline{t}) := \mathbf{D}_n/(\mathbf{D}_n\mathfrak{m}(\underline{t})\mathbf{D}_n)$$

and an injective homomorphism

$$\tilde{\eta}(\underline{t}) : A \to \varprojlim_{n\geq 1} \mathbf{D}_n(A)(\underline{t}),$$

see Laudal (2004).

2.2.1 *Formal curves of representations*

Since $\mathrm{Ext}^1_A(V,V)$ is the tangent space of the mini-versal deformation space of V as an A-module, we see that the non-commutative space $\mathrm{Ph}(A)$ also parametrizes the set of *generalized momenta*, i.e. the set of pairs of an A-module V, and a tangent vector of the formal moduli of V, at that point. Therefore, the above implies that any representation, $\rho : \mathrm{Ph}^\infty(A) \to \mathrm{End}_k(V)$, corresponds to a family of $\mathrm{Ph}^n(A)$-module-structures on V, for $n \geq 1$, i.e. to an A-module $V_0 := V$, an element $\xi_0 \in \mathrm{Ext}^1_A(V,V)$, i.e. a tangent of the deformation functor of $V_0 := V$, as A-module, an element $\xi_1 \in \mathrm{Ext}^1_{\mathrm{Ph}(A)}(V,V)$, i.e. a tangent of the deformation functor of $V_1 := V$ as $\mathrm{Ph}(A)$-module, an element $\xi_2 \in \mathrm{Ext}^1_{\mathrm{Ph}^2(A)}(V,V)$, i.e. a tangent of the deformation functor of $V_2 := V$ as $\mathrm{Ph}^2(A)$-module, etc.

All this is just $\rho_0 : A \to \mathrm{End}_k(V)$, considered as an A-module, together with a sequence $\{\xi_n\}, 0 \leq n$, of a tangent, or a *momentum*, ξ_0, an acceleration vector, ξ_1, and any number of higher-order *momenta* ξ_n. Thus, specifying a $\mathrm{Ph}^\infty(A)$-representation V implies specifying a *formal curve* through v_0, the base-point, of the *mini-versal deformation space* of the A-module V. Formally, this curve is given by the composition of the homomorphism $\epsilon(\tau) := \exp(\tau\delta)$ and ρ.

This is seen as follows. Consider the diagram,

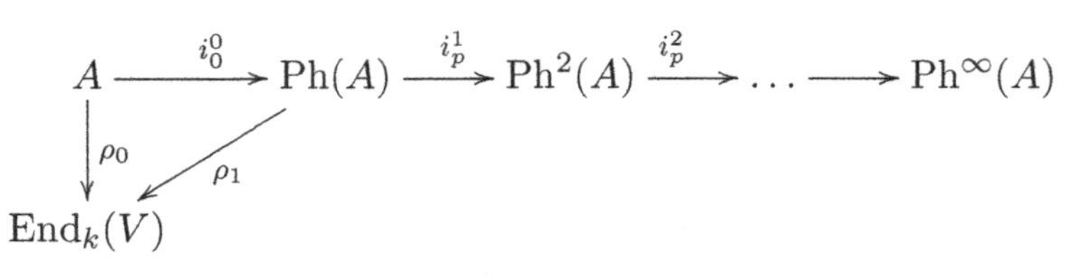

$$\begin{array}{ccccccc}
A & \xrightarrow{i_0^0} & \mathrm{Ph}(A) & \xrightarrow{i_p^1} & \mathrm{Ph}^2(A) & \xrightarrow{i_p^2} \cdots \longrightarrow & \mathrm{Ph}^\infty(A) \\
\downarrow{\scriptstyle \rho_0} & \swarrow{\scriptstyle \rho_1} & & & & & \\
\mathrm{End}_k(V) & & & & & &
\end{array}$$

where, for each integer n, the symbol i_p^n, for $p = 0, 1, \ldots, n$ signifies the family of A-morphisms between $\mathrm{Ph}^n(A)$ and $\mathrm{Ph}^{n+1}(A)$ defined above. Suppose now that we can extend ρ_0 to a morphism ρ_1, which should be seen as a momentum for the representation ρ_0, and suppose moreover that we can continue, finding morphisms $\rho_p^n : \mathrm{Ph}^p(A) \to \mathrm{End}_k(V)$, for $p = 2, \ldots, \infty$, making the diagram commute, and such that

$$i_p^{n-1}\rho_n = \rho_{n-1}$$

for all $n \geq 2$, then it is relatively easy to do the computation, and find that the map

$$[\delta] : A \to \mathrm{Ph}^*(A),$$

defined by $[\delta](a) := \sum_{n=0}^{\infty} \frac{1}{n!} d_n(\ldots(d_0))(a)$, composed with any one of the ρ_p, will be an algebra homomorphism.

If we arrange for all the $\rho_p : \mathrm{Ph}^p(A) \to \mathrm{End}_k(V) \otimes k[\tau]/(\tau)^p$ to be the (obvious) graded homomorphisms, we have in fact found a formal curve, parametrized by τ in the moduli space of $\mathrm{Rep}(A)$. See Laudal (2008).

It is, however, impossible to *prepare* a physical situation such that a measurement, i.e. an object like ρ_0, is given by an infinite sequence $\{\xi_n\}$, of dynamical data. We shall have to be satisfied with a finite number of data, and normally with just the first one, i.e. the momentum ξ_0, given by ρ_1. This is the problem of *Preparation and of the Time Evolution* of a representation ρ, to be treated in the sequel.

2.3 The Generalized de Rham Complex

Consider now the diagram

$$\begin{array}{ccccccccc}
A & \xrightarrow{i_0^0} & \mathrm{Ph}(A) & \xrightarrow{i_p^1} & \mathrm{Ph}^2(A) & \xrightarrow{i_p^2} & \mathrm{Ph}^3(A) & \xrightarrow{i_p^3} & \\
 & & \uparrow & & \uparrow & & \uparrow & & \\
 & & \mathfrak{m}_1^1 & \xrightarrow{i_p^1} & \mathfrak{m}_2^1 & \xrightarrow{i_p^2} & \mathfrak{m}_3^1 & \xrightarrow{i_p^3} &
\end{array}$$

where, for each integer n, the symbol i_p^n, for $p = 0, 1, \ldots, n$ signifies the family of A-morphisms between $\mathrm{Ph}^n(A)$ and $\mathrm{Ph}^{n+1}(A)$ defined above, and where $\mathfrak{m}_n^1$ is the ideal of $\mathrm{Ph}^n(A)$ generated by $\mathrm{im}(d)$, which is the same as the ideal generated by the family,

$\{i_p^{n-1}(i_p^{n-2}(\ldots(i_p^1(d(A))\ldots))\}$, for all possible p. Inductively, let $\mathfrak{m}_n^m$ be the ideal generated by $\mathfrak{m}_n^1\mathfrak{m}_n^{m-1}$.

We find an extended diagram

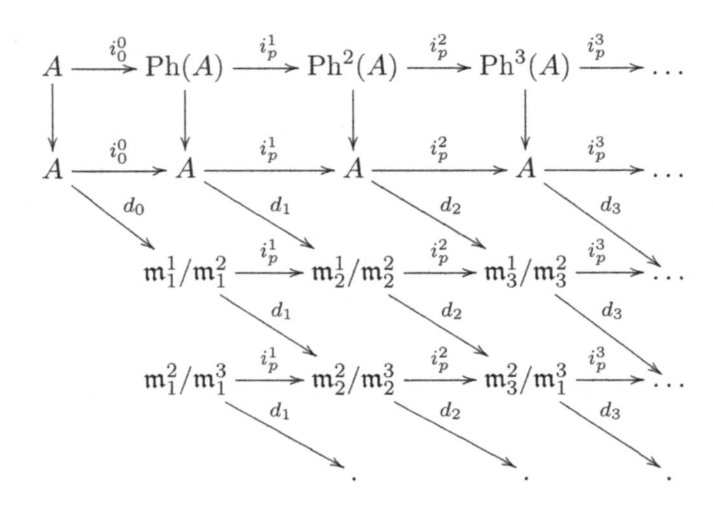

The diagonals are not necessarily complexes, but it suffices to kill d^2, to kill all d^n, $n \geq 2$, and for this it suffices to kill $d_1 d_0$, as one can easily see, operating with the edge homomorphisms $i_p^n, n \geq 2$ on the elements, $d_1(d_0(a))$ for $a \in A$. Therefore, we shall, in this general situation, make the following definition,

Definition 2.3. The curvature $R(A)$ of the associative k-algebra A is the k-linear composition of d_0 and d_1,

$$R(A) = d_0 d_1 : A \to \mathfrak{m}_2^2/\mathfrak{m}_2^3.$$

Now, kill the curvature $R(A)$, and all the terms under the first diagonal, beginning with $\mathfrak{m}_1^2/\mathfrak{m}_1^3$, together with all terms generated by the actions of the edge homomorphisms on these terms, and let $\boldsymbol{\Omega}_n^m$ be the resulting quotient of $\mathfrak{m}_n^m/\mathfrak{m}_n^{m+1}$, for $n \geq 0$. Clearly, $\Omega_n^0 = A$ for all $n \geq 0$, and we have got a graded semi-co-simplicial A-module, with a k-differential d, such that $d^2 = 0$, looking like

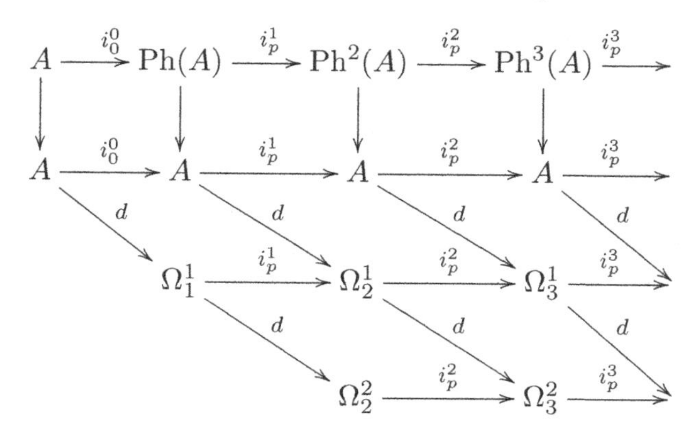

It is a graded complex, in two ways. First, as a complex induced from the semi-co-simplicial structure, with differential of bi-degree (1,0), and second, as a complex with differential d, of bi-degree (1,1).

Lemma 2.2. *Suppose A is commutative, then there is a natural morphism of complexes of A-modules,*

$$\Omega_A^* \subset \Omega_*^*,$$

with

$$\Omega_A^n := \wedge^r \Omega_A \simeq \Omega_n^n.$$

Proof. Let $a_i \in A, i = 1, \ldots, r$, and compute in Ω_*^r the value of $d^r(a_1 a_2 \ldots a_r)$. It is clear that this gives the formula

$$\sum d_{i_1}(a_1) d_{i_2}(a_2)..d_{i_r}(a_r) = 0,$$

the sum being over all permutation $(i_1, i_2, \ldots, i_r)$ of $(0, 1, \ldots, r-1)$. Here, we consider A as a sub-algebra of $\mathrm{Ph}^n(A)$ via the unique compositions of the $i_0^s : \mathrm{Ph}^s(A) \subset \mathrm{Ph}^{s+1}(A)$. In particular, we have,

$$d_0(a_1)d_1(a_2) + d_1(a_1)d_0(a_2) = 0,$$

for all $a_1, a_2 \in A$. This relation and the relation $d_0(a_2)d_1(a_1) = d_1(a_1)d_0(a_2)$, which follows from commutativity, $d(a_2)a_1 = a_1 d(a_2)$, forcing the left and right A-action on Ω_A to be equal, immediately give us $d_0(a_1)d_1(a_2) = -d_0(a_2)d_1(a_1)$.

Consider now the diagram

$$\begin{array}{ccccccccc}
A & \xrightarrow{i_0^0} & \mathrm{Ph}(A) & \xrightarrow{i_0^1} & \mathrm{Ph}^2(A) & \xrightarrow{i_0^2} & \mathrm{Ph}^3(A) & \xrightarrow{i_0^3} & \ldots \\
\downarrow & & \downarrow & & \downarrow & & \downarrow & & \\
A & \xrightarrow{i_0^0} & A \oplus \Omega_A^1 & \xrightarrow{i_0^1} & A \oplus \Omega_A^1 \oplus \Omega_A^2 & \xrightarrow{i_0^2} & A \oplus \Omega_A^1 \oplus \Omega_A^2 \oplus \Omega_A^3 & \xrightarrow{i_0^3} & \ldots,
\end{array}$$

where the bottom line is a sequence of Nagata-extensions of the k-algebra A, and the vertical homomorphisms correspond to the natural derivations among these, defined by the derivations of the de Rham complex, Ω^*, of A.

The universality of the two systems proves that there is a surjective map,

$$\alpha : \Omega_n^n \to \Omega_A^n := \wedge^n \Omega_A.$$

The map that sends the element $da_1 \wedge da_2 \wedge \cdots \wedge da_n \in \Omega_A^r$ to $d_0(a_1)d_1(a_2)..d_{n-1}(a_r) \in \Omega_n^n$ is an inverse, proving that α is an isomorphism. □

It follows from this that in the commutative case, for any scheme X considered as a covering of affine schemes in some sense, there are two spectral sequences converging to the same cohomology, first

$$E(1)_{p,q}^2 = H^p(H_{dR}^q(X, \Omega_*^*)),$$

then

$$E(2)_{p,q}^2 = H_{dR}^q(X, H^p(\Omega_*^*)).$$

Now let V be a right A-module, and assume $c(V) = 0$, such that there exists an element $\nabla' \in \mathrm{Hom}_k(V, V \otimes_A \mathrm{Ph}(A))$ with $c = \iota(\nabla')$. This implies that for $a \in A$ and $v \in V$, we have $\nabla'(va) = \nabla'(v)a + v \otimes d_0(a)$. Composing ∇' with the projection, $o : \mathrm{Ph}(A) \to A$, corresponding to the 0-derivation of A, we therefore obtain an A-linear homomorphism $P : V \to V$, *a potential.* Since $i_0^0 : A \to \mathrm{Ph}(A)$ is a section of o, we find a k-linear map

$$\nabla_0 := \nabla' - P : V \to V \otimes \mathfrak{m}_1^1.$$

Using the property

$$d_n \circ i_{j+1}^{n+1} = i_j^n \, d_{n+1},$$

we find well-defined k-linear maps

$\nabla_1 : V \to V \otimes \Omega_2^1$, $\nabla_2 : V \to V \otimes \Omega_3^1, \ldots, \nabla_n : V \to V \otimes \Omega_{n+1}^1 \; \forall n \geq 0,$

given by

$$\nabla_{n+1} := \nabla_n \circ i_1^{n+1}, \; n \geq 0,$$

such that, for all $v \in V, \omega \in \Omega_p^n$, the formula

$$\nabla_n(v \otimes \omega) = \nabla_n(v)\omega + v \otimes d_n(\omega)$$

makes sense, and defines a sequence of *derivations*,

$$\nabla_n : V \otimes \Omega_n^p \to V \otimes \Omega_{n+1}^{p+1},$$

sometimes just denoted d_n, and called a *connection* ∇, on the A-module V. We obtain a situation just like above,

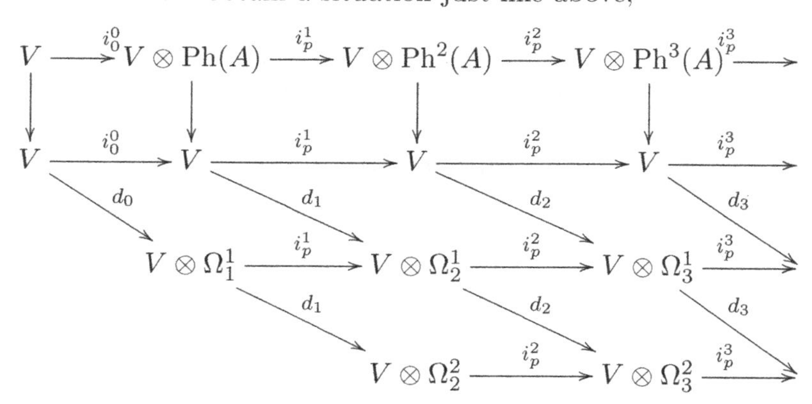

In general, there are no reasons for these derivations, $d_n := \nabla_n$, $n \geq 0$, to define complexes, and we shall make the following definition.

Definition 2.4. The curvature $R(V, \nabla)$ of the connection ∇, defined on the right A-module V, is the k-linear map, composition of d_0 and d_1,

$$R(V) = d_0 d_1 : V \to V \otimes \Omega_2^2.$$

The following lemma is then easily proved.

Lemma 2.3. *Suppose A is commutative, and assume $c(V) = 0$. Let $\nabla : \Theta_A \to \mathrm{End}_k(V)$ be the classical connection corresponding to ∇_0. Suppose moreover that the curvature R of ∇ is 0, then $R(V) = 0$, implying that $d^2 = 0$, and so the diagonals in the diagram above are all complexes.*

Proof. We may put

$$\nabla(v_i) = \sum_{j,k,} a_{i,j}^k v_j d_0(x_k)$$

and obtain

$$\nabla_1(\nabla_0(v_i)) = \sum_{j,k,l} \delta_{x_l} a_{i,j}^k v_j d_1(x_l) d_0(x_k) + \sum_{j,k,l,m} a_{i,j}^k a_{j,m}^l v_m d_1(x_l) d_0(x_k).$$

Now, the classical curvature of ∇ may be defined as

$$R_{k,l}^i = \sum_j \delta_{x_l} a_{i,j}^k v_j + \sum_{j,m} a_{i,j}^k a_{j,m}^l v_m - \sum_j \delta_{x_l} a_{i,j}^k v_j - \sum_{j,m} a_{i,j}^l a_{j,m}^k v_m,$$

so if $R = 0$, and $d_1(x_l)d_0(x_k) = -d_1(x_k)d_0(x_l)$, we find that $\nabla_1(\nabla_0(v_i)) = 0$, from which it follows that $d^2 = 0$. □

2.3.1 *Excursion into the Jacobian conjecture*

The conjecture, referred to as Jacobian, says that for algebraically closed fields k, any algebraic morphism

$$F : \mathbf{A}^n \to \mathbf{A}^n,$$

with everywhere non-trivial Jacobi determinant, $\det J(F)$, is an isomorphism. It is usually formulated by considering $F = \{f_i \in k[x_1, \ldots, x_n] : i = 1, \ldots, n\}$ as the algebraic homomorphism

$$F : k[x_1, \ldots, x_n] \to k[x_1, \ldots, x_n], F(x_i) := f_i,$$

and $J(F) = (f_{i,j})$, where we put

$$f_{i,j} = \delta_{x_i}(f_i).$$

Since $\det(J(F))$ must be a non-zero element of $k[x_1, \ldots, x_n]$, it is clear that there exists an inverse $(g_{i,j})$, of the matrix $J(F)$, with

$g_{i,j} \in k[x_1, \ldots, x_n]$, which can be written as

$$\sum_i g_{k,i} f_{i,j} dx_j = \delta_{j,k} dx_j.$$

The integral g_i of $\sum_j g_{i,j} dx_j$ is a polynomial such that

$$dg_i = \sum_j g_{i,j} dx_j.$$

We might try out the following optimistic equation:

$$g_k(f_1, \ldots, f_n) = x_k.$$

Since $dg_k(f_1, \ldots, f_n) = \sum_i g_{k,i}(f_1, \ldots, f_n) f_{i,j} dx_j$, and since we may assume that F has the origin $x_i = 0$ as fix-point, we find that

$$g_k(f_1, \ldots, f_n) = x_k(\mathrm{mod}(\underline{x})^2),$$

and we may reduce the question to the case where

$$f_i(x_1, \ldots, x_n) = x_i + \alpha_i, \alpha_i \in (\underline{x})^2,$$

which is a well-known result.

Chapter 3

Non-Commutative Algebraic Geometry

Mathematics is, since the time of Galilei, the language of physics. And since Descartes, Newton and Leibnitz, geometry and differential algebra have been our best tools for making the Universe understandable. The last centuries have seen an amazing development in science and technology due to the the parallel achievements in mathematics and physics. The theory of general relativity and the modern theory of quantum physics have transformed our world view and our daily life in a way almost unimaginable just 50 years ago. And the pace of change is, seemingly, accelerating. And so is the pace of change of the mathematical bases for these two grand theories. The differential geometry and the operator algebra have served these two fundamental sciences well for a century, but the human curiosity does not rest. The feeling that they should, somehow, be united has been there for a long time and has produced a lot of new mathematics. Algebraic geometry is as old as geometry, but has seen a formidable development the last 70 years, starting with the Grothendieck era, including a new foundation for the age-old theory of deformations. Operator algebra has in the same period, due to work of von Neumann, Gelfand and Connes, been transformed into a fascinating non-commutative geometry.

3.1 Moduli of Representations

The physicists have, of course, taken advantage of these developments and used the new mathematics to construct new models. At the moment, the situation is, nevertheless, that there are still two theories: the general relativity, treating gravitation and to some extent electroweak forces, and quantum field theory, taking care of relativistic quantum theory. The result of the latter is the Standard Model, a wonder of an effective theory for most of the forces of nature, but not including gravitation.

The hope has therefore been that, by creating some sort of fusion of classical algebraic, and the new non-commutative geometry, one would be able to create a model fusing the standard model and the theory of gravitation. This is, in our views, what a mature *Non-Commutative Algebraic Geometry* should be about.

There are many attempts to create a geometry, based on the classical algebraic geometry, modified by Serre, Chevalley and Grothendieck, but where the algebra part is extended from commutative to associative algebras. In this paper, we will give reasons for why we think this effort must include the study of non-commutative deformation of algebraic structures.

The idea is to look at the common goal of quantum theory and Grothendieck's scheme theory, which is the study of the local and global properties of the set of representations of algebras, together with their dynamical structure.

A point in scheme theory is a representation of a (commutative) ring, i.e. a morphism, $\rho : A \to R$, where R is another ring. The scheme is, in a general sense, the moduli space $\mathrm{Rep}(A)$, of such representations. The object of scheme theory is then to study the properties of these moduli spaces, their categorical relations, and eventually to classify them.

In quantum theory, the objects of interest are also representations $\rho : A \to R$ of a ring of observables A, but here $R = \mathrm{End}_k(V)$, where k is a field we may use for *measuring*, for any observable $a \in A$, the eigenvalues of $\rho(a)$ as operator on the fixed k-vector space V. The aim is to study the structure of the moduli space of such representations, $\mathrm{Rep}(A)$, and in particular to understand the dynamical properties of this space.

In both cases, the local structure of the moduli spaces are defined via deformation theory. But here is where non-commutative deformation theory enters not only because the rings we must work with are non-commutative but also because the local structure we are interested in is no longer given by commutative algebra. In fact, as we shall see, the local structure of Rep(A), in a finite family $\mathfrak{B} \subset \text{Rep}(A)$, is not the superposition of the local structures of each one of the representations. Non-commutative deformation of the family $\mathfrak{B}$ produces a homomorphism,

$$\eta : A \to O(\mathfrak{V}),$$

with the property that η is the universal "embedding" of A in an algebra O for which every representation $V \in \mathfrak{V}$ of A extends to a "simple" representation of O. This is the O-construction, the localization process in non-commutative algebra, see Eriksen *et al.* (2017).

Example 3.1 (Blowing down sub-schemes). Let us first take an easy example. Given a scheme, or just an affine scheme defined by a k-algebra A, then a sub-scheme, locally given by a quotient C of A, can be considered as a representation of A, and the Blow Down of Spec(C) in Spec(A), is given by

$$\eta : A \to O(\mathfrak{V}),$$

where $\mathfrak{V}$ is the family of points outside Spec(C) in Spec(A) plus the representation C.

3.2 Moduli of Simple Modules

The basic notions of affine non-commutative algebraic geometry related to a (not necessarily commutative) associative k-algebra, for k an arbitrary field, have been treated by many authors in several works (see e.g. Procesi (1967, 1973); Laudal (2000, 2002, 2003); Eriksen *et al.* (2017). Given a finitely generated algebra A, let $\text{Simp}_{<\infty}(A)$ be the category of simple finite-dimensional representations, i.e. right modules, of A. We show, in Laudal (2003), that any *geometric k-algebra A*, see also Laudal (2011), may be recovered from the (non-commutative) structure of $\text{Simp}_{<\infty}(A)$ and that there is an underlying quasi-affine (commutative) scheme structure on each component $\mathrm{Simp}_n(A) \subset \mathrm{Simp}_{<\infty}(A)$, parametrizing the simple representations of dimension n. In fact, we have shown the following.

Theorem 3.1. *There is a commutative k-algebra $C(n)$ with an open sub-variety $U(n) \subseteq \mathrm{Simp}_1(C(n))$, an étale covering of $\mathrm{Simp}_n(A)$, over which there exists a* versal representation $\tilde{V} \simeq C(n) \otimes_k V$, *a vector bundle of rank n defined on* $\mathrm{Simp}_1(C(n))$, *and a* versal family, *i.e. a morphism of algebras,*

$$\tilde{\rho} : A \longrightarrow \mathrm{End}_{C(n)}(\tilde{V}) \to \mathrm{End}_{U(n)}(\tilde{V}),$$

inducing all isomorphism classes of simple n-dimensional A-modules.

$\mathrm{End}_{C(n)}(\tilde{V})$ also induces a bundle, of *operators*, on the étale covering $U(n)$ of $\mathrm{Simp}_n(A)$.

3.2.1 *Evolution in the moduli of simple modules*

Assume given a derivation, $\xi \in \mathrm{Der}_k(A)$. Pick any $v \in \mathrm{Simp}_n(A)$ corresponding to the right A-module V, with structure homomorphism $\rho_v : A \to \mathrm{End}_k(V)$, then ξ composed with ρ_v, gives us an element,

$$\xi_v \in \mathrm{Ext}^1_A(V, V).$$

Therefore, ξ defines a unique one-dimensional distribution in $\Theta_{\mathrm{Simp}_n(A)}$, which, once we have fixed a versal family, defines a vector field,

$$[\xi] \in \Theta_{\mathrm{Simp}_n(A)},$$

and, in good cases, a (rational) derivation,

$$[\xi] \in \mathrm{Der}_k(C(n)).$$

Moreover, we have, in Laudal (2011), proved the following theorems.

Theorem 3.2. *Formally, at any point $v \in U(n) \subset \mathrm{Simp}(C(n))$, with local ring $\hat{C}(n)_v$, there is a derivation $[\xi] \in \mathrm{Der}_k(\hat{C}(n)_v)$, and a Hamiltonian $Q_\xi \in \mathrm{End}_{\hat{C}(n)_v}(\hat{V}_v)$, such that, as operators on $\hat{V}_v$, we have*

$$\xi = [\xi] + [Q_\xi, -].$$

This means that for every $a \in A$, considered as an element $\tilde{\rho}(a) \in \mathrm{End}_{\hat{C}(n)_v}(\hat{V}_v) \simeq M_n(\hat{C}(n)_v)$, $\xi(a)$ acts on $\hat{V}_v$ as

$$\tilde{\rho}(\xi(a)) = [\xi](\tilde{\rho}(a)) + [Q_\xi, \tilde{\rho}(a)].$$

This result will turn out to be a very general version of the Dirac equation.

Note also that we have the canonical isomorphism,

$$\mathrm{Der}_k(A, A) \simeq \mathrm{Mor}_A(\mathrm{Ph}(A), A),$$

therefore, the derivation ξ, and the A-module V, correspond to a $\mathrm{Ph}(A)$-module V_ξ.

The Schrödinger equation, for $\phi \in V_\xi$ and where time is ξ, is then

$$\xi(\phi) = Q(\phi)$$

and, see again Laudal (2011), the solutions are given as follows.

Theorem 3.3. *The* evolution operator $u(\tau_0, \tau_1)$ *that changes the state* $\phi(\tau_0) \in \tilde{V}(v_0)$ *into the state* $\phi(\tau_1) \in \tilde{V}(v_1)$*, where* τ *is a parameter of the integral curve* $\mathbf{c}$*, connecting the two points* v_0 *and* v_1*, i.e. the* time passed*, is given by*

$$\phi(\tau_1) = u(\tau_0, \tau_1)(\phi(\tau_0)) = \exp\left[\int_{\mathbf{c}} Q(\tau)d\tau\right](\phi(\tau_0)),$$

where $\exp \int_{\mathbf{c}}$ *is the non-commutative version of the ordinary action integral, essentially defined by the equation,*

$$\exp\left[\int_{\mathbf{c}} Q(\tau)dt\right] = \exp\left[\int_{\mathbf{c}_2} Q(\tau)d\tau\right] \circ \exp\left[\int_{\mathbf{c}_1} Q(\tau)d\tau\right],$$

where $\mathbf{c}$ *is* $\mathbf{c}_1$ *followed by* $\mathbf{c}_2$*.*

There is an important special case of the above results, which we shall refer to as the *Singular Case*. Suppose $[\xi]_v = 0$, then the derivation $\xi \in \mathrm{Der}_k(A, \mathrm{End}_k(V))$ maps to $0 \in \mathrm{Ext}^1_A(V, V)$. This situation deserves the status as a Corollary.

Corollary 3.1. *In the general case, let*

$$\rho : A \to \mathrm{End}_k(V)$$

be any representation, and suppose given a derivation $\xi \in \mathrm{Der}_k(A, \mathrm{End}_k(V))$*, corresponding to a* 0*-tangent vector of* V*, meaning that* ξ *maps to* $0 \in \mathrm{Ext}^1_A(V, V)$*.*

Then there exists an operator, $Q_\xi \in \mathrm{End}_k(V)$*, such that for any* $a \in A$*,* $\xi(a) = [Q_\xi, a]$*, and the* evolution operator*, in this case, is* *reduced to its first-order term, given by the* Hamiltonian Q_ξ, *in the state space* V, *as the map,*

$$\forall v \in V,\ \delta_t(v) = \xi(v) = Q_\xi(v).$$

This implies that the corresponding zeroth-order changes of the A*-module structure of* V, *can be considered, equivalently, as a* Heisenberg process, *or as a* Schrødinger process.

Proof. The fact that the condition implies that there exists an operator, $Q_\xi \in \mathrm{End}_k(V)$, such that for any $a \in A$, $\xi(a) = [Q_\xi, a]$, means that the lifting of the A-module V to the $A \otimes k[\epsilon]$-module $V_\xi := V \otimes k[\epsilon]$ with the action of A defined by ξ is trivial. The automorphism, $E_\xi := (id + Q_\xi \epsilon)$ of $V_0 := V \otimes k[\epsilon]$, induces an isomorphism between the trivial lifting V_0 and V_ξ. In fact, for any $a \in A$, as operator in $V \otimes k[\epsilon]$, we have the formula of left operators,

$$\begin{aligned}(id + Q_\xi\epsilon)(a(id - Q_\xi\epsilon)) &= (id + Q_\xi\epsilon)(a - aQ_\xi\epsilon)\\ &= a + (Q_\xi a - aQ_\xi)\epsilon = a + \xi(a)\epsilon.\end{aligned}$$

Thus, the infinitesimal action of ξ in V is the endomorphism Q_ξ, which again induces the infinitesimal action (i.e. the derivation), of $\mathrm{End}_k(V)$,

$$\mathrm{ad}(Q_\xi) \in \mathrm{Der}_k(\mathrm{End}_k(V)),$$

since, for any $\psi \in \mathrm{End}_k(V)$, we find, as above,

$$\begin{aligned}(id + Q_\xi\epsilon)(\psi(id - Q_\xi\epsilon)) &= (id + Q_\xi\epsilon)(\psi - \psi Q_\xi\epsilon)\\ &= \psi + (Q_\xi\psi - \psi Q_\xi)\epsilon = \psi + [Q_\xi, \psi]\epsilon.\end{aligned}$$ □

3.3 Non-Commutative Deformations of Swarms

To be able to treat localizations in non-commutative algebraic geometry, we introduced Laudal (2000, 2002, 2003), a theory of non-commutative deformations of families of modules of associative k-algebras and the notion of a *swarm* of right modules (or more generally of objects in a k-linear abelian category).

Let $\mathbf{a}_r$ denote the category of r-pointed not necessarily commutative k-algebras R. The objects are the diagrams of k-algebras

$$k^r \xrightarrow{\iota} R \xrightarrow{\pi} k^r,$$

such that the composition of ι and π is the identity. Any such *r-pointed k-algebra* R is isomorphic to a k-algebra of $r \times r$-matrices $(R_{i,j})$. The radical of R is the bilateral ideal $\mathrm{Rad}(R) := \ker \pi$, such that $R/\mathrm{Rad}(R) \simeq k^r$. The dual k-vector space of $\mathrm{Rad}(R)/\mathrm{Rad}(R)^2$ is called the tangent space of R.

For $r = 1$, there is an obvious inclusion of categories $\underline{l} \subseteq \mathbf{a}_1$, where $\underline{l}$, as usual, denotes the category of commutative local Artinian k-algebras with residue field k.

Fix a (not necessarily commutative) associative k-algebra A and consider a right A-module M. The ordinary deformation functor $\mathsf{Def}_M : \underline{l} \to \mathbf{Sets}$ is then defined. Assuming $\mathrm{Ext}^i_A(M, M)$ has finite k-dimension for $i = 1, 2$, it is well known, see Schlessinger (1968) or Laudal (2002) that Def_M has a pro-representing hull H, *the formal moduli of M*. Moreover, the tangent space of H is isomorphic to $\mathrm{Ext}^1_A(M, M)$, and H can be computed in terms of $\mathrm{Ext}^i_A(M, M)$, $i = 1, 2$ and their *matric* Massey products, see Laudal (2002).

In the general case, consider a finite family $\mathbf{V} = \{V_i\}_{i=1}^r$ of right A-modules and put $V := \oplus_{i=1}^r V_i$. Assume that $\dim_k \mathrm{Ext}^1_A(V_i, V_j) < \infty$. Any such family of A-modules will be called a *swarm*. We shall define a deformation functor,

$$\mathsf{Def}_{\mathbf{V}} : \mathbf{a}_r \to \mathbf{Sets},$$

generalizing the functor Def_M above. Given an object $\pi : R = (R_{i,j}) \to k^r$ of $\mathbf{a}_r$, consider the k-vector space and left R-module $(R_{i,j} \otimes_k V_j)$. It is easy to see that

$$\mathrm{End}_R((R_{i,j} \otimes_k V_j)) \simeq (R_{i,j} \otimes_k \mathrm{Hom}_k(V_i, V_j)).$$

Clearly, π defines a k-linear and left R-linear map

$$\pi(R) : (R_{i,j} \otimes_k V_j) \to \oplus_{i=1}^r V_i,$$

inducing a homomorphism of R-endomorphism rings,

$$\tilde{\pi}(R) : (R_{i,j} \otimes_k \mathrm{Hom}_k(V_i, V_j)) \to \oplus_{i=1}^r \mathrm{End}_k(V_i).$$

The right A-module structure on the V_i's is defined by a homomorphism of k-algebras,

$$\eta_0 : A \to \oplus_{i=1}^r \operatorname{End}_k(V_i) \subset (\operatorname{Hom}_k(V_i, V_j)) =: \operatorname{End}_k(V).$$

Note that this homomorphism also provides each $\operatorname{Hom}_k(V_i, V_j)$ with an A-bimodule structure. Let $\mathsf{Def}_{\mathbf{V}}(R) \in \mathbf{Sets}$ be the set of isomorphism classes of homomorphisms of k-algebras,

$$\eta' : A \to (R_{i,j} \otimes_k \operatorname{Hom}_k(V_i, V_j)),$$

such that $\tilde{\pi}(R) \circ \eta' = \eta_0$, where the equivalence relation is defined by inner automorphisms in the R-algebra $(R_{i,j} \otimes_k \operatorname{Hom}_k(V_i, V_j))$ inducing the identity on $\oplus_{i=1}^r \operatorname{End}_k(V_i)$. One easily proves that $\mathsf{Def}_{\mathbf{V}}$ has the same properties as the ordinary deformation functor and we may prove the following, see Laudal (2002).

Theorem 3.4. *The functor* $\mathsf{Def}_{\mathbf{V}}$ *has a pro-representable hull, i.e. an object* $H := H(\mathbf{V})$ *of the category of pro-objects* $\hat{\mathbf{a}}_r$ *of* $\mathbf{a}_r$ *together with a versal family*

$$\tilde{V} = (H_{i,j} \otimes V_j) \in \varprojlim_{n \geq 1} \mathsf{Def}_{\mathbf{V}}(H/\mathfrak{m}^n),$$

where $\mathfrak{m} = \operatorname{Rad}(H)$*, such that the corresponding morphism of functors on* $\mathbf{a}_r$

$$\kappa : \operatorname{Mor}(H, -) \to \mathsf{Def}_{\mathbf{V}}$$

defined for $\phi \in \operatorname{Mor}(H, R)$ *by* $\kappa(\phi) = R \otimes_\phi \tilde{V}$*, is smooth, and an isomorphism on the tangent level.* H *is uniquely determined by a set of matric Massey products defined on subspaces*

$$D(n) \subseteq \operatorname{Ext}^1(V_i, V_{j_1}) \otimes \cdots \otimes \operatorname{Ext}^1(V_{j_{n-1}}, V_k),$$

with values in $\operatorname{Ext}^2(V_i, V_k)$.

Moreover, the right action of A *on* $\tilde{V}$ *defines a homomorphism of* k*-algebras,*

$$\eta : A \longrightarrow O(\mathbf{V}) := \operatorname{End}_H(\tilde{V}) = (H_{i,j} \otimes \operatorname{Hom}_k(V_i, V_j)).$$

The k*-algebra* $O(\mathbf{V})$*, called the ring of observables of* $\mathbf{V}$*, acts on the family of* A*-modules* $\{V_i\}_{i=1}^r$*, extending the action of* A*.*

If $\dim_k V_i < \infty$*, for all* $i = 1, \ldots, r$*, the operation of associating* $(O(\mathbf{V}), \mathbf{V})$ *to* $(A, \mathbf{V})$ *is a closure operation.*

There is a crucial result, see Laudal (2000, 2003).

Theorem 3.5 (A generalized Burnside theorem). *Let A be a finite-dimensional k-algebra, k an algebraically closed field. Consider the family $\mathbf{V} = \{V_i\}_{i=1}^r$ of all simple A-modules, then*

$$\eta : A \longrightarrow O(\mathbf{V}) = (H_{i,j} \otimes \mathrm{Hom}_k(V_i, V_j))$$

is an isomorphism. Moreover, the k-algebras A and H are Morita-equivalent.

We also prove that there exists, in the non-commutative deformation theory, an obvious analogy to the notion of pro-representing (modular) substratum H_0 of the formal moduli H, see Laudal (1979). The tangent space of H_0 is determined by a family of subspaces

$$\mathrm{Ext}^1_0(V_i, V_j) \subseteq \mathrm{Ext}^1_A(V_i, V_j), \quad i \neq j,$$

the elements of which should be called the almost split extensions (sequences) relative to the family $\mathbf{V}$, and by a subspace

$$T_0(\Delta) \subseteq \prod_i \mathrm{Ext}^1_A(V_i, V_i),$$

which is the tangent space of the deformation functor of the full sub-category of the category of A-modules generated by the family $\mathbf{V} = \{V_i\}_{i=1}^r$, see Laudal (1986). If $\mathbf{V} = \{V_i\}_{i=1}^r$ is the set of all indecomposables of some Artinian k-algebra A, we show that the above notion of *almost split sequence* coincides with that of Auslander, see Reiten (1985).

Using this we consider, in Laudal (2002, 2003), the general problem of classification of iterated extensions of a family of modules $\mathbf{V} = \{V_i,\}_{i=1}^r$, and the corresponding classification of filtered modules with graded components in the family $\mathbf{V}$, and extension type given by a directed representation graph Γ. This turns out to be the starting point for our treatment of *Interactions* in physics and will be postponed, see Chapter 12.

To any, not necessarily finite, swarm $\underline{c} \subset \mathbf{mod}(A)$ of right-A-modules, we have associated two associative k-algebras, see

Laudal (2000, 2003),

$$O(|\underline{c}|,\pi) = \varprojlim_{\mathbf{V}\subset|\underline{c}|} O(\mathbf{V}),$$

and a sub-quotient $\mathbf{O}_\pi(\underline{c})$, together with natural k-algebra homomorphisms

$$\eta(|\underline{c}|) : A \longrightarrow O(|\underline{c}|,\pi)$$

and $\eta(\underline{c}) : A \longrightarrow \mathbf{O}_\pi(\underline{c})$ with the property that the A-module structure on $\underline{c}$ is extended to an $\mathbf{O}$-module structure in an optimal way. We then defined an *affine non-commutative scheme* of right A-modules to be a swarm $\underline{c}$ of right A-modules, such that $\eta(\underline{c})$ is an isomorphism. In particular, we considered, for finitely generated k-algebras, the swarm $\mathrm{Simp}^*_{<\infty}(A)$ consisting of the finite-dimensional simple A-modules, and the *generic* point A, together with all morphisms between them. The fact that this is a swarm, i.e. that for all objects $V_i, V_j \in \mathrm{Simp}_{<\infty}$ we have $\dim_k \mathrm{Ext}^1_A(V_i,V_j) < \infty$, is easily proved. We have in Laudal (2003) proved the following result, (see (4.1), loc.cit. for more on the definition of the notion of *geometric* k-algebra).

Proposition 3.1. *Let A be a* geometric *k-algebra, meaning that the natural homomorphism,*

$$\eta := \eta(\mathrm{Simp}^*(A)) : A \longrightarrow \mathbf{O}_\pi(\mathrm{Simp}^*_{<\infty}(A)),$$

is injective, then η is an isomorphism, i.e. $\mathrm{Simp}^*_{<\infty}(A)$ *is a scheme for A.*

In particular, $\mathrm{Simp}^*_{<\infty}(k\langle x_1,x_2,\ldots,x_d\rangle)$ is a scheme for $k\langle x_1,x_2,\ldots,x_d\rangle$. To analyze the local structure of $\mathrm{Simp}_n(A)$, we need the following, see Laudal (2003, 3.23).

Lemma 3.1. *Let $\mathbf{V} = \{V_i\}_{i=1,\ldots,r}$ be a finite subset of* $\mathrm{Simp}_{<\infty}(A)$, *then the morphism of k-algebras,*

$$A \to O(\mathbf{V}) = (H_{i,j} \otimes_k \mathrm{Hom}_k(V_i,V_j))$$

is topologically surjective.

Proof. Since the simple modules V_i $(i = 1, \ldots, r)$ are distinct, there is an obvious surjection

$$\eta_0 : A \to \prod_{i=1,..,r} \mathrm{End}_k(V_i).$$

Put $\mathfrak{r} = \ker \eta_0$, and consider for $m \geq 2$ the finite-dimensional k-algebra, $B := A/\mathfrak{r}^m$. Clearly $\mathrm{Simp}(B) = \mathbf{V}$, so that by the generalized Burnside theorem, see Laudal (2002, 2.6), we find

$$B \simeq O^B(\mathbf{V}) := (H^B_{i,j} \otimes_k \mathrm{Hom}_k(V_i, V_j)).$$

Consider the commutative diagram

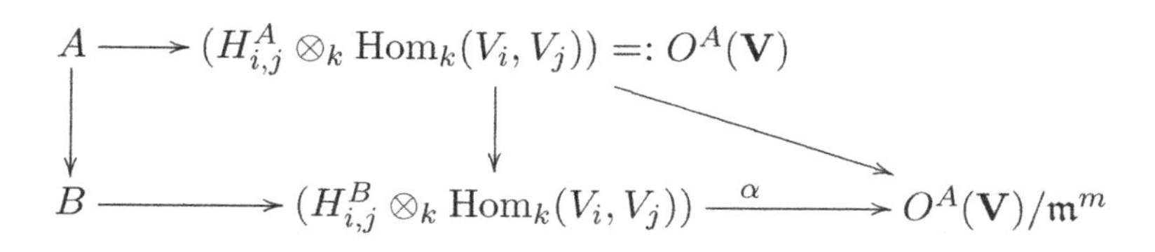

where all morphisms are natural. In particular, α exists since $B = A/\mathfrak{r}^m$ maps into $O^A\mathbf{V}/\,\mathrm{Rad}^m$ and therefore induces the morphism α commuting with the rest of the morphisms. Consequently, α has to be surjective, and we have proved the contention. □

Example 3.2. As an example of what may occur in rank infinity, we shall consider the invariant problem, $\mathbb{A}^1_{\mathbb{C}}/\mathbb{C}^*$. Here, we are talking about the algebra $A = \mathbb{C}[x](\mathbb{C}^*)$, the crossed product of $\mathbb{C}[x]$ with the group $\mathbb{C}^*$. If $\lambda \in \mathbb{C}^*$, the product in A is given by $x \times \lambda = \lambda \times \lambda^{-1}x$. There are two "points," i.e. orbits, modeled by the obvious origin $V_0 := A \to \mathrm{End}_{\mathbb{C}}(\mathbb{C}(0))$, and by $V_1 := A \to \mathrm{End}_{\mathbb{C}}(\mathbb{C}[x, x^{-1}])$. We may also choose the two points $V_0 := \mathbb{C}(0), V_1 := \mathbb{C}[x]$, in line with the definitions of Laudal (2000). Obviously, $\mathbb{C}[x]$ correspond to the closure of the orbit $\mathbb{C}[x, x^{-1}]$. This choice is the best if one wants to make visible the adjacencies in the quotient, and we shall therefore treat both cases.

We need to compute

$$\mathrm{Ext}^p_A(V_i, V_j), \quad p = 1, 2, \quad i, j = 1, 2.$$

Now,

$$\mathrm{Ext}^1_A(V_i, V_j) = \mathrm{Der}_{\mathbb{C}}(A, \mathrm{Hom}_{\mathbb{C}}(V_i, V_j))/\,\mathrm{Triv}, \quad i, j = 1, 2,$$

and since x acts as zero on V_1, and $\mathbb{C}^*$ acts as identity on V_1 and as homogeneous multiplication on V_0, we find

$$\begin{aligned}\mathrm{Der}_k(A, \mathrm{Hom}_k(V_0, V_0))/\,\mathrm{Triv} &= \mathrm{Der}_k(A, \mathrm{Hom}_k(V_0, V_0)) \\ &= \mathrm{Der}_{\mathbb{C}}(A, \mathbb{C}(0)).\end{aligned}$$

Any $\delta \in \mathrm{Der}_k(A, \mathbb{C}(0))$ is determined by its values, $\delta(x), \delta(\lambda) \in \mathbb{C}(0) | \lambda \in \mathbb{C}^*$. Moreover, since in A we have $(\lambda) \times (\lambda^{-1}x) = x \times (\lambda)$, we find

$$\delta(\lambda\mu) = \delta(\lambda) + \delta(\mu), \quad \delta((\lambda) \times (\lambda^{-1}x)) = \delta(x \times (\lambda)).$$

The left-hand side of the last equation is $\delta((\lambda^{-1}x)) = \lambda^{-1}\delta(x)$, and the right-hand side is $\delta(x)$, and since this must hold for all $\lambda \in \mathbb{C}^*$, we must have $\delta(x) = 0$. Moreover, since $\delta(\lambda\mu) = \delta(\lambda) + \delta(\mu)$, it is clear that continuity of δ implies that δ must be equal to $\alpha \ln(|.|)$ for some $\alpha \in \mathbb{C}$. (To simplify the writing, we shall put $\log := \ln(|.|)$.) Therefore,

$$\mathrm{Ext}^1_A(V_0, V_0) = \mathrm{Der}_k(A, \mathrm{Hom}_{\mathbb{C}}(V_0, V_0)) = \mathbb{C}.$$

The cup product of this class, $\log \cup \log$, sits in $\mathrm{HH}^2(A, \mathbb{C}(0)) = \mathrm{Ext}^2_a(V_0, V_0)$, and is given by the 2-co-cycle

$$(\lambda, \mu) \to \log(\lambda) \times \log(\mu).$$

This is seen to be a boundary, i.e. there exists a map $\psi : \mathbb{C}^* \to \mathbb{C}(0)$, such that for all, $\lambda, \mu \in \mathbb{C}^*$ we have

$$\log(\lambda) \times \log(\mu) = \psi(\lambda) - \psi(\lambda\mu) + \psi(\mu).$$

Just put $\psi_{1,1} := \psi_2 = -1/2 \log^2$. Therefore, the cup product is zero, and if we, in general, put

$$\psi_n := \psi_{1,1,\ldots,1} = (-1)^{n+1} 1/(n!) \log^n, \quad n \geq 1,$$

where n is the number of 1's in the first index, then computing the Massey products of the element $\log \in \mathrm{Ext}^1_A(V_0, V_0)$, we find the n-th

Massey product

$$[\log, \log, \ldots, \log] = \left\{ (\lambda, \mu) \to \sum_{p=1,\ldots,n-1} \psi_p \psi_{n-p} \right\},$$

and this is easily seen to be the boundary of the 1-co-chain

$$\psi_{n+1} = (-1)^{n+2} 1/((n+1)!) \log^{n+1}.$$

Therefore, all Massey products are zero. Of course, we have not yet proved that they could be different from zero, i.e. we have not computed the *obstruction* group $\mathrm{Ext}^2_A(V_0, V_0)$ and found it non-trivial! Now, this is unnecessary.

But, take the example $V_0 = \mathbb{C}[x, x^{-1}]$, then every

$$\delta \in \mathrm{Ext}^1_A(V_0, V_0) = \mathrm{Der}_{\mathbb{C}}(A, \mathrm{Hom}_{\mathbb{C}}(V_0, V_0)) / \mathrm{Triv}$$

is determined by the values of $\delta(x)$ and $\delta(\lambda)$, $\lambda \in \mathbb{C}^*$. Since $\mathrm{Ext}^1_{\mathbb{C}[x]}(V_0, V_1) = 0$, we may find a trivial derivation such that subtracting from δ we may assume $\delta(x) = 0$. But then the formula

$$\delta(x \times \lambda) = \delta(\lambda \times (\lambda^{-1} x))$$

implies

$$x\delta(\lambda) = \delta(\lambda)(\lambda^{-1} x),$$

from which it follows that

$$\delta(\lambda)(x^p) = (\lambda^{-1} x)^p \delta(\lambda)(1).$$

Now, since $\lambda\mu = \mu\lambda$ in $\mathbb{C}^*$, we find

$$(\lambda^{-1}\mu x)^p \delta(\lambda)(1)(\mu x) = (\lambda\mu^{-1} x)^p \delta(\lambda)(1)(\lambda x),$$

which should hold for any pair of $\mu, \lambda \in \mathbb{C}^*$, and any p. This obviously implies $\delta = 0$.

This argument shows not only that

$$\mathrm{Ext}^1_A(V_1, V_1) = \mathrm{Der}_{\mathbb{C}}(A, \mathrm{Hom}_{\mathbb{C}}(V_1, V_1))/\,\mathrm{Triv} = 0,$$

when $V_1 = \mathbb{C}[x, x^{-1}]$ but also when $V_1 = \mathbb{C}[x]$. Finally, we find that the formula above,

$$x\delta(\lambda) = \delta(\lambda)(\lambda^{-1}x),$$

shows that for

$$\delta \in \mathrm{Ext}^1_A(V_1, V_0) = \mathrm{Der}_{\mathbb{C}}(A, \mathrm{Hom}_{\mathbb{C}}(V_1, V_0))/\,\mathrm{Triv},$$

we have $\delta(\lambda)(xx^p) = 0$ for all p. Therefore,

$$\mathrm{Ext}^1_A(V_1, V_0) = \mathrm{Der}_{\mathbb{C}}(A, \mathrm{Hom}_{\mathbb{C}}(V_1, V_0))/\,\mathrm{Triv} = 0,$$

when $V_1 = \mathbb{C}[x, x^{-1}]$. However, when $V_1 = \mathbb{C}[x]$, we find that δ with $\delta(\lambda)(1) \neq 0$ and with $\delta(\lambda)(x^p) = 0$ for $p \geq 1$ survives. These will, as above, give rise to a logarithm of the real part of $\mathbb{C}^*$. Therefore, in this case, $\mathrm{Ext}^1_A(V_1, V_0) = \mathbb{C}$. The mini-versal families look like

$$H = \begin{pmatrix} \mathbb{C}[[t]] & 0 \\ 0 & \mathbb{C} \end{pmatrix}$$

when $V_1 = \mathbb{C}[x, x^{-1}]$ and like

$$H = \begin{pmatrix} \mathbb{C}[[t]] & 0 \\ \langle \mathbb{C} \rangle & \mathbb{C} \end{pmatrix}$$

when $V_1 = \mathbb{C}[x]$.

Chapter 4

The Dirac Derivation and Dynamical Structures

As we have seen in Chapter 2, the dynamics of the space of representations of our algebra A, i.e. the dynamics of the space of measurements of the family of observables ξ that A is assumed to represent, can be encoded in the category of representations of the k-algebra $\mathrm{Ph}^{\infty}(A)$, and the universal Dirac derivation δ. We would therefore like to use the tools developed above, for the k-algebra $\mathrm{Ph}^{\infty}(A)$, and with $\xi = \delta$.

However, $\mathrm{Ph}^{\infty}(A)$ is rarely of finite type, and so the moduli space of simple modules does not have a classical algebraic geometric structure. We shall therefore introduce the notion of a *dynamical structure*, to reduce the problem to a situation we can handle. This is also what physicists do. They invoke a *parsimony principle*, or an *action principle*, originally proposed by Fermat, and later by Maupertuis, with the purpose of reducing the preparation needed, to be able to see ahead.

4.1 Dynamical Structures

Definition 4.1. Let A be a k-algebra, and define a *dynamical structure*, σ, to be a two-sided δ-stable ideal $(\sigma) \subset \mathrm{Ph}^{\infty}(A)$ such that

$$A(\sigma) := \mathrm{Ph}^{\infty}(A)/(\sigma),$$

the corresponding *dynamical system* is of finite type. A dynamical structure, or system, is *of order* $\leq n$ if the canonical morphism,

$$\sigma : \mathrm{Ph}^{(n-1)}(A) \to A(\sigma),$$

is surjective. If A is generated by the *coordinate functions*, $\{t_i\}_{i=1,2,\ldots,d}$ a dynamical system of order n may be defined by a *force law*, i.e. by a system of equations,

$$\delta^n t_p = \Gamma^p(\underline{t}_i, \underline{d}t_j, \underline{d}^2 t_k, .., \underline{d}^{n-1} t_l), \quad p = 1, 2, \ldots, d.$$

Put

$$A(\sigma) := \mathrm{Ph}^{\infty}(A)/(\delta^n t_p - \Gamma^p),$$

where $\sigma := (\delta^n t_p - \Gamma^p)$ is the two-sided δ-ideal generated by the defining equations of σ. Obviously, δ induces a derivation $\delta_\sigma \in \mathrm{Der}_k(A(\sigma), A(\sigma))$, also called the Dirac derivation, and usually just denoted δ.

Clearly, if $\xi \in \mathrm{Der}_k(A)$, then A is a dynamical system of order 1, defined by the dynamical structure given by

$$\sigma = \{\delta^n(t_i) = \xi^n(t_i), \quad i = 1, 2, \ldots, d, n \geq 1\}.$$

Note that if σ_i, $i = 1, 2$, are two different order n dynamical systems, then we may well have

$$A(\sigma_1) \simeq A(\sigma_2) \simeq \mathrm{Ph}^{(n-1)}(A)/(\sigma_*)$$

as k-algebras.

Assuming that the k-algebra A is finitely generated, and given a dynamical structure σ, then by definition $A(\sigma)$ is finitely generated, and we can use the machinery of Chapter 3, with $A = A(\sigma)$ and $\xi = \delta$, the Dirac derivation. We obtain, as above, see Theorem 3.1.

Theorem 4.1. *There exists a* versal family, *i.e. a morphism of algebras,*

$$\tilde{\rho} : A(\sigma) \longrightarrow \mathrm{End}_{C(n)}(\tilde{V}) \to \mathrm{End}_{U(n)}(\tilde{V}),$$

inducing all isoclasses of simple n-dimensional $A(\sigma)$-modules.

Moreover, formally, at any point $v \in U(n) \subset \mathrm{Simp}(C(n))$, with local ring $\hat{C}(n)_v$, there is a derivation $[\delta] \in \mathrm{Der}_k(\hat{C}(n)_v)$, and a

Hamiltonian $Q \in \mathrm{End}_{\hat{C}(n)_v}(\tilde{V}_v)$, *such that, as operators on* $\tilde{V}_v$, *we have*

$$\delta = [\delta] + [Q, -].$$

This means that for every $a \in A(\sigma)$, *considered as an element* $\tilde{\rho}(a) \in M_n(\hat{C}(n)_v)$, $\delta(a)$ *acts on* $\tilde{V}_v$ *as*

$$\tilde{\rho}(\delta(a)) = [\delta](\tilde{\rho}(a)) + [Q, \tilde{\rho}(a)].$$

In line with our general philosophy, where time is a metric on an appropriate moduli space, the Dirac derivation δ is the time-propagator for representations. We shall consider $[\delta]$ as measuring *time* in $\mathrm{Simp}_n(\mathbf{A}(\sigma))$, and respectively in $\mathrm{Spec}(C(n))$. This is reasonable, since the last equation is equivalent to the following statement: The derivation δ induces an extension of $\tilde{V}_v$, as A-module, which modulo the derivation $[\delta]$, of $\hat{C}(n)_v$, is trivial. This is formally true for any derivation δ, by the definition of the versal family, i.e. the $\hat{C}(n)_v$-module $\hat{V}_v$.

Note also that $\mathrm{End}_{C(n)}(\tilde{V}) \simeq M_n(C(n))$, and be prepared in the sequel, to see this used without further warning. There are local (and even global) extensions of this result, where $[\delta]$ and Q may be assumed to be defined (rationally) on $C(n)$ (see Laudal, 2011). In this case, we may see that, provided the field k is (sufficiently) algebraically closed, any *quantum field*, $\psi \in \mathrm{End}_{C(n)}(\tilde{V})$, can be expressed as a (finite) rational polynomial of generalized *creation* and *annihilation* operators, see (4.4), loc. cit.

Assume for a while that $k = \mathbb{R}$, the real numbers, and that our constructions go through, as if k were algebraically closed. Let $v(\underline{\tau}_0) \in \mathrm{Simp}_n(\mathbf{A}(\sigma))$ be an element, an *event*. Suppose there exists an integral curve $\mathbf{c}$ of $[\delta]$ through $v(\tau_0) \in \mathrm{Simp}_1(C(n))$, ending at $v(\tau_1) \in \mathrm{Simp}_1(C(n))$, given by the automorphisms $e(\tau) := \exp(\tau[\delta])$, for $\tau \in [\tau_0, \tau_1] \subset \mathbb{R}$. The supremum of τ for which the corresponding point $v(\tau)$ of $\mathbf{c}$ is in $\mathrm{Simp}_n(\mathbf{A}(\sigma))$ should be called the *lifetime* of the event (particle). It is relatively easy to compute these lifetimes, and so to be able to talk about *decay*, when the fundamental vector field $[\delta]$ has been computed. In Laudal (2011), we have also proposed a mathematically sound way of treating interactions, purely in terms of non-commutative deformation theory.

Let $\phi(\tau_0) \in \tilde{V}(v_0) \simeq V$ be a (classically considered) state of our *quantum system* at the time τ_0 and consider the (uni)versal family,

$$\tilde{\rho} : \mathbf{A}(\sigma) \longrightarrow \mathrm{End}_{C(n)}(\tilde{V}),$$

restricted to $U(n) \subseteq \mathrm{Simp}_1(C(n))$, the étale covering of $\mathrm{Simp}_n(\mathbf{A}(\sigma))$. We shall consider $\mathbf{A}(\sigma)$ as our *ring of observables*. What happens to $\phi(\tau_0) \in \tilde{V}(v_0)$ when *time* passes from τ_0 to τ, along $\mathbf{c}$? This leads to a solution of the Schrödinger equation,

$$\frac{d\phi}{d\tau} = Q(\phi),$$

along $\mathbf{c}$, applying Theorem 4.1, proving that ϕ is completely determined, by the value of $\phi(\tau_0)$, for any $\tau_0 \in \mathbf{c}$. Here, we shall not go into the problem of *preparing* $\phi(\tau_0) \in V(\tau_0)$, i.e. of how to exactly determine *where we are*, at some chosen clock-time, τ (see Laudal, 2011).

Theorem 4.2. *The* evolution operator $u(\tau_0, \tau_1)$ *that changes the state* $\phi(\tau_0) \in \tilde{V}(v_0)$ *into the state* $\phi(\tau_1) \in \tilde{V}(v_1)$, *where* τ *is a parameter of the integral curve* $\mathbf{c}$ *connecting the two points* v_0 *and* v_1, *i.e. the* time passed, *is given by*

$$\phi(\tau_1) = u(\tau_0, \tau_1)(\phi(\tau_0)) = \exp\left[\int_{\mathbf{c}} Q(\tau)d\tau]\ (\phi(\tau_0))d\tau\right],$$

where $\exp \int_{\mathbf{c}}$ *is the non-commutative version of the ordinary action integral, essentially defined by the equation*

$$\exp\left[\int_{\mathbf{c}} Q(\tau)dt\right] = \exp\left[\int_{\mathbf{c}_2} Q(\tau)d\tau\right] \circ \exp\left[\int_{\mathbf{c}_1} Q(\tau)d\tau\right],$$

where $\mathbf{c}$ *is* $\mathbf{c}_1$ *followed by* $\mathbf{c}_2$.

In the situation of Theorem 4.2, we observe that, since $\delta = [\delta] + [Q, -]$ is a derivation defined in the algebra $\mathrm{End}_{C(n)}(\tilde{V})$, the eigenvalues $\Lambda := \{\lambda\}$ of the eigenvectors a_λ of δ will have a structure as an additive sub-monoid of the reals. Assume now that $[[\delta], Q] = 0$, (which will be the case for trivial metrics in Section 4.4), and suppose that $[\delta](\psi) = \nu\psi$ and $[Q, \psi] = \epsilon\psi$, then

$$\begin{aligned}\exp(\delta)(\psi) &= \exp(\mathrm{ad}(Q))\exp([\delta])(\psi) = \exp(\mathrm{ad}(Q))(\psi(t+\nu)) \\ &= \exp(\epsilon)\exp(\nu)(\psi),\end{aligned}$$

which means that if Λ has a generator h_0, then we have a Heisenberg relation,

$$\Delta E \times \Delta t \geq h := \exp(h_0),$$

where $\exp(\epsilon) = \Delta E$, $\exp(\nu) = \Delta t$. Compare with Laudal (2011, (4.4)), where we consider the singular situation, corresponding to $\delta = \mathrm{ad}(Q)$.

4.2 Gauge Groups and Invariant Theory

We may use the above in an attempt to make precise the notion of *gauge group*, gauge fields, and gauge invariance, and thus to be able to understand why the physicists define their objects, the *fields* and *particles*, the way they do.

Suppose, in line with our philosophy, that we have uncovered the moduli space $\mathbf{M}$ of the mathematical model X, of our phenomena $\mathbf{P}$, and that A is the *affine k-algebra* of (an affine open subset of) this space, assumed to contain all the parameters of our interest of the states of X.

4.2.1 *The global gauge group and invariant theory*

Suppose, furthermore, that we have identified a k-Lie algebra $\mathfrak{g}_0 \subset \mathrm{Der}_k(A)$ of infinitesimal automorphisms, i.e. of derivations of A, a *global gauge group*, leaving invariant the physical properties of our phenomena $\mathbf{P}$. We would then be led to consider the *quotient space* $\mathbf{M}/\mathfrak{g}_0$, which in our non-commutative algebraic geometry is equivalent to restricting our representations, $\rho : A \to \mathrm{End}_k(V)$, to those representations (ρ, V) for which

$$\mathfrak{g}_0 \subset \mathfrak{g}_V = \mathfrak{g}_\rho = \{\gamma \in \mathrm{Der}_k(A) |\ ks(\gamma) = \kappa(\delta\rho) = 0\},$$

see Section 2.1, Definition 2.1, and Lemma 2.1. This would then imply that the corresponding *Hamiltonians*, Q_γ, define a $\mathfrak{g}_0$-connection on V,

$$Q : \mathfrak{g}_0 \longrightarrow \mathrm{End}_k(V),$$

such that for all $a \in A$, and for all $\xi \in \mathfrak{g}_0$, $\rho(\xi(a)) = [Q_\xi, \rho(a)]$. This is usually written as,

$$\rho(\xi(a)) = [\xi, \rho(a)].$$

The curvature, i.e. the obstruction for Q to be a Lie algebra homomorphism,

$$R(\xi_1, \xi_2) := [Q_{\xi_1}, Q_{\xi_2}] - Q_{[\xi_1, \xi_2]} \in \mathrm{End}_A(V),$$

corresponds to a *global force* acting on the representation ρ. These forces, *mediated* by the *gauge-particles*, $\xi \in \mathfrak{g}_0$, will be the first to be studied in some detail, see the next sub-section. Recall also from Section 2.1.1, the definition of the map,

$$\begin{aligned}\kappa : \mathrm{Der}_k(A, \mathrm{Hom}_k(U,V)) &\to \mathrm{Der}_k(A, \mathrm{Hom}_k(U,V))/\,\mathrm{Triv}\\ &= \mathrm{Ext}^1_A(U,V),\end{aligned}$$

and put

$$\begin{aligned}\mathrm{Rep}(A, \mathfrak{g}_0) :&= \{\rho \in \mathrm{Rep}(A) | \kappa(\xi\rho) = 0, \forall \xi \in \mathfrak{g}_0\}\\ &= \{\rho \in \mathrm{Rep}(A) | \mathfrak{g}_0 \subset \mathfrak{g}_\rho\},\end{aligned}$$

where $\mathrm{Rep}(A)$ is the category of all representations of A, and note that, in the commutative situation, if we consider the case where the gauge group $\mathfrak{g}_0 = \mathrm{Der}_k(A)$, then $\mathrm{Rep}(A, \mathfrak{g}_0)$ is the category of *A-Connections*, for which the space of isomorphism classes is discrete with respect to time. Note that this is also the situation in classical quantum theory, where the *Hilbert Space* is always considered as the unique state space of interest.

Definition 4.2. An object $V \in \mathrm{Rep}(A, \mathfrak{g}_0)$ is called simple if there are no non-trivial sub-objects of V in $\mathrm{Rep}(A, \mathfrak{g}_0)$. The generalized quotient $\mathrm{Simp}(A)/\mathfrak{g}_0$, is by definition, the set, $\mathrm{Simp}(A, \mathfrak{g}_0)$, of iso-classes of simple objects in $\mathrm{Rep}(A, \mathfrak{g}_0)$.

If the curvature also vanishes, there is a canonical homomorphism,

$$\phi : U(\mathfrak{g}_0) \to \mathrm{End}_k(V),$$

where $U(\mathfrak{g}_0)$ is the universal algebra of the Lie algebra $\mathfrak{g}_0$.

In the general case, let

$$A'(\mathfrak{g}_0) \subset \mathrm{End}_k(A)$$

be the sub-algebra generated by A and $\mathfrak{g}_0$. Then we put, for all $a \in A, \xi \in \mathfrak{g}_0$,

$$A(\mathfrak{g}_0) = A'(\mathfrak{g}_0)/(\xi a - a\xi - \xi(a)),$$

and we have an identification between the set of $\mathfrak{g}_0$-connections on V, and the set of k-algebra homomorphisms,

$$\rho_{\mathfrak{g}} : A(\mathfrak{g}_0) \to \mathrm{End}_k(V),$$

since any such would respect the relation above, such that, for $a \in A$, $\xi \in \mathfrak{g}_0$,

$$\rho_{\mathfrak{g}_0}(\xi a) = \rho_{\mathfrak{g}_0}(\xi)\rho_{\mathfrak{g}}(a) = \rho_{\mathfrak{g}_0}(a)\rho_{\mathfrak{g}_0}(\xi) + \rho_{\mathfrak{g}_0}(\xi(a)).$$

Therefore, $\mathrm{Rep}(A)/\mathfrak{g}_0 := \mathrm{Rep}(A, \mathfrak{g}_0) \simeq \mathrm{Rep}(A(\mathfrak{g}_0))$, and we note, for memory, the trivial.

Lemma 4.1. *In the above situation, we have the following isomorphisms:*

$$\mathrm{Rep}(A)/\mathfrak{g}_0 := \mathrm{Rep}(A, \mathfrak{g}_0) \simeq \mathrm{Rep}(A(\mathfrak{g}_0)),$$
$$\mathrm{Simp}(A)/\mathfrak{g}_0 := \mathrm{Simp}(A, \mathfrak{g}_0) \simeq \mathrm{Simp}(A(\mathfrak{g}_0)).$$

Note that the commutant in $A(\mathfrak{g}_0)$, of A and $\mathfrak{g}_0$, is the sub-ring

$$A^{\mathfrak{g}_0} := \{a \in A | \forall \xi \in \mathfrak{g}_0, \xi(a) = 0\} \subset A.$$

Note also that the commutativization, $A(\mathfrak{g}_0)^{\mathrm{com}}$, of $A(\mathfrak{g}_0)$ is the quotient of $A(\mathfrak{g}_0)$ by an ideal containing $\{\xi(a) | a \in A, \xi \in \mathfrak{g}_0\}$. Therefore, there is a natural map

$$A^{\mathfrak{g}_0} \to A(\mathfrak{g}_0)^{\mathrm{com}}.$$

However, this map may not be injective, so we cannot, in general, identify the rank 1 points of $\mathrm{Simp}(A, \mathfrak{g}_0)$, with $\mathrm{Simp}_1(A^{\mathfrak{g}_0})$.

If the k-algebra C is assumed commutative, the classical invariant theory identifies the two schemes, $\mathrm{Spec}(C)/\mathfrak{g}_0$ and $\mathrm{Spec}(C^{\mathfrak{g}_0})$, which in the above light, is not entirely kosher. However, if $\mathfrak{a} \subset C$ is an ideal, stable under the action of $\mathfrak{g}_0$, then since any derivation γ of C acts on the multiplicative operators $a \in C$ as $\gamma(c) = \gamma c - c\gamma$, it is clear that the quotient $C/\mathfrak{a}$ is a $C(\mathfrak{g}_0)$-representation. Moreover, as representation of C, we have

$$C/\mathfrak{a} \in \mathrm{Simp}_1(C)/\mathfrak{g}_0$$

if and only if the subset $\mathrm{Simp}_1(C/\mathfrak{a}) \subset \mathrm{Simp}_1(C)$ is the closure of a *maximal integral sub-variety* for $\mathfrak{g}_0$. The *space* of such integral sub-varieties is what we, in Laudal (2003), have termed the *non-commutative quotient*, $\mathrm{Spec}(C)/\mathfrak{g}_0$.

4.2.2 *The local gauge group*

Suppose now, in the general case, that there is an A-Lie algebra $\mathfrak{g}_1$, acting A-linearly on those A-modules V, which we would consider of *physical* interest. $\mathfrak{g}_1$ should be called a *local gauge group*. One may then want to know whether the given action of $\mathfrak{g}_0$ moves $\mathfrak{g}_1$ in its formal moduli as A-Lie algebra. If so, the action of $\mathfrak{g}_1$ would not be invariant under the gauge transformations induced by $\mathfrak{g}_0$, and we should not consider $(\rho, \mathfrak{g}_1)$ as physically kosher. If, on the other hand, the action of $\mathfrak{g}_0$ does not move $\mathfrak{g}_1$ in its formal moduli, it should follow that there is a relation between the $\mathfrak{g}_0$-action (i.e. the connection) on V, and the action of $\mathfrak{g}_1$. Now, since $\mathfrak{g}_0 \subset \mathrm{Der}_k(A)$, it follows from the Kodaira–Spencer map,

$$\mathfrak{ks} : \mathrm{Der}_k(A) \to A^1(A, \mathfrak{g}_1 : \mathfrak{g}_1),$$

see Bjar and Laudal (1990, Lemma 2.3), that we have the following result.

Lemma 4.2. *Let* $c_{i,j}^k \in A$ *be the structural constants of* $\mathfrak{g}_1$ *with respect to some* A*-basis* $\{x_i\}$, *and let* $\pi : \mathfrak{F} \to \mathfrak{g}_1$ *be a surjective morphism of a free* A*-Lie algebra* $\mathfrak{F}$ *onto* $\mathfrak{g}_1$, *mapping the generators* $\mathfrak{x}_i$ *of* $\mathfrak{F}$ *onto* x_i. *Let* $\mathfrak{F}_{i,j} = [\mathfrak{x}_i, \mathfrak{x}_j] - \sum_k c_{i,j}^k \mathfrak{x}_k \in \ker(\pi)$, *and let* $\xi \in \mathrm{Der}_k(A)$. *Then,* $\mathfrak{ks}(\xi)$ *is the element of* $A^1(A, \mathfrak{g}_1 : \mathfrak{g}_1)$ *determined by* *the element of* $\mathrm{Hom}_{\mathfrak{F}}(\ker(\pi), \mathfrak{g}_1)$, *given by the map*

$$\mathfrak{F}_{i,j} \to -\sum_k \xi(c_{i,j}^k)x_k.$$

For $\mathfrak{ks}(\xi)$ *to be 0, there must exist a derivation* $D_\xi \in \mathrm{Der}_A(\mathfrak{F}, \mathfrak{g}_1)$, *such that*

$$D_\xi(\mathfrak{F}_{i,j}) = D_\xi\left([\mathfrak{x}_i, \mathfrak{x}_j] - \sum_k c_{i,j}^k \mathfrak{x}_k\right) = -\sum_k \xi(c_{i,j}^k)x_k.$$

Consider this result for all $\xi \in \mathfrak{g}_0$, then we find.

Theorem 4.3. *The action of* $\mathfrak{g}_0$ *does not move* $\mathfrak{g}_1$ *in its formal moduli, if and only if there exists a connection,*

$$\mathfrak{D} : \mathfrak{g}_0 \to \mathrm{Der}_k(\mathfrak{g}_1),$$

i.e. a map such that, for $\xi \in \mathfrak{g}_0,\ a \in A,\ \gamma \in \mathfrak{g}_1$,

$$\mathfrak{D}_\xi(a \cdot \gamma) = a\mathfrak{D}_\xi(\gamma) + \xi(a) \cdot \gamma.$$

Proof. Assume the action of $\mathfrak{g}_0$ does not move $\mathfrak{g}_1$, then there exists by the lemma an A-derivation D such that $D_\xi(\mathfrak{F}_{i,j}) = D_\xi([\mathfrak{x}_i, \mathfrak{x}_j] - \sum_k c_{i,j}^k \mathfrak{x}_k) = -\sum_k \xi(c_{i,j}^k)x_k$. Define $\mathfrak{D}_\xi$ by its actions on the base elements x_i, by

$$\mathfrak{D}_\xi(x_i) = D_{-\xi}(\mathfrak{r}_i).$$

Then, since D_ξ is a π-derivation, we find

$$D_\xi([\mathfrak{r}_i, \mathfrak{r}_j]) = [D_\xi(\mathfrak{r}_i), x_j] + [x_j, D_\xi(\mathfrak{r}_j)] = \sum_k c_{i,j}^k D_\xi(\mathfrak{r}_k) - \sum_k \xi(c_{i,j}^k)x_k$$

from which it follows that

$$\begin{aligned}\mathfrak{D}_\xi([x_i, x_j]) = \mathfrak{D}_\xi\left(\sum_k c_{i,j}^k x_k\right) &= \sum_k c_{i,j}^k \mathfrak{D}_\xi(x_k) + \sum_k \xi(c_{i,j}^k)x_k \\ &= [\mathfrak{D}_\xi(x_i), x_j] + [x_i, \mathfrak{D}_\xi(x_j)],\end{aligned}$$

which means that $\mathfrak{D}_\xi$ is a connection. Inversely, if the connection $\mathfrak{D}$ exists, let D be defined by the values $D_\xi(\mathfrak{r}_i) := \mathfrak{D}_{-\xi}(x_i)$. $\square$

If the curvature

$$R(\xi_1, \xi_2) := [\mathfrak{D}_{\xi_1}, \mathfrak{D}_{\xi_2}] - \mathfrak{D}_{[\xi_1, \xi_2]},$$

representing the second-order action of $\mathfrak{g}_0$ on $\mathfrak{g}_1$, vanishes, the map

$$\mathfrak{D} : \mathfrak{g}_0 \to \mathrm{Der}_k(\mathfrak{g}_1),$$

is a Lie-algebra morphism, and $\mathfrak{D}$ defines a Lie algebra structure on the sum

$$\mathfrak{g} := \mathfrak{g}_0 \oplus \mathfrak{g}_1.$$

The Lie product of the sum is defined as the product in each Lie algebra, with the cross-products defined for $\xi \in \mathfrak{g}_0, \gamma \in \mathfrak{g}_1$ as

$$[\xi, \gamma] = \mathfrak{D}_\xi(\gamma).$$

The situation above comes up when we have chosen a dynamical structure σ for A, with Dirac derivation δ. Assume that there exists, as above, a global gauge group $\mathfrak{g}_0 \subset \mathrm{Der}_k(A)$, and suppose moreover that there is an A-Lie algebra $\mathfrak{g}_1$ that acts as a local gauge group on the A-module V. Then we would be lead to consider the quotient of A by both $\mathfrak{g}_0$ and $\mathfrak{g}_1$, i.e. the representations

$$\rho : A \to \mathrm{End}_k(V),$$

for which there exists a diagram

$$\begin{array}{ccc} \mathfrak{g}_0 & \xrightarrow{\mathfrak{D}} \mathrm{Der}_k(\mathfrak{g}_1) \longleftarrow & \mathfrak{g}_1 \\ \downarrow \nabla & & \downarrow \nabla_1 \\ \mathrm{End}_k(V) & \longleftarrow & \mathrm{End}_A(V) \end{array}$$

where $\mathfrak{D}$, which we might call a *generalized spin structure*, and ∇ are connections, and ∇_1 is the action of $\mathfrak{g}_1$ on V. If $\mathfrak{D} = \mathrm{ad}(\nabla)$ has vanishing curvature, then there is a connection of the Lie algebroid $\mathfrak{g}$,

$$\nabla_2 : \mathfrak{g} \to \mathrm{End}_k(V).$$

The category of representations $\rho : A \to \mathrm{End}_k(V)$ with this property vis a vis the Lie algebra $\mathfrak{g}$, and simple as such, will be written as,

$$\mathrm{Simp}(A)/\mathfrak{g}.$$

According to our philosophy, this should be the object of study in mathematical physics. See Section 10.5 *for a possible relation to the notion of bi-cross products of Hadfield and Majid* (2007).

Note, for later use, that if $A = k[t_1, \dots, t_d]$ is a polynomial algebra and

$$\rho : A \to \mathrm{End}_k(V)$$

is an object of $\mathrm{Simp}(A)/\mathfrak{g}$, with an extension to

$$\rho_\xi : \mathrm{Ph}(A) \to \mathrm{End}_A(V)$$

corresponding to a derivation $\xi \in \mathrm{Der}_k(A, \mathrm{End}_k(V))$, then, since in $\mathrm{Ph}(A)$ we have the relations $[dt_i, t_j] = [dt_j, t_i]$, we must have

$$\rho_\xi(dt_i) \in \mathrm{End}_A(V), \rho_\xi([dt_i, t_j]) =: g_\xi^{i,j} = g_\xi^{j,i} \in \mathrm{End}_k(V).$$

Any element $\xi \in \mathfrak{g}_0$ composed with ρ defines a morphism,

$$\rho_\xi : \mathrm{Ph}(A) \to \mathrm{End}_k(V)$$

with $\rho_\xi(dt_i) = \xi(t_i) \in A$, and so

$$g_\xi^{i,j} = [\rho_\xi(dt_i), t_j] = 0$$

for all $\xi \in \mathfrak{g}_0$.

Note also that physicists have a way of classifying or naming *states*, i.e. the elements of the representation vector space V, according to certain numbers associated to them, like spin, charge, hyperspin, etc. We find this in the situation above, as follows.

Consider the Cartan sub-algebra $\mathfrak{h} \subset \mathfrak{g}_1$. It will operate on the above representation space V as diagonal matrices, and the eigenvectors may be labeled by the corresponding eigenvalues.

Note also that if $V_i \in \mathrm{Rep}(A, \mathfrak{g}), i = 1, 2$, then it follows that an extension of the $A(\mathfrak{g})$-module V_1 with V_2 will also sit in $\mathrm{Rep}(A, \mathfrak{g})$.

Note, finally that, given an action of the Lie algebra $\mathfrak{g}_0$ on A, since Ph( ) is a functor in the category of algebras and algebra morphisms, the action of $\mathfrak{g}_0$ extends to $\mathrm{Ph}(A)$, but not necessarily to a dynamical system of the type $A(\sigma)$.

This will turn out to be important for our version of the Standard Model. There we will also meet the following extension of the situation above.

4.3 The Generic Dynamical Structures Associated to a Metric

It turns out that the geometry of algebraic schemes, commutative or not, is best described by introducing some kind of metrics, i.e. local length functions. These should be related to the universal Dirac derivation, and give rise to dynamical structures with which we may work along the lines of Chapter 3. The first, and simplest, case is the case of a classical affine scheme, and we shall in this section reduce to the smooth case, and assume that A is a commutative polynomial k-algebra.

4.3.1 *The commutative case, metrics, and gravitation*

Now, let $k = \mathbb{R}$ be the real numbers, and consider a commutative polynomial k-algebra, $C = k[t_1, \ldots, t_d]$. Moreover, let

$$g = 1/2 \sum_{i,j=1,..,r} g_{i,j} dt_i dt_j \in \mathrm{Ph}(C),$$

correspond to a (non-degenerate) Riemannian metric, i.e. such that $g_{i,j} = g_{j,i}$. The determinant $\mathfrak{m}$, of the matrix $(g_{i,j})$, is non-zero on an affine open subspace of $\mathbf{A}^d := \mathrm{Spec}(C)$, and there we put

$$(g^{i,j}) := (g_{i,j})^{-1}.$$

In the sequel, we shall consider the localization, $C \to C_{\mathfrak{m}}$ of C, and the diagram of canonical morphisms,

$$\begin{array}{ccc} C & \longrightarrow & C_{\mathfrak{m}} \\ \downarrow & & \downarrow \\ \mathrm{Ph}(C) & \longrightarrow & \mathrm{Ph}(C_{\mathfrak{m}}), \end{array}$$

but work with C instead of $C_{\mathfrak{m}}$, making sure that every construction made for $\mathrm{Ph}^*(C)$ goes through for $\mathrm{Ph}^*(C_{\mathfrak{m}})$. In particular, all representations $\rho : C \to \mathrm{End}_k(V)$ are supposed to be extendible to $C_{\mathfrak{m}}$, which simply means that $\rho(\mathfrak{m})$ is invertible and $V = V_{\mathfrak{m}}$.

Recall that the Levi–Civita connection,

$$\nabla : \Theta_C \to \mathrm{End}_k(\Theta_C)$$

is defined to be killing the metric, i.e. it satisfies the equation

$$\nabla(g) = 0.$$

Moreover, it is without torsion, i.e.

$$\nabla_\xi(\kappa) - \nabla_\kappa(\xi) = [\xi, \kappa], \forall \xi, \kappa \in \Theta_C.$$

Recall also that, for $f \in C$,

$$\nabla_\xi(f\kappa) = \xi(f)\kappa + fD_\xi(\kappa),$$

where $D_\xi \in \mathrm{End}_C(\Theta_C)$ is given in terms of the Christoffel symbols Γ, defined by

$$\Gamma^k_{i,j} = 1/2 \sum_l g^{k,l}(\delta_{t_j} g_{l,i} + \delta_{t_i} g_{l,j} - \delta_{t_k} g_{i,j}),$$

as

$$D_{\delta_i}(\delta_j) = \sum_k \Gamma^k_{i,j}\delta_k,$$

where we have put $\delta_i := \delta_{t_i}$. The dual action of D,

$$D' : \Theta_C \to \mathrm{End}_k(\Omega_C),$$

normally just called D, comes out as

$$D_{\delta_i}(dt_j) = -\sum_k \Gamma^j_{i,k} dt_k.$$

A short computation then proves that, for all $\xi \in \Theta_C$, we have

$$\nabla_\xi(g) = \sum_{i,j} \nabla_\xi(g_{i,j}dt_i dt_j) = \sum_{i,j} \xi(g_{i,j})dt_i dt_j - \sum_{i,j,k,l} g_{l,j}\Gamma^l_{k,i}\xi_k dt_i dt_j$$
$$- \sum_{i,j,k,l} g_{i,l}\Gamma^l_{k,j}\xi_k dt_i dt_j.$$

Now, in $\mathrm{Ph}^\infty(C)$, we have the formula

$$\delta(g) = \sum_{i,j,k=1,..,r} \delta_{t_k}(g_{i,j})dt_k dt_i dt_j + \sum_{i,j,=1,..,r} g_{i,j}(\delta^2 t_i dt_j + dt_i \delta^2 t_j),$$

in which we may plug in the formula

$$\delta^2 t_l = -\Gamma^l := -\sum \Gamma^l_{i,j} dt_i dt_j$$

on the right-hand side, and see that we, in the *commutative situation*, i.e. for the dynamical structure generated by

$$(\sigma_0) := \{[dt_i, t_j], [dt_i, dt_j] | t_i, t_j \in C\},$$

have got the *Lagrange equation*

$$\delta(g) = 0.$$

We shall consider it as a *force law*,

$$d^2 t_l = -\Gamma^l := -\sum \Gamma^l_{i,j} dt_i dt_j,$$

of the dynamical structure (σ_0). The corresponding dynamical system is then, of course, the commutative algebra,

$$C(\sigma_0) = k[\underline{t}, \underline{\xi}],$$

where ξ_j is the class of dt_j. The Dirac derivation now takes the form

$$[\delta] = \sum_l (\xi_l \delta_{t_l} - \Gamma^l \delta_{\xi_l}),$$

coinciding with the fundamental vector field $[\delta]$ in $\mathrm{Simp}_1(C(\sigma_0)) = Spec(k[t_i, \xi_j])$.

The equation

$$[\delta](g) = 0$$

implies that g is constant along the integral curves of $[\delta]$ in $\mathrm{Simp}_1(\mathrm{Ph}(C))$, and these integral curves project into $\mathrm{Simp}_1(C)$ to give the geodesics of the metric g, with the equations

$$\ddot{t}_l = -\sum_{i,j} \Gamma^l_{i,j} \dot{t}_i \dot{t}_j.$$

For a general treatment of Force Laws, see Theorem 4.5. But we may also look at this from another point of view. Suppose given any dynamical structure with Dirac derivation δ on $\mathrm{Ph}(C)$. Consider $\mathrm{Simp}_1(\mathrm{Ph}(C))$. It is obviously represented by $C(1) := k[\underline{t}, \underline{\xi}]$, and the Dirac derivation induces a derivation $[\delta] \in \mathrm{Der}_k(C(1))$, and the Hamiltonian must vanish. Therefore, we have two options for *the same notion of time* in the picture, g and $[\delta]$. The last derivation must therefore be a *Killing vector field*, i.e. we must have a solution for the Lagrange equation,

$$[\delta](g) = 0,$$

and we are left with the above solution for δ.

Since the metric is related to the gravitational force, the group of isometries, $O(g)$, of the metric g, i.e. the group of algebraic automorphisms of C leaving the metric g invariant would, in line with our philosophy, be an obvious global gauge group. We shall refer to its Lie algebra as $\mathfrak{o}(g)$. Since $\mathrm{Ph}(*)$ is a functor, $O(g)$ would also act on $\mathrm{Ph}(C)$, and would induce an action of $\mathfrak{o}(g)$ on $\mathrm{Ph}(C)$, and so also on $\mathrm{Ph}^\infty(C)$.

4.3.2 *The Lie algebra of isometries*

Consider the metric

$$g = 1/2\sum_{i=1}^{d} g_{i,j}dt_i dt_j \in \mathrm{Ph}(C),$$

and a derivation $\eta = \sum_i \eta_i \delta_{t_i} \in \mathrm{Der}_k(C)$ acting on C, and so on $\mathrm{Der}_k(C)$, by the Lie product, and by functoriality on $\mathrm{Ph}(C)$. In particular, η acts on Ω_C such that $\eta(dt_i)$ is defined by

$$\eta(dt_j)(\delta_k)) = dt_j([\eta, \delta_k]) = -\delta_k(\eta_j).$$

The Lie algebra of *Killing vectors* is the Lie algebra of derivations,

$$\mathfrak{o}(g) = \{\eta \in \mathrm{Der}_k(C) | \eta(g) = 0\},$$

where the equation $\eta(g) = 0$ is equivalent to

$$\sum_{i,j} \eta(g_{i,j})dt_i dt_j - \sum_{i,j,k} g_{i,j}\delta_{t_k}(\eta_i)dt_k dt_j - \sum_{i,j,k} g_{i,j}\delta_{t_k}(\eta_j)dt_i dt_k = 0,$$

implying, for all $i, j = 1, \ldots, d$,

$$\eta(g_{i,j}) - \sum_k \delta_{t_i}(\eta_k)g_{k,j} - \sum_k \delta_{t_j}(\eta_k)g_{i,k} = 0.$$

Compare this with the formulas defining the Levi-Civita connection,

$$D : \Theta_C \to \mathrm{End}_k(\Theta_C),$$

see above,

$$\sum_{i,j} D_\eta(g_{i,j}dt_i dt_j) = \sum_{i,j} \eta(g_{i,j})dt_i dt_j + \sum_{i,j,k,l} g_{l,j}\Gamma^l_{k,i}\eta_k dt_i dt_j$$
$$+ \sum_{i,j,k,l} g_{i,l}\Gamma^l_{k,j}\eta_k dt_i dt_j = 0$$

and see that in this case the condition for $\eta \in \mathfrak{o}(g)$ is the usual

$$g(D_{\delta_i}(\eta), \delta_j) + g(D_{\delta_j}(\eta), \delta_i) = 0.$$

There are two fundamental examples, the Euclidean and the Minkowski metrics. First, suppose all $g_{i,j}$ are constants, and we are interested in the linear derivations. We obtain that the derivations are given in terms of matrices, $(\gamma_{i,j}) := (\delta_{t_i})(\gamma_j)$, where $\gamma_{i,j}g_{j,j} = -\gamma_{j,i}g_{i,i}$. This gives in dimension 2, for the Euclidean, respectively, for the Minkowski metric,

$$(\gamma_{i,j}) = \begin{pmatrix} 0 & 1 \\ -1 & 0 \end{pmatrix}, \quad (\gamma_{i,j}) = \begin{pmatrix} 0 & 1 \\ 1 & 0 \end{pmatrix}.$$

The corresponding one-dimensional (rotation) Lie groups acting on C, with coordinates (t_1, t_2), are given by the exponential

$$O(g) = \exp\left(\tau \begin{pmatrix} 0 & 1 \\ -1 & 0 \end{pmatrix}\right) = \begin{pmatrix} \cos(\tau) & \sin(\tau) \\ -\sin(\tau) & \cos(\tau) \end{pmatrix},$$

respectively,

$$O(g) = \exp\left(\tau \begin{pmatrix} 0 & 1 \\ 1 & 0 \end{pmatrix}\right) = \begin{pmatrix} \cosh(\tau) & \sinh(\tau) \\ \sinh(\tau) & \cosh(\tau) \end{pmatrix},$$

as we know.

4.4 Metrics, Gravitation, and Energy

In the commutative case (and for the corresponding one-dimensional representation) treated above, the Hamiltonian was trivial, and the notion of time was taken care of by a vector field $[\delta]$. Let us now consider representations ρ, for which the Dirac derivation $[\delta]$ of Theorem 4.5 vanish, and the notion of time is taken care of by the Hamiltonian Q.

This is accomplished by introducing another dynamical structure related to the metric g. Note first that a non-degenerate metric, $g = 1/2 \sum_{i,j=1}^{d} g_{i,j} dt_i dt_j \in \mathrm{Ph}(C)$, induces a duality, i.e. an isomorphism of C-modules

$$\Theta_C = \mathrm{Hom}_C(\Omega_C, C) \simeq \Omega_C \simeq \mathrm{Hom}_C(\Theta_C, C),$$

such that

$$\delta_i(dt_j) = g_{i,j}, dt_i(\delta_j) = g^{i,j}.$$

Recall the relations $[dt_i, t_j] = [dt_j, t_i]$ in $\mathrm{Ph}(C)$, and consider the bilateral ideal (σ_g) of $\mathrm{Ph}(C)$ generated by

$$(\sigma_g) = ([dt_i, t_j] - g^{i,j}),$$

and put

$$C(\sigma_g) := \mathrm{Ph}(C)/(\sigma_g).$$

Let, moreover,

$$T := \sum_j T_j dt_j = -1/2 \sum_{i,j,l} \delta_{t_l}(g_{i,j}) g^{l,i} dt_j = 1/2 \sum_{i,j,l} \delta_l(g^{l,i}) g_{i,j} dt_j.$$

An easy computation shows that

$$T_l = -1/2 \left(\sum_k \Gamma_{k,l}^k + \sum_{k,p,q} g^{k,q} \Gamma_{k,q}^p g_{p,l} \right) = -1/2 \left(\sum_j \Gamma_{j,l}^j + \bar{\Gamma}_{j,l}^j \right),$$

where $\bar{\Gamma}_{j,l}^j := \sum_{p,q} g^{j,q} \Gamma_{j,q}^p g_{p,l}$.

Consider the inner derivation of $C(\sigma_g)$, defined by

$$\delta := \mathrm{ad}(Q),\ \ Q = g - T.$$

After a dull computation, we obtain, in $C(\sigma_g)$,

$$\delta(t_i) = [Q, t_i] = dt_i,\ \ \delta^2(t_i) = [Q, dt_i], \quad i = 1, \ldots, d.$$

Therefore, by universality, we have a well-defined dynamical structure (σ_g), with Dirac derivation, $\delta = \mathrm{ad}(g - T)$.

It is clear that if $\phi : C \to C'$ is an isomorphism of k-algebras, and $g' = \phi(g)$, then $C(\sigma_g)$ is isomorphic to $C'(\sigma_{g'})$, so the construction is *covariant* in the language of physicists. In particular, (σ_g) is invariant w.r.t. isometries, implying that $\mathfrak{o}(g)$ is a sub-Lie algebra of the global gauge group $\mathfrak{g}_0$ of $C(\sigma_g)$.

Any representation

$$\rho : C(\sigma_g) \to \mathrm{End}_k(V),$$

extending a representation

$$\rho_0 : C \to \mathrm{End}_k(V),$$

and defining a representation

$$\rho_1 : \mathrm{Ph}(C) \to \mathrm{End}_k(V),$$

is defined by

$$\rho(t_i) = \rho_0(t_i),\ \ \rho(dt_i) = [\rho_1(Q), \rho_0(t_i)] = \rho_1([Q, t_i]),$$

and since $\rho([dt_i, t_j]) = [\rho(dt_i), t_j] = g^{i,j}$, we find, putting

$$\delta_i := \delta_{t_i},\ \ \xi_i := \sum_{j=1}^{d} g^{i,j}\delta_j,$$

that

$$\rho(dt_i) = \nabla_{\xi_i}$$

defines a connection,

$$\nabla : \mathrm{Der}_k(C) =: \Theta_C \to \mathrm{End}_k(E),$$

since, for non-degenerated metrics, the ξ_i's generate $\mathrm{Der}_k(C)$ as a C-module.

Since any connection on a free C-module E is given as

$$\nabla_{\delta_{t_i}} = \delta_{t_i} + \psi_i,$$

where $\psi_i \in \mathrm{End}_C(E)$, we find in this case

$$\rho_1(dt_i) = \nabla_{\xi_i} = \xi + \psi_i,$$

for some (arbitrary) *potential* $\psi = \{\psi_i\} \in \mathrm{End}_C(V)^d$, see Section 4.5.

In the general case, the metric is non-degenerate in $C_{\mathfrak{m}}$, and the derivation $\{\xi_i, i = 1, \ldots, d\}$ forms a C-basis for $\mathrm{Der}_k(C_{\mathfrak{m}})$, and therefore any representation $\rho : C(\sigma_g) \to \mathrm{End}_k(V)$ induces a C-connection,

$$\nabla : \mathrm{Der}_k(C_{\mathfrak{m}}) \to \mathrm{End}_k(V).$$

Fixing one, then any other connection is given by

$$\nabla'_{\xi_i} = \nabla_{\xi_i} + \psi_i,$$

Remark 4.1 (Notations). From now on we shall use the notations

$$\nabla_{\delta_i},\ \nabla_{\xi_i},$$

when there exists a C-basis $\{v_l\}$ of V for which $\rho(dt_i)(v_l) = 0$ for all v_l.

Any other morphism $\rho' : C(\sigma_g) \to \mathrm{End}_k(V)$ extending ρ_0 may also be defined by

$$\rho'(dt_i) = \rho(dt_i) + \psi_i,$$

for some potential $\psi \in \mathrm{End}_C(V)^d$. Put $\rho_\psi := \rho' = \rho + \psi$. Since the derivation $\eta \in \mathrm{Der}_k(C, \mathrm{End}_k(V))$ induced by ρ_ψ is given as $\eta(t_i) = [\rho_\psi(g - T), t_i]$. Put $Q_\eta := Q_\psi = \rho_\psi(g - T)$, then it follows from Section 2.1.1, that the image of η in $\mathrm{Ext}^1_C(V, V)$ must be 0. Now,

$$Q_\psi = \rho(g - T) + [\psi] + 1/2 g_{i,j}\psi_i\psi_j + 1/2 \left(\sum_l \Gamma^j_{j,l} + \bar{\Gamma}^j_{j,l} \right) \psi_l,$$

where

$$[\psi] = \sum_i \psi_i \nabla_{\delta_{t_i}} =: \sum_i \psi_i \delta_{t_i}.$$

Note that if we consider, for a given ψ, the representation

$$\psi : \mathrm{Ph}(C) \to \mathrm{End}_C(V),$$

given by $\psi(t_i) = t_i, \psi(dt_i) = \psi_i$, the formula above would look like

$$Q_\psi = \rho(g - T) + [\psi] + \psi(g - T).$$

We may consider this as an action in the state space V, a sum of a *horizontal*, a *vertical*, and between them a *mixed* component, $[\psi]$. Denote the *horizontal* and the *vertical* terms,

$$Q_h := Q = \rho(g - T), \; Q_v := \psi(g - T).$$

We obtain the following formula for the *time development* of ρ_ψ, i.e. for the action $[Q] := [Q, -]$, of the Dirac derivation δ on $\mathrm{End}_k(V)$,

$$[Q] := \mathrm{ad}(Q_h + [\psi] + Q_v), i.e. \; \rho_\psi(d^{n+1}t_i) = [Q_h + [\psi] + Q_v, \rho_\psi(d^n t_i)].$$

Compare with Lemma 2.1, and see that what we there termed $[\xi]$, replaced by δ here, vanishes. Therefore, we might be tempted to write

$$\delta = [Q].$$

However, the Dirac derivation acts on the algebra of observables, $C(\sigma_g)$, and our $[Q]$ acts trivially on the moduli space of representations, but as the time development in each representation. In Lemma 2.1, it was reasonable to write $\delta = [\delta] + [Q, -]$, since for geometric algebras $A(\sigma)$ that are determined by its simple finite-dimensional representations, one might expect $[\delta] + [Q, -]$ to determine δ. At the end of this section, and in the Scholie Theorem 4.7, we shall see that we may consider the algebras $C(\sigma_g)$ as fibers of a family of algebras parametrized by the metrics of C, then the distinctions between δ, $[\delta]$, and $[Q] = [Q, -]$ become more serious, and we shall therefore choose to reserve the notation δ for the derivation of the algebra of observables, and fuse $[\delta]$ and $[Q] = [Q, -]$.

The formula above then, finally, reads as

$$[\delta] := \mathrm{ad}(Q_h + [\psi] + Q_v),$$

meaning that for any $n \geq 1$, we have

$$\rho_\psi(d^{n+1}t_i) = [Q_h + [\psi] + Q_v, \rho_\psi(d^n t_i)].$$

Remark 4.2 (The global gauge group of $C(\sigma)$). Since any representation (ρ_1, V) of $C(\sigma_g)$ induces a C-connection on V, it is reasonable to accept $\mathfrak{g}_0 = \mathrm{Der}_k(C)$ as the global gauge group for $C(\sigma_g)$. We should therefore be interested in the invariant theory of $C(\sigma_g)$ modulo $\mathfrak{g}_0$. Now, $\mathrm{Der}_k(C)$, is generated by the derivations $\{\xi_i\}, i = 1, \ldots, d$, and clearly $\mathfrak{o}(g) \subset \mathfrak{g}_0$, we should therefore also try to express $\mathfrak{o}(g)$ in terms of the $\xi_i' s$.

Any Killing vector must have the form

$$\eta = \sum_i \alpha_i \xi_i.$$

Put this into the equation of Definition 2.2,

$$\eta(g_{i,j}) - \sum_k \delta_{t_i}(\eta_k) g_{k,j} - \sum_k \delta_{t_j}(\eta_k) g_{i,k} = 0,$$

and use the well-known formulas,

$$\delta_{t_l}(g_{i,k}) = \sum_p (g_{p,k} \Gamma^p_{i,l} + g_{i,p} \Gamma^p_{k,l}),$$

$$\delta_{t_q}(g^{r,m}) = -\sum_k (g^{r,k} \Gamma^m_{k,q} + g^{k,m} \Gamma^r_{k,q}).$$

We find that $\eta \in \mathfrak{o}(g)$ if and only if

$$\delta_{t_j}(\alpha_i) + \delta_{t_i}(\alpha_j) = 2 \sum_k \Gamma^k_{i,j} \alpha_k, \forall i, \quad j = 1, \ldots, d.$$

Moreover, the Lie structure is given by

$$[\xi_i, \xi_j] = \sum_k c^k_{!,j} \xi_k, \; c^k_{!,j} = \Gamma^{j,i}_k - \Gamma^{i,j}_k,$$

where we have put

$$\Gamma^{j,i}_p := \sum_k g^{j,k} \Gamma^i_{k,p}, \; \Gamma^{i,j}_p = \sum_k g^{i,k} \Gamma^j_{k,p}.$$

Note that the representation $\rho = \rho_{lc}$ of $C(\sigma_g)$, defined on Θ_C, by the Levi-Civita connection, has a Hamiltonian,

$$Q := \rho(g - T) = 1/2 \sum_{i,j} g^{ij} \nabla_{\delta_i} \nabla_{\delta_j},$$

i.e. the generalized Laplace–Beltrami operator, which is also invariant w.r.t. isometries, although the proof demands some algebra.

For the representation ρ_Θ, and for an element (a state) $\phi \in \Theta_C$, we would, in line with classical Quantum Theory, assume the dynamics given by the Schrödinger equation

$$\frac{d\phi}{d\tau} = Q(\phi),$$

where τ would be an ad hoc chosen time parameter. But again, see the discussion above, we have only one option for the notion of time, namely the metric g. We must therefore wait, until Chapter 10, for a relation to the classical Schrödinger equation.

We find a general equation of motion for the representations of $C(\sigma_g)$, formulated as in the already proved statement.

Theorem 4.4 (The generic equation of motion). *Assume the metric g is non-degenerate. Then we know that the derivations ξ_i generate $\mathrm{Der}_k(C)$. Put $\delta_i := \delta t_i$.*

Let $\rho_0 : C \to \mathrm{End}_k(V)$ be a representation, and let $\rho : C(\sigma_g) \to \mathrm{End}_k(V)$ be an extension, considered as a preparation, i.e. as fixing a momentum *of ρ. Put*

$$\rho(dt_i) = \nabla_{\xi_i} \in \mathrm{End}_C(V),$$

and let now ρ_ψ be the representation given by

$$\rho_\psi(dt_i) = \nabla_{\xi_i} + \psi_i, \;\; \psi_i \in \mathrm{End}_C(V).$$

The corresponding ρ_0-derivation,

$$\eta \in \mathrm{Der}_k(C, \mathrm{End}_k(V)),$$

maps to $0 \in \mathrm{Ext}^1_C(V, V)$ since we have

$$\eta(t_i) = \rho_\psi(dt_i) = [\rho_\psi(g - T), t_i] =: [Q_\eta, \rho_0(t_i)].$$

This implies that the time development in $\mathrm{End}_k(V)$ is given by the derivation

$$[\delta] = \mathrm{ad}(Q_h + [\psi] + Q_v),$$

where

$$Q_h := \rho(g - T) = Q := 1/2 \sum_{i,j} g^{i,j} \nabla_{\delta_i} \nabla_{\delta_j},$$

$$[\psi] := \sum_i \psi_i \nabla_{\delta_i},$$

$$Q_v := \psi(g - T) = 1/2 \sum_{i,j} g_{i,j} \psi_i \psi_j + 1/2 \left(\sum_{j,l} \Gamma^j_{j,l} + \bar{\Gamma}^j_{j,l} \right) \psi_l.$$

In particular,

$$\eta(t_i) = \rho_\psi(dt_i) = \rho(dt_i) + \psi_i.$$

Therefore, the time development induced in $\mathrm{End}_k(V)$ *by* ρ_ψ *is, infinitesimally, given by* $[\delta]$ *and the corresponding first-order time development in the state space is given by the operator*

$$Q_\psi = Q + [\psi] + Q_v : V \to V.$$

The total energies measured by the representation should be the eigenvalues of $[\delta]$ *and* Q_ψ.

This turns out to be a general version of the Dirac equation, and a result closely related to Lemma 2.1 that we shall return to frequently in the sequel. Note that the potentials ψ will pop up as the tangent directions in the space of connections **P**, see Section 4.5. The "classical" Dirac equation appears when we put $\psi_k = \sum_i \gamma_i g^{i,k}$ with $\gamma_i \in \mathfrak{g}_1$, and find

$$[\psi] = \sum_i \gamma_i \xi_i =: [\delta].$$

This choice of ψ_i and, in particular, the choice of the $\gamma_i' s$ will follow from Chapters 9 and 10.

Remark 4.3. Since

$$T = \sum_l T_l dt_l, \text{ with } T_l = -1/2 \left(\sum_j (\Gamma^j_{j,l} + \bar{\Gamma}^j_{j,l}) \right)$$
$$= -1/2(Tr\nabla_l + Tr\bar{\nabla}_l),$$

and since

$$Q = \rho(g - T) = 1/2 \sum g^{i,j} \nabla_{\delta_i} \nabla_{\delta_j},$$

we find that the purely GR-Schrödinger energy equation

$$Q(\phi) = E\phi,$$

takes the form, in C,

$$1/2 \sum g^{i,j} \nabla_{\delta_i} \nabla_{\delta_j}(\phi) = E\phi.$$

Picking out the energy E of "states" $\phi \in C$, we find solutions of the form

$$\phi = \exp(-\underline{e}.\underline{t}) \text{ where } \underline{e} := (e_1, \ldots, e_d), \underline{t} := (t_1, \ldots, t_d),$$

with energy given by

$$E = 1/2 \sum_{i,j} e_i g^{i,j} e_j.$$

Note that the energy E is $1/2$ of the square length of the momentum $\omega = \rho(\sum_l e_l dt_l) = \sum_l e_l \xi_l$, as one might have expected, since we classically have

$$E = 1/2m|\omega|^2,$$

where m=mass. Note also that the classical curvature in this case is

$$\rho(F_{i,j}) = [\rho(dt_i), \rho(dt_j)] - \sum_p (\Gamma^{j,i}_p - \Gamma^{i,j}_p)\rho(dt_p) = [\nabla_{\xi_i}, \nabla_{\xi_j}] - \nabla_{[\xi_i, \xi_j]},$$

and that we usually write

$$R_{i,j} := [dt_i, dt_j], \; F_{i,j} := R_{i,j} - \sum_p (\Gamma^{j,i}_p - \Gamma^{i,j}_p) dt_p$$

for the elements in $C(\sigma_g)$.

We shall, however, when it is clear what we are talking about, write $F_{i,j}$ for $\rho(F_{i,j})$.

Obviously, $F_{i,j}$ is the obstruction for ρ inducing a representation of the Lie algebra $\Theta_C = \mathrm{Der}_k(C)$.

Now, to be able to handle this time development, we need to know formulas, for $d^l t_i,\ l \geq 1,\ i = 1, \ldots, n$, in $C(\sigma_g)$. To this end, put

$$\bar{\nabla}_l := (\bar{\Gamma}^j_{i,l}),$$

$$\bar{\Gamma}^i_{p,q} := \sum_{l,r} g^{r,i}\Gamma^l_{r,p} g_{l,q}.$$

Computing, we find (see Laudal, 2011) for a proof.

Theorem 4.5 (The generic force laws). *In $C(\sigma_g)$, where the Dirac derivation δ is defined, we have the following force law, expressed in two different ways in* $\mathrm{Ph}(C)$,

$$(1)\ d^2t_i := \delta^2(t_i) = [g - T, dt_i] = -1/2\sum_{p,q}(\bar{\Gamma}^i_{p,q} + \bar{\Gamma}^i_{q,p})dt_p dt_q$$

$$+ 1/2\sum_{p,q} g_{p,q}(R_{p,i}dt_q + dt_p R_{q,i}) + [dt_i, T],$$

$$(2)\ d^2t_i = -\sum_{p,q}\Gamma^i_{p,q}dt_p dt_q - 1/2\sum_{p,q} g_{p,q}(F_{i,p}dt_q + dt_p F_{i,q})$$

$$+ 1/2\sum_{l,p,q} g_{p,q}[dt_p, (\Gamma^{i,q}_l - \Gamma^{q,i}_l)]dt_l + [dt_i, T],$$

$$= -\sum_{p,q}\Gamma^i_{p,q}dt_p dt_q - \sum_{p,q} g_{p,q}F_{i,p}dt_q + 1/2\sum_{p,q} g_{p,q}[F_{i,q}, dt_p]$$

$$+ 1/2\sum_{l,p,q} g_{p,q}[dt_p, (\Gamma^{i,q}_l - \Gamma^{q,i}_l)]dt_l + [dt_i, T].$$

Remark 4.4. We shall consider the above formulas as general Force Laws, in $\mathrm{Ph}(C)$, induced by the metric g. This means the following.

First, assume given a representation

$$\rho_0 : C \to \mathrm{End}_k(V),$$

and pick any tangent vector (momentum) of the formal moduli of the C-module V, i.e. an extension of ρ_0,

$$\rho_1 : \mathrm{Ph}(C) \to \mathrm{End}_k(V),$$

then, if ρ_1 can be extended to a representation

$$\rho_2 : \mathrm{Ph}^2(C) \to \mathrm{End}_k(V)$$

with $\rho(d_2(d_1 t_i)) = \rho_1(d^2 t_i)$ given by the formula of the Force Law, this means that the force law has induced a second-order momentum in the formal moduli space of the representation ρ_1, usually called $E \cdot \mathbf{a}(\rho_0)$, where E is the energy of the object in movement, and $\mathbf{a}$ is the acceleration, explaining the name Force Law.

We might also consider $(\mathfrak{c}_g)$, the δ-stable ideal generated by any one of these equation in $\mathrm{Ph}^\infty(C)$. Since the force laws above hold in the dynamical system defined by (σ_g), we obviously have $(\mathfrak{c}_g) \subset (\sigma_g)$, and we might hope these new dynamical systems might lead to new Quantum Field Theories, as defined above, with equally new and interesting properties.

One immediate result is that the restriction of the force laws (Section 4.3) to the commutative case reduces to the General Relativity, as we have seen above, since we find the same geodesics, see Section 4.3.1, and also Laudal (2011).

For a connection ∇, on a free C-module E, the second Force Law above will now take the form, in $\mathrm{End}_C(E)$,

$$\begin{aligned}\rho_E(d^2 t_i) + \sum_{p,q} \Gamma^i_{p,q} \nabla_{\xi_p} \nabla_{\xi_q} &= 1/2 \sum_p F_{p,i} \nabla_{\delta_p} + 1/2 \sum_p \nabla_{\delta_p} F_{p,i} \\ &\quad + 1/2 \sum_{l,q} \delta_q (\Gamma^{i,q}_l - \Gamma^{q,i}_l) \nabla_{\xi_l} + [\nabla_{\xi_i}, \rho_E(T)],\end{aligned}$$

where we, as above, have put $\rho(dt_i) = \nabla_{\xi_i} = \sum_j g^{i,j} \nabla_{\delta_j}$.

Note that considering the representation ρ_Θ, corresponding to the Levi-Civita connection, the above translates into,

$$\rho(dt_i) = [Q, t_i], \quad \rho(d^2 t_i) = \sum_{j=1}^{d} [Q, \rho(dt_i)],$$

where Q is the Laplace–Beltrami operator.

Given any *observable* $a \in \mathrm{Ph}(C)$, we would expect that the dynamics of the future *values* of a would be the spectrum of the operator,

$$f(\tau) := \exp(\tau \cdot \mathrm{ad}(Q))(a).$$

Collecting the above results, and definitions, we may, in the light of the general philosophy of this work, express what we have done as follows. Given the moduli space of "models," our "universe," assumed to be given as an affine space, $\underline{C} := \mathrm{Spec}(C)$, "time" is defined by a metric g. The "furniture," i.e. the non-gravitational material content in the universe, is identified with the category of representations $\mathrm{Rep}(C)$. The dynamical properties of any such representation ρ_0 is "controlled" by the possible extensions of ρ_0 to $\mathrm{Ph}^\infty(C)$, and the time operator, given by the Dirac derivation δ. The results of this sub-section fuse the two notions of time. This fusion is given by the generic dynamical structure, $C(\sigma_g)$, of the k-algebra C, induced by a metric g, the Dirac Derivation δ in $C(\sigma_g)$, the Time-evaluation, and Energy operator, $Q = g - T$. This we see is implying General Relativity and Quantum Theory, nicely related.

4.4.1 *The case of subspaces, spectral triples*

What happens if we have a subspace of a space with a given metric, and want to compare the two possible dynamical structures? This is a problem that has been central to the development both of general relativity and cosmology, and we shall take a look at the general problem, but stick to the case where the spaces are affine algebraic varieties, and, in fact, affine spaces, $\underline{C} := \mathrm{Spec}(C) \subset \underline{B} := \mathrm{Spec}(B)$, with

$$\Phi : B = k[\underline{x}] \to C = k[\underline{t}],$$

the morphism of k-algebras corresponding to the inclusion. We then have polynomial functions defining Φ,

$$x_i = x_i(t_1, \ldots, t_d), \quad i = 1, \ldots, n.$$

Assume there is a metric $h = 1/2 \sum h_{i,j} dx_i dx_j \in \mathrm{Ph}(B)$, so that we may consider the dynamical system $B(\sigma_h)$. A natural problem would be to try and find a metric $g = 1/2 \sum g_{i,j} dt_i dt_j \in \mathrm{Ph}(C)$, such that Φ induces a homomorphism of dynamical systems,

$$\phi : B(\sigma_h) \to C(\sigma_g).$$

It is easy to see that in $C(\sigma_g)$ we have the formulas

$$dx_k = \sum_j \delta_{t_j}(x_k) dt_j + 1/2 \sum_{i,j} \delta_{t_i}(\delta_{t_j})(x_k) g^{i,j}$$

such that

$$[dx_k, x_l] = \sum_{i,j} \frac{\partial x_k}{\partial t_i} \frac{\partial x_l}{\partial t_j} g^{i,j}.$$

Therefore, the condition for, given the metric h of $\underline{B}$, the existence of a compatible metric g in $\underline{C}$ is

$$h^{k,l} = \sum_{i,j} \delta_{t_i}(x_k) \delta_{t_j}(x_l) g^{i,j}.$$

The analogy with the structure of the three fundamental forms in the study of unique time coordinates in general relativity, associated to a Cauchy hypersurface, is obvious.

Example 4.1. Consider the special case where $C = k[t]$ and $\mathrm{Spec}(C)$ is a curve γ, in $\mathrm{Spec}(B)$, then the condition for the metric h of $\underline{B}$, to be compatible with the metric g in $\underline{C}$, is

$$h^{k,l} = \frac{\partial x_k}{\partial t} \frac{\partial x_l}{\partial t} g^{1,1}.$$

So,

$$\sum_{k,l} h_{k,l} \frac{\partial x_k}{\partial t} \frac{\partial x_l}{\partial t} g^{1,1} = 1.$$

For any $f \in B$, the image of $df \in \mathrm{Ph}(B)$ in $\mathrm{Ph}(C)_{\mathrm{com}}$ is

$$df = \frac{\partial f}{\partial t} dt = \sum_i \frac{\partial f}{\partial x_i} \frac{\partial x_i}{\partial t} dt$$

the square norm of which is

$$\|df\|^2 = \sum_{i,j} \frac{\partial f}{\partial x_i} \frac{\partial f}{\partial x_j} g^{1,1} \frac{\partial x_i}{\partial t} \frac{\partial x_j}{\partial t} = \sum_{i,j} h^{i,j} \frac{\partial f}{\partial x_i} \frac{\partial f}{\partial x_j}.$$

Consider the subset $B_{<h} \subset B$ defined by

$$B_{<h} = \left\{ f \in B \middle| \left\| \sum_{i,j} \frac{\partial f}{\partial x_i} \frac{\partial f}{\partial x_j} h^{i,j} \right\| \leq 1 \right\}.$$

Let now $o, p \in \mathrm{Spec}(C)$, then for any curve γ linking o and p, in $\mathrm{Spec}(B)$, of length $|\gamma|$,

$$f(p) - f(o) = \int_\gamma df \leq |\gamma|,$$

so it is reasonable that the distance in $\mathrm{Spec}(B)$ should be given by

$$d(o, p) = \sup_{f \in B_{<h}} |f(p) - f(o)|.$$

This may be generalized to define metrics in "spaces" provided with a "spectral triangle" (see Connes, 2018), and the comments in Chapter 14.

4.4.2 *Relations to Clifford algebras*

In the situation above, where we are given a polynomial algebra $C = k[t_1, \ldots, t_d]$, and a non-singular metric g, we know that the C-module of differentials Ω_C, generated by dt_i, is provided with the metric g^{-1}. Therefore, we find that for every point in $\underline{t} \in \underline{C}$, there is a quadratic form on $T_{\underline{t}} := \Omega_{C,\underline{t}}$, given by $g^{-1}(\underline{t})$. We might then consider the Clifford algebra $Cl(T_{\underline{t}}, g^{-1})$. This can now be generalized, to construct a generalized Clifford algebra

$$Cl(C, g) := \mathrm{Ph}(C)/(dt_i dt_j + dt_j dt_i - 2g^{i,j} - g^{i,i} - g^{j,j}).$$

Given a point $\underline{t} \in \underline{C}$, we have canonical homomorphisms of k-algebras,

$$\mathrm{Ph}(C) \to Cl(C, g) \to Cl(T_{\underline{t}}, g^{-1}).$$

However, there are no decompositions of this composed morphism, into something like

$$\mathrm{Ph}(C) \to C(\sigma_g) \to Cl(T_{\underline{t}}, g^{-1}).$$

Nevertheless, as some physicists have remarked, the algebra $Cl(T_{\underline{t}}, g^{-1})$ may be of interest in quantum theory, in particular, in relation to the notion of rotation (in quantum mechanics) (see Lasenby *et al.*, 2004). Its use in the theory of gravity seems to me, however, to be very unnatural. We shall come back to this in Chapter 10, where we shall try to make the Toy Model, treated in the following chapter, more palatable for us, the human observers.

4.5 Potentials and the Classical Gauge Invariance

The space of representations, ρ of $C(\sigma_g)$, on a free (or projective) C-module V, is given as above by

$$\rho_\psi(t_i) = t_i, \; \rho_\psi(dt_i) = \sum_{l=1}^{n} g^{il}\delta_l + \psi_i,$$

where $\psi_i \in \mathrm{End}_C(V)$. The set of isoclasses of these representations is identified with the space of equivalence classes of the corresponding *Potentials*,

$$\underline{\psi} = (\psi_1, \psi_2, ..\psi_n) \in \mathrm{End}_C(V)^n.$$

It does not form an algebraic variety, but it has a nice structure, which we will work out in the next sub-section.

But before this, take a look at the diagram,

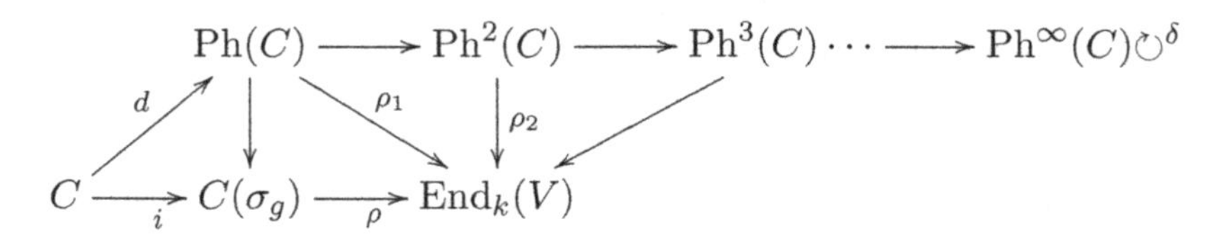

where $\rho_0 := i \circ \rho$ is a representation of a (commutative polynomial) k-algebra C, g a metric, and ρ a momentum of ρ_0 defined on the corresponding $C(\sigma_g)$. Consider a derivation, $\zeta \in \mathrm{Der}_k(C(\sigma_g), \mathrm{End}_k(V))$, and its composition, $\eta := i \circ \zeta$ with $i : C \to C(\sigma_g)$. Obviously, η will induce an element in $\mathrm{Ext}^1_C(V, V)$. If this element does not vanish, it would not be reasonable to consider ζ as change of the momentum ρ, of the object ρ_0, since ρ_0 representing (a point of) space should be left unchanged. Let us therefore introduce the following notation.

Given homomorphisms of k-algebras,

$$A \xrightarrow{i} B \xrightarrow{\rho} \mathrm{End}_k(V), \rho_0 := i \circ \rho,$$

put

$$\mathrm{Ext}^i_{\rho,\rho_0}(V,V) := \ker(\mathrm{Ext}^i_\rho(V,V) \to \mathrm{Ext}^i_{\rho_0}(V,V)).$$

This comes up next in Section 4.5.1.

4.5.1 *Infinitesimal structure on* $\mathbf{Rep}(C(\sigma_g))$

The set of *Potentials* Ψ is a *torsor* under

$$\mathcal{P} := (\mathrm{End}_C(V))^n,$$

and, of course, $\mathrm{End}_C(V)$ means the $\mathrm{End}_{\rho_0}(V)$, i.e. where C acts via the representation $\rho_0 := i \circ \rho$.

Note that if $V = \Theta_C$, the Levi-Civita connection provides a natural isomorphism $\Psi \simeq (\mathrm{End}_C(V))^n$. Now, consider two representations,

$$\rho_l : C(\sigma_g) \to \mathrm{End}_k(V), \quad l = 1, 2,$$

represented by elements $\psi(l) \in \mathcal{P}, l = 1, 2$, as

$$\rho_l(t_i) = \rho_0(t_i), \rho_l(dt_i) = \xi_i + \psi_i(l), \quad l = 1, 2, \ \xi_i := \sum_{l=1}^{n} g^{il}\delta_l.$$

The tangent space $\mathbf{T}_{\rho_1,\rho_2}$, between ρ_l, $l = 1, 2$, may be identified with a quotient of $\mathcal{P}$. We have

$$\mathbf{T}_{\rho_1,\rho_2} = \mathrm{Ext}^1_{C(\sigma_g)}(\rho_1, \rho_2) = \mathrm{Der}_k(C(\sigma_g), \mathrm{End}_k(V))/\,\mathrm{Triv},$$

where Triv is the set of trivial derivations, i.e. those of the form $\zeta(a) = \Omega\rho_1(a) - \rho_2(a)\Omega)$, where $\Omega \in \mathrm{End}_C(V)$. Any derivation $\zeta \in \mathrm{Der}_k(C(\sigma_g), \mathrm{End}_k(V))$ maps the relations $([dt_i, t_j] - g^{i,j})$ of $C(\sigma_g)$ (as quotient of $\mathrm{Ph}(C)$) to zero, so we shall have

$$[\zeta(dt_i), t_j] + \rho_1(dt_i)\zeta(t_j) - \zeta(t_j)\rho_2(dt_i) = \zeta(g^{i,j}).$$

Now, for the purpose of application to physics and elementary particles, (see above, and in later chapters, in particular, in Chapter 10), we shall consider the derivations ζ for which the restrictions $i \circ \zeta : C \to \mathrm{End}_k(V)$ vanish as elements in $\mathrm{Ext}^1_C(V, V)$. They form the subspace

$$\mathrm{Ext}^1_{C(\sigma_g),\rho_0}(\rho, \rho) := \ker\{\mathrm{Ext}^1_{C(\sigma_g)}(\rho, \rho) \to \mathrm{Ext}^1_C(\rho, \rho)\}.$$

The point is that given a representation ρ_0, we may be interested in the changes of a chosen momentum ρ, therefore to the possible accelerations, or forces, in a given space defined by ρ_0, but without altering ρ_0.

For any $\zeta \in \mathrm{Ext}^1_{C(\sigma_g),\rho_0}(\rho, \rho)$, there must exist a linear map $\Phi_0 \in \mathrm{End}_k(V)$, such that $\zeta(t_j) = t_j\Phi_0 - \Phi_0 t_j$, for all j. The derivations ζ and $\zeta - \mathrm{ad}(\Phi_0)$ are equivalent as elements of $\mathrm{Ext}^1_{C(\sigma_g)}(\rho_1, \rho_2)$. We may, therefore, for a chosen ζ, assume all $\zeta(t_i) = 0$, and it follows from the above equation that

$$\zeta([dt_i, t_j] - g^{i,j}]) = [\zeta(dt_i), t_j] = 0,$$

for all i, j, so the derivation ζ is determined by the family of elements, $\zeta(dt_i) \in \mathrm{End}_C(V), i = 1, \ldots, n$.

Therefore, any potential $\psi \in \mathfrak{P}$ will define a derivation

$$\psi \in \mathrm{Der}_k(C(\sigma_g), \mathrm{End}_k(V))/\,\mathrm{Triv},$$

with

$$\psi(t_i) = 0, \psi(dt_j) = \psi_j,$$

since for any $x([dt_i, t_j] - g^{i,j})y \in \sigma_g, x, y \in \mathrm{Ph}(C)$, we have

$$\psi(x([dt_i, t_j] - g^{i,j}))y = \psi(x)\rho(([dt_i, t_j] - g^{i,j}))y) \\ + \rho(x)\psi(([dt_i, t_j] - g^{i,j}))y) = 0.$$

In particular, for any representation $\rho : C(\sigma_g) \to \mathrm{End}_k(V)$, we find

$$\mathbf{T}_\rho := \mathbf{T}_{\rho,\rho} := \mathrm{Ext}^1_{C(\sigma_g),\rho_0}(\rho, \rho) = \mathcal{P}/\,\mathrm{Triv},$$

where the trivial derivations $W \in \mathrm{Triv}$, mapping t_i to 0, are exactly those given by the n-tuples,

$$\left(\left(\sum_j^n g^{1,j}\left(\frac{\partial\Phi}{\partial t_j}\right) + [\psi_1, \Phi]\right), \ldots, \left(\sum_{j=1}^n g^{n,j}\left(\frac{\partial\Phi}{\partial t_j}\right) + [\psi_n, \Phi]\right)\right),$$

for some $\Phi \in \mathrm{End}_C(V)$ by

$$W(t_i) = 0, \; W(dt_i) = \left(\sum_j^n g^{i,j} \left(\frac{\partial \Phi}{\partial t_j} \right) + [\psi_i, \Phi] \right).$$

The expression

$$\Phi(\psi) := (\xi_1(\Phi) + [\psi_1, \Phi], \ldots, \xi_n(\Phi) + [\psi_n, \Phi]),$$

therefore, corresponds to an infinitesimal gauge transformation,

$$\Phi \in \mathrm{Der}_k(\mathcal{P}),$$

of the space $\mathcal{P}$.

The *physical* relevant tangent space is therefore the quotient

$$\mathbf{P} = \mathcal{P}/\mathfrak{h},$$

of $\mathcal{P}$ with respect to the action of the abelian Lie algebra $\mathfrak{h} := \mathrm{End}_C(V)$.

As in the finite-dimensional situation, the Dirac derivation, here $\delta = \mathrm{ad}(g - T)$, induces a vector field,

$$[\delta] \in \Theta_{\mathcal{P}},$$

so long as we, by vector field, understand any map, which to an element ψ in $\mathcal{P}$ associates an element in its tangent space, i.e. in $\mathrm{Ext}^1_{C(\sigma)}(V_{\rho_\psi}, V_{\rho_\psi})$. It must, however, vanish at ρ_ψ for $\psi = 0$, since the Dirac derivation $\delta = \mathrm{ad}(g - T)$ necessarily must be mapped to a trivial derivation in $\mathrm{Der}_k(C(\sigma_g), \mathrm{End}_k(V))$, therefore, to 0 in $\mathrm{Ext}^1_{C(\sigma_g)}(V_\rho, V_\rho)$. But then it corresponds to an infinitesimal transformation of V as we have seen in Section 4.2

$$[\delta] = \mathrm{ad}(Q_h + [\psi] + Q_v),$$

meaning that

$$\rho_\psi(d^{n+1}t_i) = [Q_h + [\psi] + Q_v, \rho_\psi(d^n t_i)].$$

This may be interpreted as saying that time, defined by the dynamical structure (σ_g), acts, in all orders, within each representation, $\rho_\psi : C(\sigma_g) \to \mathrm{End}_k(V)$!

Remark 4.5. The formula

$$\mathrm{Ext}^1_{C(\sigma_g),\rho_0}(\rho, \rho) = \mathcal{P}/\,\mathrm{Triv}$$

will be used in later chapters, see Section 10.3, where we treat the classical Maxwell equation, in Section 12.2, the Weak Interaction, and in Chapter 12, where we explain how to produce new physical objects from simpler ones, using the properties of the Ext^1 functor in homological algebra. The identification of the set of Potentials with elements of this Ext-group, may explain the notion of "force carrying bosons," mediating force between particles.

4.5.2 *Physics and the Chern–Simons class*

Physicists usually write $\delta\phi := \Phi(\phi)$, not caring to mention Φ, taking for granted that $\delta\phi := \delta_\Phi(\phi)$ stands for an infinitesimal movement of ϕ in the direction of Φ, and call the transformation above an *infinitesimal gauge transformation.* The literature on gauge theory, and its relation to non-commutativity of space, and to quantization of gravity, is huge. I think that the introduction of the *non-commutative phase space*, and in the metric case, the *generic* dynamical system,

$$(\sigma_g) = ([dt_i, t_j] - g^{i,j}),$$

can, to some degree, elucidate the philosophy behind this effort. See, e.g. Sachs and Wu (1977) and Cerchiai *et al.* (2000), where the authors initially introduce non-commutativity in the ring of *observables* generated by *coordinates*, $\hat{x}^\nu$, by imposing

$$[\hat{x}^\nu, \hat{x}^\mu] = \Theta^{\nu,\mu},$$

where $\Theta^{i,j}$ are constants.

The above treatment of the notion of gauge groups and gauge transformations may also explain why, in physics, one considers *potentials* as *interaction carriers*, thus as particles *mediating force* upon other particles. And maybe one can also see why the notion of Ghost Fields, or Particles, of Faddeev and Popov, comes in. It seems to me that the introduction of ghost particles is linked to working with a particular section of the quotient map, $\mathcal{P} \to \mathbf{P}$.

The Dirac derivation, which is entirely dependent upon the notion of a non-commutative phase space, is not (explicitly) found in present day physics, although it is found hidden in the spectral paradigm of Connes as the self-adjoint operator $\mathfrak{D}$, which together with the involutive algebra $\mathfrak{A}$, and the $\mathfrak{A}$-representation on a Hilbert space $\mathfrak{H}$, constitute the spectral triple $(\mathfrak{A},\mathfrak{H},\mathfrak{D})$. Here, the relation to our treatment of time and metric is that $\mathfrak{D}$ "plays the role of the inverse line element," see Connes (2018), and Chapter 14.

The parsimony principle is therefore, normally, introduced via the construction of a Lagrangian, and an Action Principle, i.e. a function of the (assumed physically significant) variables, the fields and their derivatives, defined in $\mathcal{P}$, assumed to be invariant under the gauge transformations, so really defined in $\mathbf{P}$, and supposed to stay stable during *time* development, (see Laudal, 2011). A non-trivial element in the toolbox of the physicists, helping them to guess the Lagrangian, is the Chern–Simons functional, touched upon in (2.3), and that we now spell out in some generality. Since everything we have done above is functorially, we may work on non-singular schemes, instead of commutative k-algebras. Let us, as above, assume given a metric g on some scheme X, and that we have an affine covering, given in terms of a family of commutative k-algebras C_α and a bundle V, defined on X, corresponding to a set of representations, $\rho_0 : C = C_\alpha \to \mathrm{End}_k(V)$. Let $\rho : O_X(\sigma_g) \to \mathrm{End}_k(V)$ be a momentum at ρ_0, i.e. an extension of ρ_0 to $O_X(\sigma_g)$. Then we know that ρ induces a connection on V, and we have denoted by $\mathcal{P} := \mathrm{End}_{O_X}(V)^d$ the set of such connections, or if we want to, the set of representations $\rho : O_X(\sigma_g) \to \mathrm{End}_k(V)$. Recall that, given the metric, there is a unique 0-*object* in $\mathcal{P}$, defined by $\rho_1(dt_i) = \xi_i$, such that all other representations ρ are ρ_1 plus a potential $\psi \in \mathrm{End}_{O_X}(V)^d$.

Consider now the class $\mathrm{ch}^n(\rho) \in \mathrm{HH}^n(O_X, \mathrm{End}_k(V))$, see Section 2.1.4, Chern classes, defined by the Hochschild co-chain, the k-linear map,

$$\mathrm{ch}^n : C_\alpha^{\otimes n} \to \mathrm{End}_k(V),$$

defined by

$$\mathrm{ch}^n(c_1 \otimes c_2, \dots, \otimes c_n) = \rho_1(dc_1 dc_2 \dots dc_n) \in \mathrm{End}_k(V).$$

We know that ch^n is a co-cycle, since, see Section 2.1.4,

$$\begin{aligned}\delta(\mathrm{ch}^n)((c_1 \otimes c_2 \cdots \otimes c_{n+1})) &= c_1\rho_1((dc_2 dc_3 \dots dc_{n+1})) \quad (4.1)\\ &+ \sum_1^n (-1)^i \rho_1((dc_1 \dots d(c_i c_{i+1}) \dots dc_{n+1}))\\ &+ (-1)^{n+1}\rho_1(dc_1 \dots dc_n)c_{n+1} = 0. \quad (4.2)\end{aligned}$$

The *Generalized Chern–Simons Class* of ρ is then the class in the obvious double complex, defined by the covering $\{C_\alpha\}$, and the classes, for every $C := C_\alpha$,

$$\text{csh}^n(\rho) \in \text{HH}^n(C, \text{End}_k(V)),$$

defined by $\text{csh}^n = 1/n!\ \text{ch}^n$.

Let $\Phi \in \mathfrak{h} := \text{End}_C(V)$, and consider Φ as a Hochschild 0-co-cycle,

$$\Phi : k = C^0 \to \text{End}_k(V), \Phi(\alpha) = \alpha\Phi,$$

then

$$\delta\Phi(c) = \rho(dc)\Phi - \Phi\rho(dc).$$

In particular,

$$\delta\Phi(t_i) = [\xi_i + \psi_i, \Phi] = \xi_i(\Phi) + [\psi_i, \Phi].$$

This proves that

$$\text{csh}^1(\rho_1 + \psi)\,\text{csh}^1(\rho_1 + \psi + \Phi(\psi)),$$

i.e. the *Ghost Fields* have well-defined Chern–Simons class, so that the Chern–Simons class is a well-defined functional,

$$\text{csh}^* : \mathbf{P} \to \text{HH}^*(C, \text{End}_k(V)).$$

4.6 A Generalized Yang–Mills Theory

Given a metric $g \in \text{Ph}(C)$ as above, and assume there is a *gauge group*, i.e. a k-Lie algebra $\mathfrak{g}_0$, operating on $C = k[t_1, \ldots, t_d]$ with extension to $C(\sigma_g)$, where σ is a dynamical structure. Normally, σ would be σ_g, and $\mathfrak{g}_0 = \text{Der}_k(C)$, since all representations of $C(\sigma_g)$ have a C-connection. Clearly, therefore, the Lie algebra of Killing vectors $\mathfrak{o}(g)$ is contained in $\mathfrak{g}_0$.

In the general situation, we have, together with the global gauge group $\mathfrak{g}_0$, also a local gauge group, i.e. a C-Lie algebra $\mathfrak{g}_1$ acting insensitively upon the representations V that pop up in our theory, and also insensitive to the action of $\mathfrak{g}_0$. We, therefore, have a $C(\sigma)$-connection

$$\mathfrak{D} : \mathfrak{g}_0 \to \text{Der}_k(\mathfrak{g}_1),$$

which is a kind of a general spin structure, or *Coupling Morphism*. Recall that when the curvature of $\mathfrak{D}$ vanishes, $\mathfrak{g} := \mathfrak{g}_0 \oplus \mathfrak{g}_1$ is a Lie algebroid.

Let us pause to consider a non-trivial example of this situation.

Theorem 4.6. *Assume $\mathfrak{g}_1^0$ is a finite-dimensional k-Lie algebra and consider the C-Lie algebra $\mathfrak{g}_1 := \mathfrak{g}_1^0 \otimes_k C$. Consider a representation,*

$$\rho : C(\sigma_g) \to \mathrm{End}_k(V)$$

with global gauge group $\mathfrak{g}_0 = \mathrm{Der}_k(C)$ and local gauge group $\mathfrak{g}_1 \subset \mathrm{End}_C(V)$.

Let the $\mathfrak{g}_0$-connection,

$$\mathfrak{D} : \mathfrak{g}_0 \to \mathrm{Der}_k(\mathfrak{g}_1),$$

be defined as

$$\mathfrak{D}(\xi_i) = \mathrm{ad}(\nabla_{\xi_i} + \psi_i), \textit{with } \psi_i = g^{i,l}\gamma_l, \ \gamma_l \in \mathfrak{g}_1^0.$$

Then we find that the curvature of $\mathfrak{D}$ is

$$F_{i,j}^{\psi} = \mathrm{ad}([\psi_i, \psi_j]).$$

It vanishes if and only if

$$[[\gamma_i \gamma_j], \gamma] = 0,$$

for all $\gamma \in \mathfrak{g}_1, \forall i, j = 1, \ldots, n$.

Put now

$$\rho_{\psi,A} : C(\sigma_g) \to \mathrm{End}_k(V),$$

defined by $\rho_1(dt_i) = \nabla_\xi + \psi_i + A_i$ where $A_i \in C$ is a usual potential, then the curvature of $\rho_{\psi,A}$ is equal to

$$F_{i,j}^{\rho_{\psi},A} = \mathrm{ad}([\psi_i, \psi_j]) + \xi_i(A_j) - \xi_j(A_i).$$

Proof. By definition, the curvature of $\mathfrak{D}$ is

$$F_{i,j} := [\mathfrak{D}(\xi_i), \mathfrak{D}(\xi_j)] - \mathfrak{D}([\xi_i, \xi_j]),$$

where

$$\begin{aligned}[\mathfrak{D}(\xi_i), \mathfrak{D}(\xi_j)] &= [\mathrm{ad}(\xi_i) + \mathrm{ad}(\psi_i), \mathrm{ad}(\xi_j) + \mathrm{ad}(\psi_j)] \\ &= \mathrm{ad}([\xi_i, \xi_j]) + \mathrm{ad}([\xi_i, \psi_j]) + \mathrm{ad}([\psi_i, \xi_j]) + \mathrm{ad}([\psi_i, \psi_j]),\end{aligned}$$

and

$$\mathfrak{D}([\xi_i, \xi_j]) = \sum_k c_{i,j}^k \mathfrak{D}(\xi_k) = \sum_k c_{i,j}^k \,\mathrm{ad}(\xi_k) + \sum_k c_{i,j}^k \,\mathrm{ad}(\psi_k),$$

where $c_{i,j}^k = (\Gamma_k^{j,i} - \Gamma_k^{i,j})$. Moreover,

$$\begin{aligned}\mathrm{ad}([\xi_i, \psi_j]) + \mathrm{ad}([\psi_i, \xi_j]) &= \mathrm{ad}(\xi_i(\psi_j)) - \mathrm{ad}(\xi_j(\psi_i)) \\ \xi_i(\psi_j) - \xi_j(\psi_i) &= \left(g^{i,l}\frac{\partial g^{j,k}}{\partial t_l} - g^{j,l}\frac{\partial g^{i,k}}{\partial t_l}\right)\gamma_k \\ &= \left(-g^{i,l}\sum (g^{j,t}\Gamma_{t,l}^k + g^{t,k}\Gamma_{t,l}^j\right) \\ &\quad + g^{j,l}\sum (g^{i,t}\Gamma_{t,l}^k + g^{t,k}\Gamma_{t,l}^i))\gamma_k \\ &= (\Gamma_t^{j,i} - \Gamma_t^{i,j})g^{t,k}\gamma_k \\ &= \sum_k c_{i,j}^k \psi_k,\end{aligned}$$

where we have used the standard formulas for the derivatives of $g^{i,j}$. Therefore,

$$\begin{aligned}F_{i,j}^{\psi} :&= [\mathfrak{D}(\xi_i), \mathfrak{D}(\xi_j)] - \mathfrak{D}([\xi_i, \xi_j]) \\ &= \mathrm{ad}([\xi_i, \xi_j]) + \sum_k c_{i,j}^k \,\mathrm{ad}(\psi_k) + \mathrm{ad}([\psi_i, \psi_j]) \\ &\quad - \sum_k c_{i,j}^k \,\mathrm{ad}(\xi_k) - \sum_k c_{i,j}^k \,\mathrm{ad}(\psi_k) \\ &= \mathrm{ad}([\psi_i, \psi_j]).\end{aligned}$$

The rest is easily seen. $\square$

This result makes us think of Dirac spinors, to which we return. Anyway, in case $\mathfrak{g} = \mathfrak{g}_0 \oplus \mathfrak{g}_1$ is a Lie algebroid, the action of $\mathfrak{g}_1$ and the connection, $\mathfrak{D}$, extends to a connection

$$\mathfrak{D} : \mathfrak{g} \to \mathrm{End}_k(V).$$

According to our philosophy, we should therefore consider (as moduli space, for our models), the invariant (or quotient) space,

$$\mathrm{Simp}(C)/\mathfrak{g} \simeq \mathrm{Simp}(C(\mathfrak{g})).$$

Theorem 4.5, can in the light of the last sub-section be reformulated to take the following form.

Theorem 4.7 (Scholie). *Consider a representation,*

$$\rho_0 : C \to \mathrm{End}_k(V),$$

with gauge group $\mathfrak{g}$, *as above, and an extension* $\rho : C(\sigma_g) \to \mathrm{End}_k(V)$, *considered as a* momentum *of* ρ_0, *and as a reference point in the set of connections* $\mathcal{P}$ *on* (V, ρ_0).

The tangent space of $\mathcal{P}$ *at any* ρ *is of the same form, i.e. a quotient* $\mathbf{P}$ *of* $\mathcal{P}$, *by the action of the Lie algebra* $\mathrm{End}_C(V)$. *However, we know that the first-order time-development, the Dirac derivation,* $\delta = \mathrm{ad}(g-T)$, *induces the trivial vector field on* $\mathcal{P}$, *so the first-order time-development of the* ρ_0, *given by any* $\rho_\psi := \rho + \psi$, $\psi \in \mathcal{P}$, *does not have a time-development as a representation given by the Dirac derivative* δ *of* $C(\sigma_g)$. *But we know that the time development on* V *is given by the formula*

$$[\delta] = \mathrm{ad}(Q_h + [\psi] + Q_v),$$

meaning that

$$\rho_\psi(d^{n+1}t_i) = [(Q_h + [\psi] + Q_v), \rho_\psi(d^n t_i)].$$

This takes care of the complete time development of an endomorphism of V, *as well as for a state vector* $\phi \in V$.

The second-order time-development is also given in terms of the Force Law in $\mathrm{Ph}(C)$,

$$d^2 t_i = -\sum_{p,q} \Gamma^i_{p,q} dt_p dt_q - \sum_{p,q} g_{p,q} F_{i,p} dt_q + 1/2 \sum_{l,p,q} g_{p,q} [F_{i,q}, dt_p]$$
$$+ 1/2 \sum_{l,p,q} g_{p,q} [dt_p, (\Gamma^{i,q}_l - \Gamma^{q,i}_l)] dt_l + [dt_i, T].$$

Given the gauge group $\mathfrak{g} = \mathfrak{g}_0 \oplus \mathfrak{g}_1$, *there are tangent directions,* $\mathfrak{v} = \{\mathfrak{v}_i\}$, *with* $\mathfrak{v}_i = \sum_j \gamma_j g^{j,i}$, *and* $\gamma_i \in \mathfrak{g}_1$, *"normal" to* C *in* V, *naturally related to derivations* $\mathfrak{D} : \mathfrak{g}_0 := \mathrm{Der}_k(C) \to \mathfrak{g}_1 := \mathfrak{g}_1^0 \otimes C$. *Put*

$$[\psi] := \sum_i \gamma_i \nabla_{\xi_i} = \sum_{i,j} \gamma_j g^{j,i} \frac{\partial}{\partial t_i} = \sum_i \mathfrak{v}_i \frac{\partial}{\partial t_i} = \mathfrak{v},$$

then we may formally write the time operator in the same form as in our general Quantum Field Theory, see Theorem 4.7,

$$[\delta] = \mathrm{ad}(Q_h + [\psi] + Q_v) \in \mathrm{Der}(\mathrm{End}_k(\tilde{V})),$$

where

$$Q_h : = \rho(g - T) = Q := 1/2 \sum_{i,j} g^{i,j} \nabla_{\delta_i} \nabla_{\delta_j},$$

$$[\psi] : = \sum_i \psi_i \nabla_{\delta_i},$$

$$Q_v : = \psi(g - T) = 1/2 \sum_{i,j} g_{i,j} \psi_i \psi_j + 1/2 \left(\sum_{j,l} \Gamma^j_{j,l} + \bar{\Gamma}^j_{j,l} \right) \psi_l + 1/2 \nabla . \psi,$$

with the corresponding first-order time-action in the state-space defined by,

$$(Q_h + [\psi] + Q_v) \in \mathrm{End}_k(V).$$

Given a classical potential, $A = \{A_i\}, A_i \in C$, *put*

$$\rho_{\psi,A} : C(\sigma_g) \to \mathrm{End}_k(V),$$

defined by $\rho_1(dt_i) = \nabla_{\xi_i} + \psi_i + A_i$, *then the curvature of* $\rho_{\psi,A}$ *is equal to*

$$F_{i,j} = \mathrm{ad}([\psi_i, \psi_j]) + \xi_i(A_j) - \xi_j(A_i),$$

where we, as before, write for any $a \in A, \nabla_{\xi_i}(a) = \xi_i(a)$.

Remark 4.6. As we shall see, the gauge groups of the Standard Model, i.e. the Lie algebras,

$$\mathfrak{g}' := \mathfrak{u}(1) \times \mathfrak{sl}(2) \subset \mathfrak{g}_1' := \mathfrak{u}(1) \times \mathfrak{su}(2) \times \mathfrak{su}(3),$$

which is part of our Toy Model (see Laudal, 2011), also pops up in our cosmological model, $\underline{\tilde{H}}$, and now as a real local gauge group in the above sense. The elementary particles in that model should, therefore, in line with the usage of present quantum theory, be the points of $\mathrm{Simp}(\mathrm{Ph}(\tilde{H}))/\mathfrak{g} = \mathrm{Simp}(\mathrm{Ph}(\tilde{H})(\mathfrak{g}))$. In fact, we shall see that an impressive part of the structure of the Standard Model, is contained in the structure of this *non-commutative quotient* (*or invariant*) *space.*

The above is a generalization of the physicists treatment of the type of representations

$$\rho_W : \mathrm{Ph}(C) \to \mathrm{End}_k(V),$$

parametrized by what they call *Gauge Fields*, the $W_i^l \in C$, in the following formula. Put $W_i := \sum_{l=1}^r W_i^l \gamma_l$, and let as above

$$[\gamma_l, \gamma_m] = \sum_{p=1}^r C_{l,m}^p \gamma_p.$$

Now, recall that $[\xi_i, \xi_j] = \sum_k (\Gamma_k^{j,i} - \Gamma_k^{i,j})\xi_k$, and consider the curvature of this representation,

$$F_{i,j} = [\rho(dt_i), \rho(dt_j)] - \sum_p (\Gamma_p^{j,i} - \Gamma_p^{i,j})\rho(dt_p) = [\nabla_{\xi_i}, \nabla_{\xi_j}] - \nabla_{[\xi_i,\xi_j]}.$$

$$\begin{aligned} F_{i,j} &= \rho_W([dt_i, dt_j]) - \sum_p (\Gamma_p^{j,i} - \Gamma_p^{i,j})\rho_W(dt_p) \\ &= \sum_{l=1}^r (\nabla_{\xi_i}(W_j^l \gamma_l) - \nabla_{\xi_j}(W_i^l \gamma_l) + \sum_{l,m,p=1}^r C_{l,m}^p W_i^l W_j^m \gamma_p \\ &\quad - \sum_p (\Gamma_p^{j,i} - \Gamma_p^{i,j}) W_p^l \gamma_l. \end{aligned}$$

Put $F_{i,j} = \sum F_{i,j}^l \gamma_l$, then we obtain the equation

$$F_{i,j}^l = (\xi_i(W_j^l) - \xi_j(W_i^l)) + \sum_{p,m=1}^{r} C_{p,m}^l W_i^p W_j^m - \sum_p (\Gamma_p^{j,i} - \Gamma_p^{i,j}) W_p^l,$$

which if $\sum_p (\Gamma_p^{j,i} - \Gamma_p^{i,j}) W_p^l = 0$, is the classical expression for the curvature in this case.

The Euler–Lagrange equations of the Lagrangian,

$$\mathbf{L}_{gf} = -1/4 F^{\mu\nu\alpha} F_{\mu\nu}^{\alpha},$$

used by physicists gives us the corresponding equation of motion,

$$\xi_\mu F_{\mu\nu}^a + c_{ab}^c W^{\mu b} F_{\mu\nu}^c = 0,$$

the *Yang–Mills equation*, corresponding to the vanishing of

$$1/2 \sum_{j=1}^{n} [F_{i,j}, \xi_j + W_j].$$

With a *Source* added, it looks like

$$\xi_\mu F_{\mu\nu}^a + c_{ab}^c W^{\mu b} F_{\mu\nu}^c = -J_\nu^a.$$

In the general metric case, with non-abelian gauge group, it is difficult to find gauge invariant Lagrangians of reasonable physical relevance, so we have to operate differently. Here is where the Generic Equation of Motion above comes in and give us equations of motion, quite generally. Recall that the Force Law is given by

$$d^2 t_i = -\sum_{p,q} \Gamma_{p,q}^i dt_p dt_q - 1/2 \sum_{p,q} g_{p,q}(F_{i,p} dt_q + dt_p F_{i,q})$$
$$+1/2 \sum_{l,p,q} g_{p,q}[dt_p, (\Gamma_l^{i,q} - \Gamma_l^{q,i})] dt_l + [dt_i, T].$$

When the metric is Euclidean, or Minkowski, this reduces to

$$d^2 t_i = -1/2 \sum_p g_{p,p}(F_{i,p} dt_p + dt_p F_{i,p})$$
$$= -\sum_p g_{p,p} F_{i,p} dt_p - 1/2 \sum_p g_{p,p}[dt_p, F_{i,p}].$$

Therefore, where we, as above, let the curvature conserve its name, $F_{i,j} := \rho(F_{i,j})$. The Yang–Mills equation above is now seen to imply

$$\rho(d^2 t_i) = -\sum_p g_{p,p} F_{i,p} \rho(dt_p).$$

This, however, indicates that the tangent to the representation $\dot{\rho}$ given by

$$\psi = (\psi_i) = \left(-1/2 \sum_{p,m} g_{p,p} \frac{\partial}{\partial t_p} (F_{i,p}^m) \gamma_m - 1/2 \sum_{p,l,m} g_{p,p} W_p^l F_{i,p}^m c_{l,m}^q \gamma_q \right),$$

is a classical 0-tangent, so the Yang–Mills equation should just tell us that there exists a potential $\Phi \in \mathrm{End}_k(V)$ such that

$$(\psi_i) = \left(\left(\sum_j^n g^{1,j} \left(\frac{\partial \Phi}{\partial t_j} \right) + [\phi_1, \Phi] \right), \ldots, \right.$$
$$\left. \left(\sum_{j=1}^n g^{n,j} \left(\frac{\partial \Phi}{\partial t_j} \right) + [\phi_n, \Phi] \right) \right).$$

Compare with the Lorentz Force Law, classically, and for an electric field,

$$\mathbf{a}_i = -\sum_{p=1}^n F_{i,p} v_p.$$

Interpreting $\rho(d^2 t_i) = m\mathbf{a}_i$ and $\rho(dt_p) = mv_p$, we recover the classical equation of movement in a field (see Laudal, 2011, p. 115).

4.7 Reuniting GR, YM, and General QFT

Let us have a look at the significance of the conclusion of this chapter, and the last theorems. We have, for every polynomial algebra C, outfitted with some metric g, proved that there exists a derivation,

$$\delta := \mathrm{ad}(g - T) \in \mathrm{Der}_k(\mathrm{Ph}(C)),$$

such that it coincides with the canonical derivation $d : C \to C(\sigma_g)$, in the generic dynamical system (GDS). The corresponding Force Laws in $\mathrm{Ph}(C)$, see Theorem 4.5, generate equations of motions in General Relativity (GR), as well as in the generalized Yang–Mills (YM) theory introduced above. In fact, we find a very satisfying identity between the notions of *Time* in GDS, GR, and in YM. The *Dirac derivation* $\delta = \mathrm{ad}(g - T)$ in GDS, inducing a quantum field theory (QFT) on the space $\mathbf{P}$ where $[\delta]$ and the corresponding Hamiltonian operator Q in any representation space are deduced from the general Laplace–Beltrami operator, $[\delta] := \mathrm{ad}(g - T)$.

Since T vanishes in $\mathrm{Ph}(C)^{\mathrm{com}}$, the time in GR reduces to the Dirac derivation,

$$[\delta] = \sum_l \left(\xi_l \frac{\partial}{\partial t_l} - \Gamma^l \frac{\partial}{\partial \xi_l}\right),$$

and a trivial Hamiltonian. The Schrödinger equation in GDS is given as

$$(Q - E)(\psi) = 0.$$

We shall come back to our generalization of QFT.

To unify QFT, GDS, and GR, we might have started with the dynamical system, $C(g)$, generated by, say, the Force Laws of type (1) or (2). Unluckily, the structure of this system, in general, seems to be very complicated. It is, for example, not easy to decide whether or not $C(g)$ has finite-dimensional representations at all.

Anyway, it seems that we may be content with the above structures, all deduced from GDS, since physicists probably do not know how to include the second-order momentum in the preparation of their experiments. That shortcoming in the GDS is therefore not yet a big theoretical problem, and we shall see, at the end of Chapter 11, that the results are models that satisfy Wightman's axioms for QFTs.

However, we may study an obvious unification of GR, GDS, and GQT, where the algebra of observables is $\mathrm{Ph}^{\infty}(C)$, and we shall show that we are able to classify, i.e. compute the moduli space of, the finite dimensional representations of $\mathrm{Ph}^{\infty}(C)$, even though this space will turn out to be of infinite dimension. Thus we may hope to extend the method of Chapter 3, and obtain a unified theory. There are, however, lots of problems involved in this scheme, one is the action of the gauge groups that turns up. Another is the philosophically, maybe reasonable, but very unpopular, consequence of this restriction of the theory, to just the finitely defined measurable entities: Our Space, and everything else modeled by such a theory, would be discrete, simple objects would have point-like structures, etc.

The computation of the moduli space of finite-dimensional representations of $\mathrm{Ph}^{\infty}(C)$, mentioned above, and the further analysis of the resulting "QFT" will, for economical reasons, be postponed and fused with the results above on generic action of equation of time, in Chapter 10, in particular, Section 10.5.

To go further, we should have to go back to our philosophy and ask ourselves why we are capable of identifying, and communicating, our sense of natural objects, or rather the impressions that we have about such objects, with their relations, like distances between them, and the like. One obvious answer, which is at the base of this chapter, is that we assume we have the notion of Time as a metric on our moduli space of the events we think we have identified, and that this clock is running smoothly, such that coupled with the universal constancy of the velocity of light, this makes most objects look the same today as yesterday up to obvious shifts corresponding to symmetries that we have called gauge symmetries. This is the reason for studying the dynamic structure (σ_g) corresponding to one fixed metric g. In a sense, we assume that most of the Furniture, which we have identified as representations of the algebra (or if one wishes, of the scheme) of observables, stays constant up to an understandable gauge. This is tantamount to the assumption that our world stays reasonably constant, even though we know, and clearly see, that there are cataclysms in the Universe, completely changing the objects involved.

Now this is also the reason why we have to go into a more technical relationship between our Time g, and the Furniture, i.e. the representations, V. First, let us admire a commutative diagram that

will help us through the arguments,

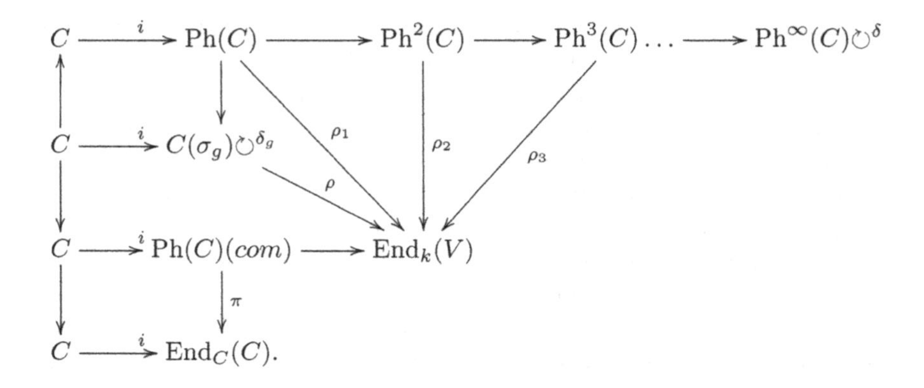

Here, $\delta_g = \mathrm{ad}(g - T)$, and π is an arbitrary representation,

$$\mathrm{Ph}(C)(\mathrm{com}) := \mathrm{Ph}(C)/([dt_i, t_j], [dt_i, dt_j]) \to \mathrm{End}_C(C).$$

The morphisms ρ_p, $p = 1, 2, \ldots$, are uniquely defined by

$$\begin{aligned}
\mathrm{ad}(g - T) \circ \rho_1 &= d \circ \rho_2 : \mathrm{Ph}(C) \to \mathrm{End}_k(V,),\\
\mathrm{ad}(g - T) \circ \rho_2 &= d \circ \rho_3 : \mathrm{Ph}^2(C) \to \mathrm{End}_k(V),\\
\mathrm{ad}(g - T) \circ \rho_n &= d \circ \rho_{n+1} : \mathrm{Ph}^n(C) \to \mathrm{End}_k(V), n \geq 1.
\end{aligned}$$

We shall now be able to prove a theorem that makes a mathematically reasonable relationship between Time and Furniture, or rather between the infinitesimal changes of one, and the infinitesimal changes of the other. Until now, we have had essentially two different notions of time, the metric of our moduli space of our models, and the Dirac derivative of $\mathrm{Ph}^{\infty}(\ )$ of the same. Now we find that the relationship between these notions is tight.

But first, let us recall that we have taken the liberty of working with two notions of metrics. First, the classical metric $\overline{g} := (g_{i,j})$, and then the element $g := 1/2\sum_{i,j} g_{i,j} dt_i dt_j \in \mathrm{Ph}(C)$. Obviously, $\overline{g} = \mathrm{sym}(g)$, the symmetrization of g. The context will clearly show which one we are talking about, so we shall continue talking about the metric g.

Theorem 4.8. *Let $\mathbf{M}$ be the space (of isomorphism classes) of metrics on C. For every point $g \in \mathbf{M}$, consider the diagram*

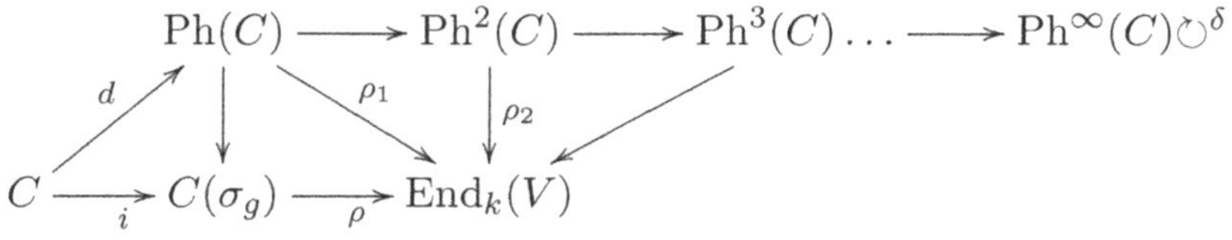

$$
\begin{array}{ccccccc}
 & & \mathrm{Ph}(C) & \longrightarrow & \mathrm{Ph}^2(C) & \longrightarrow & \mathrm{Ph}^3(C)\ldots \longrightarrow \mathrm{Ph}^\infty(C)\circlearrowright^{\delta} \\
 & \nearrow^{d} & \downarrow & \searrow^{\rho_1} & \downarrow{\scriptstyle\rho_2} & \swarrow & \\
C & \xrightarrow[i]{} & C(\sigma_g) & \xrightarrow[\rho]{} & \mathrm{End}_k(V) & &
\end{array}
$$

where $\rho_0 := i \circ \rho$ is a representation of C, and ρ a momentum of ρ_0. Let ρ_1 be the induced representation of $\mathrm{Ph}(C)$. Consider the family of k-algebras

$$\mu : \mathbf{C}(\sigma) \to \mathbf{M},$$

indexed by the possible metrics of C, such that $C(\sigma_g)$ corresponds to $g \in \mathbf{M}$.

Let

$$\mathbf{T}_{\mathbf{M},g} = \{(h_{i,j})\}, \;\; h_{i,j} = h_{j,i} \in C$$

be the tangent space to $\mathbf{M}$ at g. Define $h^{i,j}$ by

$$h_{i,j} = -\sum_{p,q} g_{i,p} h^{p,q} g_{q,j}, \;\; h = \{h_{i,j}\} \in \mathbf{T}_{\mathbf{M}},$$

and put

$$T = -1/2 \sum_{i,j,l} \frac{\partial g_{i,j}}{\partial t_l} g^{l,i} dt_j$$

and

$$T' = -1/2 \sum_{i,j,l} \left(\frac{\partial (g_{i,j} + h_{i,j}\epsilon)}{\partial t_l} \right) (g^{l,i} + h^{l;i}\epsilon) dt_j.$$

Consider now the first-order deformation of the metric $g \in \mathrm{Ph}(C)$, $g + \epsilon h \in \mathrm{Ph}_{k[\epsilon]}(C \otimes k[\epsilon]) = \mathrm{Ph}(C) \otimes k[\epsilon]$ and the corresponding

Dirac derivation $\mathrm{ad}(g + \epsilon h - T')$ *in* $\mathrm{Ph}_{k[\epsilon]}(C \otimes k[\epsilon])$. *Put* $d't_i :=$ $\mathrm{ad}(g + \epsilon h - T')(t_i)$. *Then we find, in* $C(\sigma_g) \otimes k[\epsilon]$,

$$[d't_i, t_j] = g^{i,j} - h^{i,j}\epsilon.$$

Moreover, the ρ_1*-derivation* $\eta : \mathrm{Ph}(C) \to \mathrm{End}_k(V)$ *defined by*

$$\eta(t_i) = 0,\ \eta(dt_i) = \sum_{l,q} h^{i,l}\rho_1(g_{l,q}dt_q) = \sum_{l} h^{i,l}\nabla_{\delta_l} \in \mathrm{Diff}^1(V, V),$$

corresponds to a first-order derivative of ρ_1, *i.e. to the morphism*

$$\eta(\rho_1) := \rho_2 : \mathrm{Ph}^2(C) \to \mathrm{End}_k(V),$$

for which, $\rho_2(d^2t_i) = \eta(dt_i)$. *This induces an element,*

$$\eta(h) \in \mathrm{Ext}^1_{\mathrm{Ph}(C)}(V, V),$$

where V *is the representation* ρ_1, *producing an injective map*

$$\eta : \mathbf{T}_{\mathbf{M},g} \to \mathrm{Ext}^1_{\mathrm{Ph}(C)}(V, V)$$

onto the linear subspace,

$$\mathrm{Ext}^1_{\mathrm{Ph}(C)}(V, V)^{(1)} \subset \mathrm{Ext}^1_{\mathrm{Ph}(C)}(V, V)$$

of first-order *non-trivial tangent space of the* $\mathrm{Ph}(C)$*-representation* (ρ_1, V), *defined by the derivations* $\eta : \mathrm{Ph}(C) \to \mathrm{End}_k(V)$, *where for all* i, $\eta(dt_i) \in \mathrm{Diff}^1(V, V)$.

Thus, any non-trivial deformation of the metric g *induces a non-trivial deformation of every* $\mathrm{Ph}(C)$*-representation* (ρ_1, V), *defined by a representation* $\rho : C(\sigma_g) \to \mathrm{End}_k(V)$, *and any* first-order *non-trivial deformation of the* $\mathrm{Ph}(C)$*-representation* (ρ_1, V) *induces a non-trivial deformation of the metric. (See Chapter* 10 *for the relationship to the notion of negative/positive energy).*

Proof. Consider the first-order deformation $g + \epsilon h$, of the metric g. The corresponding derivation $\mathrm{ad}(g - T)$ of $\mathrm{Ph}(C)$, defines the Dirac derivation in $C(\sigma_g)$, and the derivation $(\mathrm{ad}(g + \epsilon h - T')$ defines a derivation in $\mathrm{Ph}(C) \otimes k[\epsilon]$.

Here,

$$T = -1/2 \sum_{i,j,l} \frac{\partial g_{i,j}}{\partial t_l} g^{l,i} dt_j$$

and

$$T' = -1/2 \sum_{i,j,l} \left(\frac{\partial (g_{i,j} + h_{i,j}\epsilon)}{\partial t_l} \right) (g^{l,i} + h^{l;i}\epsilon) dt_j,$$

where $h^{i;i}$ is defined by

$$h_{i,j} = -\sum_{p,q} g_{i,p} h^{p,q} g_{q,j}.$$

We find in the quotient, $C(\sigma_g) \otimes k[\epsilon]$,

$$\begin{aligned} d't_i = [(g + h\epsilon) - T', t_i] = &\left[g - T + 1/2\epsilon \sum_{p,q,l} h_{p,q} dt_p dt_q \right. \\ &\left. + 1/2\epsilon \sum_{p,q,l} \left(\frac{\partial g_{p,q}}{\partial t_l} h^{l,p} dt_q + \frac{\partial h_{p,q}}{\partial t_l} g^{l,p} dt_q \right), t_i \right] \\ = dt_i + 1/2\epsilon \sum_{p,q,l} &\left(h_{p,q} g^{p,i} dt_q + h_{p,q} dt_p g^{q,i} + \frac{\partial g_{p,q}}{\partial t_l} h^{l,p} g^{q,i} \right. \\ &\left. + \frac{\partial h_{p,q}}{\partial t_l} g^{l,p} g^{q,i} \right) \\ = dt_i + \epsilon \sum_{l,q} &(-h^{i,l} g_{l,q}) dt_q + \epsilon P, \ P \in C, \end{aligned}$$

where

$$P = 1/2 \left(h_{p,q} \xi_p (g^{q,i}) + \frac{\partial g_{p,q}}{\partial t_l} h^{l,p} g^{q,i} + \frac{\partial h_{p,q}}{\partial t_l} g^{l,p} g^{q,i} \right).$$

Here, we have used

$$h_{p,q} dt_p g^{q,i} = h_{p,q} g^{q,i} dt_p + h_{p,q} [dt_p, g^{q,i}],$$

$$h_{p,q} [dt_p, g^{q,i}] = h_{p,q} \xi_p (g^{q,i}) = \sum_l h_{p,q} g^{p,l} \frac{\partial g^{q,i}}{\partial t_l}.$$

From this follows

$$[d't_i, t_j] = g^{i,j} - h^{i,j}\epsilon.$$

Now, consider the derivation

$$\eta \in \mathrm{Der}_k(\mathrm{Ph}(C), \mathrm{End}_k(V)),$$

defined in $\mathrm{End}_k(V)$ by

$$\eta(t_i) = 0,\ \eta(dt_i) = \sum_j h^{i,j}\nabla_{\delta_j} \in \mathrm{Diff}^1(V) \subset \mathrm{End}_k(V).$$

It is well-defined since

$$\eta[dt_i, t_j] = [\eta(dt_i), t_j] = h^{i,j} = h^{j,i} = [\eta(dt_j), t_i] = \eta[dt_j, t_i].$$

It therefore defines an element

$$\eta(h) \in \mathrm{Ext}^1_{\mathrm{Ph}(C)}(V, V)^{(1)}.$$

Moreover, $\eta(h) \in \mathrm{Ext}^1_{\mathrm{Ph}(C)}(V, V)$ is 0 only if there exists an $S \in \mathrm{End}_k(V)$ such that

$$0 = \eta(t_i) = [S, t_i]\ \forall t_i \in C,$$

implying that $S \in \mathrm{End}_C(V)$, so $S = (S_{r,s})$ is a matrix with entries in C. Furthermore, we must have

$$\eta(dt_i) = [S, \rho_1(dt_i)] = (-\xi_i(S_{k,l})) \in \mathrm{End}_C(V).$$

Since $\eta(dt_i) = \sum_l h^{i,l}\nabla_{\delta_l} \in \mathrm{Diff}^1(V, V)$, with $h^{i,j} = h^{j,i} \in C$, not all vanishing, this contradiction proves that $\eta(h) \in \mathrm{Ext}^1_{\mathrm{Ph}(C)}(V, V)$, defined by the tangent h at $g \in \mathbf{M}$, is not zero. □

Remark 4.7. Given any non-zero tangent $h \in \mathbf{T}_{\mathbf{M},g}$, we have seen that there corresponds to any representation ρ_0, with momentum ρ in terms of g, a non-zero second-order tangent ρ_2 of ρ_0, given by the element $\eta(h) \in \mathrm{Ext}^1_{\mathrm{Ph}(C)}(V, V)$, determined by the ρ_1-derivation $\eta =: \eta(h)$. The corresponding extension of V with itself, as $\mathrm{Ph}(C)$-module, is given by how t_i, dt_j operates on $V \oplus \epsilon V$, where

$$\begin{aligned} t_i(v_1, v_2) &= (t_i v_1, t_i v_2), \rho_1(dt_j)(v_1, v_2) \\ &= (\rho_1(dt_j)(v_1), (\rho_1(dt_j)(v_2) + \eta(h)(dt_j)(v_1)). \end{aligned}$$

Given a state $v \in V$, the two states $(v, 0) \in V \oplus V$ and $(0, v) \in V \oplus V$ are identical as states over $C(\sigma_g)$, but different as states over $\mathrm{Ph}(C)$, one might call them "entangled."

Thus, $\eta(h)$ is provoking an acceleration ρ_2 of the representation ρ_0, as a representation that should correspond to a "cataclysmic change" of any massy "particle" $\rho_0 : C \to \mathrm{End}(V)$ with momentum ρ. The acceleration is defined by $\rho_2(d^2 t_i) = \eta(h)(dt_i)$. Since we have

$$\eta(h)([dt_i, t_j]) = h^{i,j} \in \mathrm{End}_C(V),$$

the Hamiltonian comes out as in Theorem 3.2,

$$Q_h = \rho((g - T) = 1/2 \sum_{i,j} g^{i,j} \nabla_{\delta_i} \nabla_{\delta_j},$$

and the acceleration, given by $\eta(h)$,

$$\eta(h)(g - T) = \sum_{i,j} h^{i,j} \nabla_{\delta_i} \nabla_{\delta_j} + 1/2 \sum \frac{\partial h_{i,j}}{\partial t_i} \nabla_{\delta_j},$$

which means that the difference between the two "entangled" states in an extension of the state space is given by the difference of the "gravitational mass density," or the difference of the metrics of the two "space–times, containing these states."

This is remarkably similar to Penrose's OR proposal, as we shall see at the end of the Chapter 12, (see also Penrose and Hameroff, 2011).

The solutions of the Hamiltonian equations $Q(\phi) = E\phi$, in an appropriate one-dimensional deformation $\tilde{V}_\tau$ of ρ_1, should be a "wave" $\phi(\tau) \in V$.

Combined with the structure of our "Toy Model," see Section 4.8, this fits well with the present understanding of gravitational waves. It also fits reasonably well with the present cosmological theory, as we shall see later (see also Sormani, 2017). Since a "cataclysmic change" of any massy "particle $\rho_0 : C \to \mathrm{End}(V)$" will influence the metric, our Time, and vice versa, we find a mathematical reason for taking seriously the so-called Mach principle, that "everything depends upon everything."

Moreover, let us, for every metric $g \in \mathbf{M}$, consider the scalar curvature R as a function on the space $\underline{C} := \mathrm{Spec}(C)$, and fix a "compact" subset $\Omega \subset \underline{C}$. Since $R \sim 1/r^d$, where r is the "radius of curvature" of Ω at the corresponding point, it is not unreasonable to consider the Hilbert action,

$$S(g) := \int_{\Omega} R dv_g,$$

where dv_g is the volume element in $\underline{C}$ defined by the metric g, as related to the "gravitational mass" content of the part of space, Ω. But then one could look at S, as a functional,

$$S : \mathbf{M} \to \mathbb{R},$$

the stability of which would give us a unique vector field on $\mathbf{M}$, the Hilbert–Einstein tensor,

$$\mathfrak{G} \in \Theta_{\mathbf{M}},\ \mathfrak{G}(g) =: G := \{G_{i,j}\} = \{\mathrm{Ric}_{i,j} - 1/2 R g_{i,j}\} \in T_{\mathbf{M},g}.$$

Recall also that we have put $g = 1/2 \sum_{i,j} g_{i,j} dt_i dt_j$ and $\overline{g} = \mathrm{sym}(g)$, so one should have written $G = \mathrm{Ric} - 1/2 R\overline{g}$.

For any representation (ρ_0, V) with momentum $\rho : C(\sigma_g) \to \mathrm{End}_k(V)$, the injective map, $\eta : \mathbf{T}_{\mathbf{M},g} \to \mathrm{Ext}^1_{\mathrm{Ph}(C)}(V, V)$, would then give us a unique element,

$$[\mathfrak{G}(g)] = [G] := \eta(G) \in \mathrm{Ext}^1_{\mathrm{Ph}(C)}(V, V),$$

which would be a kind of *Universal Field Equation* of Einstein–Hilbert type, for the Universe, with respect to the Furniture, i.e. the family of representations, with prescribed momenta, $\{V, \rho_1\}$.

Anyway, an "increment" h, of the metric g in $\Theta_{\mathbf{M}}$, would correspond to the element $\eta(h) \in \mathrm{Ext}^1_{\mathrm{Ph}(C)}(V, V)$, a first-order increment of ρ_1, which again corresponds to an increment of energy of the furniture (V, ρ_1), given by the representation ρ of $C(\sigma_g)$. So, again we find a Schrödinger-type equation, equating derivation with respect to time, i.e. with respect to the action of $[G]$,

$$\frac{\partial}{\partial t}(\rho_1) = \rho_2,$$

where ρ_2 corresponds to an Hamiltonian operator Q on V.

We shall return to this in Chapter 6 in relation to the notion of entanglement in physics. For this reason, it will be important to recall that for any representation (ρ_0, V) of C, with momentum $\rho : C(\sigma_g) \to \mathrm{End}_k(V)$, all extensions of ρ with itself are trivial, i.e. $\mathrm{Ext}^1_{C(\sigma_g)}(V, V) = 0$.

4.8 Family of Representations versus Family of Metrics

Consider again the diagram

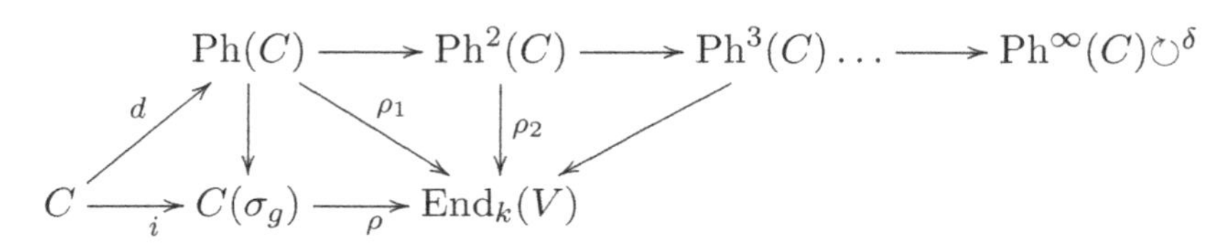

where $\rho_0 := i \circ \rho$ is a representation of a commutative polynomial k-algebra C, g a metric, and ρ a momentum of ρ_0 defined on the corresponding $C(\sigma_g)$.

Let ρ_1 be the induced representation of $\mathrm{Ph}(C)$, and consider a deformation of ρ_1 over the one-dimensional polynomial algebra $k[\tau]$,

$$\tilde{\rho}_1 : \mathrm{Ph}(C) \to \mathrm{End}_{k[\tau]}(\tilde{V}),$$

with $\rho_1 = \tilde{\rho}(0)$.

For any τ_0, we may look at the composition

$$k[\tau] \to k[\epsilon] \simeq k[\tau]/(\tau - \tau_0)^2 \to k(\tau_0) := k[\tau]/(\tau - \tau_0).$$

The corresponding extensions

$$\mathrm{Ph}(C) \overset{\tilde{\rho}_1}{\to} \mathrm{End}_{k[\tau]}(\tilde{V}) \to \mathrm{End}_{k[\epsilon]}(V \otimes k[\epsilon]) \to \mathrm{End}_k(V),$$

define the following notation:

$$\frac{\partial}{\partial \tau}(\tilde{\rho}_1)(\tau_0) \in \mathrm{Ext}^1_{\mathrm{Ph}(C)}(V, V).$$

Given a curve g_τ in $\mathbf{M}$, and a representation ρ of $C(\sigma_{g_{\tau_0}})$, we know what to understand by

$$\frac{\partial}{\partial \tau}(g_\tau)(\tau_0) \in \mathrm{Ext}^1_{\mathrm{Ph}(C)}(V, V)^{(1)}.$$

If the curve g_τ is an integral curve of the vector field $\mathfrak{G}$, then

$$\frac{\partial}{\partial \tau}(g_\tau)(\tau_0) = [\mathfrak{G}(g_{\tau_0})] \in \mathrm{Ext}^1_{\mathrm{Ph}(C)}(V, V)^{(1)}.$$

We may formulate at least three problems:

First, given a curve g_τ in $\mathbf{M}$, and a representation $\rho_{\tau_0} : C(\sigma_{g_{\tau_0}}) \to \mathrm{End}_k(V)$, does there exist a one-dimensional family of representations,

$$\tilde{\rho}_1 : \mathrm{Ph}(C) \to \mathrm{End}_{k[\tau]}(\tilde{V}),$$

with $\tilde{V} = V \otimes_k k[\tau]$, $\rho_{\tau_0 1} = \tilde{\rho}_1(\tau_0)$, such that, for all τ_1,

$$\frac{\partial}{\partial \tau}(g_\tau)(\tau_1) = \frac{\partial}{\partial \tau}(\tilde{\rho}_1)(\tau_1) \in \mathrm{Ext}^1_{\mathrm{Ph}(C)}(V_{\tau_1}, V_{\tau_1})^{(1)} \ ?$$

Second, given a metric g, and a representation $\rho : C(\sigma_g) \to \mathrm{End}_k(V)$, inducing a representation ρ_1 of $\mathrm{Ph}(C)$, let

$$\tilde{\rho}_1 : \mathrm{Ph}(C) \to \mathrm{End}_{k[\tau]}(\tilde{V})$$

be a one-dimensional family of representations with $\rho_1 = \tilde{\rho}_1(0)$, such that for all τ_0,

$$\frac{\partial}{\partial \tau}(\tilde{\rho}_1)(\tau_0) \in \mathrm{Ext}^1_{\mathrm{Ph}(C)}(V_{\tau_0}, V_{\tau_0})^{(1)},$$

does there exist a curve of metrics $g_\tau \in \mathbf{M}$, such that $g_0 = g$ and

$$\frac{\partial}{\partial \tau}(\tilde{\rho}_1)(\tau_0) = \frac{\partial}{\partial \tau}(g_\tau)(\tau_0) \in \mathrm{Ext}^1_{\mathrm{Ph}(C)}(V_{\tau_0}, V_{\tau_0})^{(1)} \ ?$$

Third, does there exist an (algebraic) integral curve, g_τ, in $\mathbf{M}$ of the vector field $\mathfrak{G}$, such that

$$\frac{\partial}{\partial \tau}(g_\tau)(\tau) = [\mathfrak{G}](g_\tau) \ ?$$

Recall that in our model the Dirac derivation δ induces the dynamical structure $\delta := \delta_g = \mathrm{ad}(g - T)$ for every metric, i.e. for every point in the space of metrics $\mathbf{M}$. It also creates the time-like vector field $\mathfrak{G}$ in $\mathbf{M}$, together with the observable time in our space $\mathrm{Spec}(C)$. So, it is reasonable to promote δ to our Chronos=Time, and maybe call $\delta_g = \mathrm{ad}(g - T)$, his Demiurge, taking care of time for us, Earth-dwellers.

If we may answer affirmatively to the first and the third questions, it seems that we have taken care of the problem of understanding the relations between the metrics and the furniture of our space Spec(C). However, suppose we find that some interesting representation, in our furniture, is the iterated extensions of sub-representations of Θ_C, with non-trivial action of a local gauge group $\mathfrak{g}_1$. Then this may impose new conditions for our time model, represented by the metric g. It is reasonable to think that g must be $\mathfrak{g}$-invariant. This would restrict our choice of metrics, and thereby also of our stock of furniture, satisfying the conditions above.

Before we introduce a more concrete model for space and time, in Chapter 5, let us sum up the essentials of our general model. Going back to Section 4.5, we proposed to study the structure of the *space*,

$$\mathbf{U} := \mathrm{Ph}^{\infty}(\mathfrak{M})/\sigma,$$

corresponding to an open affine covering of $\mathfrak{M}$, by subspaces of type $U(\sigma) := \mathrm{Simp}(A(\sigma))$, for $A := C$, a commutative polynomial algebra, and for any metric g, defined in Spec(C), we have seen that $\sigma_g = (g^{i,j} - [dt_i, t_j]) \subset \mathrm{Ph}(A)$ furnish models for a "relativistic quantum theory," where time is encoded in the metric g. Moreover, we have that these (quantum) algebras of observables $c(\sigma_g)$ form a differentiable family over the space of metrics $\mathbf{M}$.

The fact that infinitesimally changing the metric implies a universal change of all Ph(C)-representations defined over $c(\sigma_g)$ resembles the idea of positive and negative energies, the sum of which should be constant.

There are symmetries, gauge groups, i.e. Lie algebras $\mathfrak{g}$, of different kinds, and the Space we are interested in should be given by the category of representations of $C(\sigma_g)$ with $\mathfrak{g}_0$-connections, where the "states" of interest are those marked with the eigenvalues of the Cartan sub-algebra $\mathfrak{h}$ of $\mathfrak{g}_1$.

Moreover, note that distances in space are, today, measured by the time light needs to connect points. If space is empty, i.e. if the only representations we consider are the trivial representation $\mathrm{Simp}_1(C)$, then space becomes continuous. However, if we consider some furniture, given by the representation V, the eigenvalues of the time operator $\mathrm{ad}(g-T)$, operating as derivation on $\mathrm{End}_k(V)$, may be discrete. The semi-group of real eigenvalues may then have a least positive element, the Planck constant, $\hbar$ (see Laudal, 2011, Section 1.7). If the representation V is related to light, say photons, then measurable space must also be discontinuous, or quantized as it is called in the physics literature.

The role of the metric, in the construction of the quantum theory above, was based on the fact that C was a commutative polynomial algebra so that, in particular, $[dt_i, t_j] = [dt_j, t_i]$. To generalize this to the case of any associative algebra A substituting the classical metric by some non-commutative "metric" in order to obtain a finitely generated dynamical structure $\mathrm{Ph}^*(A)/\sigma$, should be a goal, but it may be difficult to achieve.

Finally, we have seen, in Section 4.5, that the *Ghost Fields* have well defined Chern–Simons class, meaning that the Chern–Simons class is a well-defined functional,

$$\mathrm{csh}^* : \mathbf{P} \to \mathrm{HH}^*(C, \mathrm{End}_k(V)).$$

Chapter 5

Time–Space and Space–Times

In a first paper on the problem of defining Time, (see Laudal, 2003), we sketched a Toy Model in physics, where the space–time of classical physics became a section of a universal fiber space $\tilde{E}$, defined on the moduli space, $\mathbb{H} := \mathrm{Hilb}^{(2)}(\mathbb{A}^3)$, of the physical systems we chose to consider. In this case, the systems are composed of an observer and an observed, both sitting in the affine real 3-space, $\mathbb{A}^3$ (see Section 1.2 for why to choose the affine space $\mathbb{A}^3$).

This moduli space, the Hilbert scheme of sub-schemes of length 2 in $\mathbb{A}^3(k)$, is easily computed and has the form $\mathbb{H} = \underline{\tilde{H}}/\mathbb{Z}_2$, where $\underline{H} = \mathbb{A}^3 \times \mathbb{A}^3$ is the space of all ordered pairs of points in $\mathbb{A}^3$, $\underline{\tilde{H}}$ is the blow-up of the diagonal $\underline{\Delta} \subset \underline{H}$, and $\gamma \in \mathbb{Z}_2$ is the obvious group-action interchanging the observer and the observed. Put $H := k[t_1, t_2, t_3, t_4, t_5, t_6]$, then $\underline{H} = \mathrm{Simp}_1(H)$.

Measurable time, in this mathematical model, turned out to be a metric g on the time–space, measuring all possible infinitesimal changes of *the state* of the objects in the family we are studying. This implies that the notion of relative velocity may be interpreted as an oriented line in the tangent space $T_{\tilde{H},(o,p)}$ of a point $(o,p) \in \underline{\tilde{H}}$. Thus, the space of velocities is compact.

This leads to a *physics* where there are no infinite velocities, and where the principle of relativity comes for free. The abelian Lie algebra of translations in $\mathbb{A}^3$ defines a three-dimensional distribution $\tilde{\Delta}$ in the tangent bundle of $\underline{\tilde{H}}$, corresponding to zero-velocities. We shall see that the natural metrics on $\underline{\tilde{H}}$ will make the distribution $\tilde{c}$, corresponding to light-velocities, normal to the distribution $\tilde{\Delta}$ corresponding to the zero-velocities.

We explained how the classical *space-time* can be thought of as a universal subspace $\underline{\tilde{M}}(l)$, of $\underline{\tilde{H}}$ defined by a fixed line $l \subset \mathbb{A}^3$.

We shall also show, see Section 5.1, how the generator $\gamma \in \mathbb{Z}_2$, here just denoted P, and called the *Parity* operator, is linked to the usual operators C, P, T in classical physics, where C, the Charge Inversion, is the identity on $\tilde{c}$, and $(-\operatorname{id})$ on $\tilde{\Delta}$, and P is the identity on $\tilde{\Delta}$, and $(-\operatorname{id})$ on $\tilde{c}$, such that for T, the Time Inversion operator $\gamma^2 = CPT = \operatorname{id}$. Moreover, we observed that the three fundamental gauge groups of current quantum theory U(1), SU(2), and SU(3) are part of the structure of the fiber space,

$$\tilde{E} \longrightarrow \underline{\tilde{H}}.$$

In fact, we might introduce bases in affine 3-space so that we may talk about the Euclidean 3-space $\mathbb{E}^3$. This is not necessary, as we shall make clear in Chapter 8, but it makes it a bit easier to see the geometry of $\underline{\tilde{H}}$.

For any point $\underline{t} = (o, p)$ in $\underline{H}$, outside the diagonal $\underline{\Delta}$, we may consider the line l in $\mathbb{E}^3$ defined by the pair of points $(o, p) \in \mathbb{E}^3 \times \mathbb{E}^3$. We may also consider the action of U(1) on the normal plane $B_o(l)$, of this line, oriented by the normal (o, p), and on the same plane $B_p(l)$, oriented by the normal (p, o). Using parallel transport in $\mathbb{E}^3$, we find an isomorphism of bundles,

$$P_{o,p} : B_o \to B_p,\ P : B_o \oplus B_p \to B_o \oplus B_p,$$

the *partition isomorphism.* Using P we may write (v, v) for $(v, P_{o,p}(v)) = P((v, 0))$. We have also seen, in loc.cit., that the line l defines a unique sub-scheme $\underline{H}(l) \subset \underline{H}$. The corresponding tangent space at (o, p), is called $A_{(o,p)}$. Together, these define a decomposition of the tangent space of $\underline{H}$,

$$T_{\underline{H}} = B_o \oplus B_p \oplus A_{(o,p)}.$$

If $\underline{\lambda} = (o, o) \in \underline{\Delta}$, and if we consider a point $\underline{\omega}$ in the exceptional fiber $E_{\underline{\lambda}}$ of $\underline{\tilde{H}}$, we find that the tangent bundle decomposes into

$$T_{\underline{\tilde{H}},\underline{\omega}} = C_{\underline{\omega}} \oplus A_{\underline{\omega}} \oplus \tilde{\Delta},$$

where $C_{\underline{\omega}}$ is the tangent space of $E_{\underline{\lambda}}$, $A_{\underline{\omega}}$ is the light velocity defining $\underline{\omega}$, and $\tilde{\Delta}$ is the zero-velocities. Both B_o and B_p as well as the bundle

$C_{(o,p)} := \{(\psi, -\psi) \in B_o \oplus B_p\}$ become complex line bundles on $\underline{H} - \underline{\Delta}$. $C_{(o,p)}$ extends to all of $\underline{\tilde{H}}$, and its restriction to E_o coincides with the tangent bundle. Tensorizing with $C_{(o,p)}$, we complexify all bundles. In particular, we find complex dimension 2, bundles $\mathbb{C}B_o$ and $\mathbb{C}B_p$, on $\underline{H} - \underline{\Delta}$, and we obtain a canonical decomposition of the complexified tangent bundle.

As we shall see, the natural metrics g on $\underline{\tilde{H}}$ will decompose the tangent space into the light-velocities $\tilde{\mathbf{c}}$ normal to the zero-velocities, $\tilde{\Delta}$ (but be prepared to change definition, when we come to the cosmological situation, and the Big Bang). This decomposition can, of course, also be extended to the complexified tangent bundle of $\underline{\tilde{H}}$,

$$T_{\underline{H}} = \tilde{\mathbf{c}} \oplus \tilde{\Delta}, \;\; \mathbb{C}T_{\underline{H}} = \mathbb{C}\tilde{\mathbf{c}} \oplus \mathbb{C}\tilde{\Delta}.$$

Clearly, U(1) act on $T_{\underline{H}}$, and SU(2) and SU(3) act naturally on $\mathbb{C}B_o \oplus \mathbb{C}B_x$ and $\mathbb{C}\tilde{\Delta}$, respectively. Moreover, SU(2) acts also on $\mathbb{C}C_{o'}$, in such a way that the actions should be *physically* irrelevant. The groups, U(1), SU(2), SU(3) are our elementary *gauge groups*, and we shall consider the corresponding Lie algebra,

$$\mathfrak{g}^* := \mathfrak{u}(1) \oplus \mathfrak{s}u(2) \oplus \mathfrak{s}u(3)$$

as the *gauge group*, in the sense of Chapter 4. We shall also see, in Chapter 8, that the gauge group structure introduced here actually follows from purely algebraic geometry, no need to introduce an Euclidean metric!

From now on, there will be no warnings when we suddenly go from a real situation, i.e. from $k = \mathbb{R}$, to a complexified situation, where $k = \mathbb{C}$. There are few problems involved and one should, unless there are mistakes, easily understand what is going on.

5.1 The Cylindrical Coordinates, Newton, and Kepler

Let us study the geometry of $\mathbb{H}$. Recall that $\underline{\tilde{H}} \to \underline{H}$ is the (real) blow-up of the diagonal $\underline{\Delta} \subset \underline{H}$, where $\underline{H}$ is the space of pairs of points in $\mathbb{E}^3$. Clearly, any point $\underline{t} \in \underline{H}$ outside the diagonal determines a vector $\xi(o,p)$ and an oriented line $l(o,p) \subset \mathbb{A}^3$, on which both the observer o and the observed p sit. This line also determines a subscheme $\underline{H}(l) \subset \underline{H}$, see above and Laudal (2005), and in $\underline{H}(l)$ there is a unique *light velocity curve* $\underline{l}(\underline{t})$, through $\underline{t}$, an integral curve of the distribution $\tilde{c}$, and this curve cuts the diagonal $\underline{\Delta}$ at a unique point $c(o,p)$, the *center of "gravity" of the observer and the observed*, and thus defines a unique point $\xi(\underline{t})$, of the blow-up of the diagonal, in the fiber of $\underline{\tilde{H}} \to \underline{H}$, above $c(o,p)$.

Recall that the subspace $\underline{\tilde{M}}(l) \subset \underline{\tilde{H}}$, corresponding to a line $l \subset \mathbb{A}^3$, referred to above, consists of all points $(o,p) \in \underline{\tilde{H}}$ for which $c(o,p) \in l$.

Let $\tilde{\Delta}$ be the sub-bundle of the tangent bundle $\Theta_{\tilde{H}}$ of tangent vectors $\eta := (\eta_1, \eta_2), \eta_2 = \eta_1$ at the points $\underline{t} = (o,x)$. We have that $\tilde{\Delta}$ corresponds to the zero-velocities. Then $\tilde{c}$, normal to $\tilde{\Delta}$ in some chosen metric, corresponds to the sub-bundle of light velocities.

There is a convenient parametrization of $\underline{\tilde{H}}$.

Definition 5.1. The cylindrical coordinates of $\underline{\tilde{H}}$ are defined by: Given a point $\underline{\lambda} \in \underline{\Delta}$, and a point $\underline{\omega} \in E(\underline{\lambda}) = \pi^{-1}(\underline{\lambda})$, the fiber of

$$\pi : \underline{\tilde{H}} \to \underline{H},$$

at the point $\underline{\lambda}$, for $o = p$. Since $E(\underline{\lambda})$ is homomorphic to S^2, parametrized by $\underline{\omega} = (\theta, \phi)$, any element of $\underline{\tilde{H}}$ is now uniquely determined in terms of the triple $\underline{t} = (\underline{\lambda}, \underline{\omega}, \rho)$, such that $c(\underline{t}) := c(o,p) = \underline{\lambda}$, and such that $\underline{\omega}$ is defined by the line $\overline{op}$, and the action of θ keeps $\underline{\omega}$ fixed. Here, $\rho \geq 0$, see also Chapter 7. Note also that, at the exceptional fiber, i.e. for $\rho = 0$, the momentum corresponding to $d\rho$ is not defined. This shows that topologically,

$$\underline{\tilde{H}} \simeq \underline{\Delta} \times S^2 \times \mathbb{A}^1_+.$$

This is remarkably close to the notion of *Future null infinity*, $\mathfrak{I}^+$, introduced by Penrose, and used in the theory of gravitational waves, see, e.g. Sormani (2017), where the S^2 factor is a consequence of the Minkowski structure of space time, in particular, the Null Cone. Note also, for later use, that in the same theory physicists define a "Trautman–Bondi mass," $M(u)$, assigned to any null surface, C_u, in their space, the intersection of which with $\mathfrak{I}^+$ would be isomorphic to $E(\underline{\lambda}) \simeq S^2$ (I propose that $M(u)$ is related to our $h(\underline{\lambda})$, which we shall meet shortly).

In 2014, see Manin and Marcolli (2014), Manin and Marcolli observed that the physicists future and past infinities could be explained as blow-ups in algebraic geometry, and they used this to comment on Penrose's theory of a cyclic universe, where eons follow eons. See Chapter 14, where I take up these notions, and their theological background!

Remark 5.1 (Reminder on the metric of a sphere). We may consider the sphere as the real affine scheme, defined by the algebra

$$A = k[x_1, x_2, x_3]/\left(\sum_i x_i^2 = 1\right), \quad k = \mathbb{R}, \;\; i = 1, 2, 3.$$

Consider the commutative version of $\mathrm{Ph}(A)$, given by

$$\mathrm{Ph}(A) = k[x_1, x_2, x_3, dx_1, dx_2, dx_3]/\left(\sum_i x_i^2 = 1, \sum_i x_i dx_i = 0\right),$$

and consider the metric

$$g = \sum_i dx_i{}^2.$$

Now, use the polar coordinates, (ρ, θ, ϕ), with

$$\rho = 1, (\theta, \phi) \in S^1 \times S^1$$

with

$$x_1 = \sin(\theta)\cos(\phi), \;\; x_2 = \sin(\theta)\sin(\phi), \;\; x_3{}^2 = 1 - \sin^2(\theta) = \cos^2(\theta).$$

Then

$$\begin{aligned} dx_1 &= \cos(\theta)\cos(\phi)d\theta - \sin(\theta)\sin(\phi)d\phi, \\ dx_2 &= \cos(\theta)\sin(\phi)d\theta + \sin(\theta)\cos(\phi)d\phi, \\ dx_3^2 &= \sin^2(\theta)d\theta^2, \end{aligned}$$

and the metric comes out as

$$g := d^2\underline{\omega} = d\theta^2 + \sin^2(\theta)d\phi^2.$$

Of course, we then would have $A \subset \tilde{A} := k\{\theta, \phi\}$, the analytical extension of A. In $\mathrm{Ph}(\tilde{A})$, the Force Laws would look like

$$\begin{aligned} d^2\theta &= \sin(\theta)\cos(\theta)d\phi^2, \\ d^2\phi &= -2\frac{\cos(\theta)}{\sin(\theta)}d\theta d\phi, \\ \Gamma^{\theta}_{\phi,\phi} &= -\sin(\theta)\cos(\theta), \\ \Gamma^{\phi}_{\phi,\theta} = \Gamma^{\phi}_{\theta,\phi} &= \frac{\cos(\theta)}{\sin(\theta)}, \end{aligned}$$

with all other Christoffel coefficients vanishing. The geodesics are solutions of the equations

$$d^2\theta = -\Gamma^{\theta}_{\phi,\phi} d\phi^2, \; d^2\phi = -\Gamma^{\phi}_{\phi,\theta} d\phi d\theta - \Gamma^{\phi}_{\theta,\phi} d\theta d\phi,$$

and are seen to be the plane sections through the center of the sphere, as they should. To accept this, consider a non-zero vector $l := (\alpha, \beta, \gamma)$ in $\mathbb{A}^3$, and let the point $\underline{\omega}(\tau)$ describe a curve on the sphere in the plane normal to l, i.e. such that

$$\alpha \sin(\theta) \cos(\phi) + \beta \sin(\theta) \sin(\phi) + \gamma \cos(\theta) = 0.$$

Now, derive the left-hand side twice, and introduce

$$\ddot{\theta} = \sin(\theta) \cos(\theta) \dot{\phi}^2,$$

$$\ddot{\phi} = -2 \frac{\cos(\theta)}{\sin(\theta)} \dot{\theta} \dot{\phi},$$

and see that it comes out right.

The Ricci tensor is given by

$$R_{\theta,\theta} = 1, R_{\phi,\phi} = \sin^2(\theta)$$

and the scalar curvature comes out as

$$R = 2.$$

Moreover,

$$\nabla_{\delta_\phi}(\delta_\phi) = -\sin(\theta) \cos(\theta) \delta_\theta,$$

$$\nabla_{\delta_\phi}(\delta_\theta) = \cot(\theta) \delta_\phi,$$

$$\nabla_{\delta_\theta}(\delta_\phi) = \cot(\theta) \delta_\phi,$$

$$\nabla_{\delta_\theta}(\delta_\theta) = 0.$$

Consider now any metric on $\underline{\tilde{H}}$. Later, in Chapter 8, we shall come back to the conditions that a metric should satisfy, given the fact that there are gauge groups acting. A first reasonable condition is that it must be of the form

$$g = h_\rho(\underline{\lambda}, \underline{\omega}, \rho) d\rho^2 + h_{\underline{\omega}}(\underline{\lambda}, \underline{\omega}, \rho) d\underline{\omega}^2 + h_{\underline{\lambda}}(\underline{\lambda}, \underline{\omega}, \rho) d\underline{\lambda}^2, \tag{5.1}$$

where $d\underline{\omega}^2 = d\theta^2 + \sin^2 \theta d\phi^2$ is the natural metric in $S^2 = E(\underline{\lambda})$, see the remark above, and $d\underline{\lambda}$ is the metric on the three-dimensional space $\underline{\Lambda}$.

Moreover, the component $h_{\underline{\omega}}(\underline{\lambda}, \underline{\phi}, \rho)$ should not depend on θ. In fact, it seems that all $h_\rho, h_{\underline{\omega}}, h_{\underline{\lambda}}$ should depend just on ρ and $\lambda = \lambda_3$.

We shall also see that to be able to model black holes, and certainly the Big Bang event itself, we shall have to accept that the metric may degenerate in a certain subspace, from now on called the *Horizon*, which turns out to be given by $\rho = h(\lambda)$, where h is a function to be chosen. We shall demand that everywhere else it is non-degenerate and, since the space $\underline{\tilde{H}}$ is smooth everywhere, locally isomorphic to an affine space, the spectrum of a polynomial algebra C, everything we have done in Chapter 4 is valid outside the Horizon in this model.

It is reasonable to believe that the geometry of $(\underline{\tilde{H}}, g)$ might explain the notions like gravitational *energy, mass, charge*, etc. In fact, we tentatively propose that the source of mass is located in the *black holes* $E(\underline{\lambda})$. This would imply that mass, charge, etc. are properties of the five-dimensional superstructure of our usual three-dimensional Euclidean space, essentially given by a *density*, $h(\underline{\lambda})$.

This might bring to mind Kaluza–Klein theory. However, it seems to me that there are important differences, making comparison very difficult.

Let us first treat the following simple case:

$$h_\rho = \left(\frac{\rho - h}{\rho}\right)^2, \quad h_{\underline{\omega}} = (\rho - h)^2, \; h_{\underline{\lambda}} = 1,$$

where h is a positive real number. This metric is everywhere defined, but note that for $\rho = 0$, there are no tangent vectors in $d\rho$ direction. As we have seen, the topology of this space is

$$\mathbb{R}^3 \times S^2 \times \mathbb{R}_+,$$

and it is clear that the metric reduces to the Euclidean metric far away from $\underline{\Delta}$, and it is singular on the *horizon* of the black hole, given by $\rho = h$, which in $\underline{H}$ is simply a sphere in the *light-space*, of radius h. Moreover, it is clear that h is also the *radius of the exceptional fiber*, since the length of the circumference of $\rho = 0$ is $2\pi h$. Clearly, the exceptional fiber, the black hole itself, is not visible, and does not bound anything. However, if $h \geq 0$, the horizon bounds a piece of space. If $h = 0$, the horizon is reduced to a point in $\underline{H}$, then the circumference, or area of the exceptional fiber, as measured using the above metric, reduces to zero, and the metric becomes the usual Euclidean metric. So, this is in tune with the proposal of a Cosmic Censorship, i.e. the non-existence of *naked singularities*.

Now, recall that the mass of a black hole, in cosmology, has been related to the area of the horizon, therefore to h^2. Thus, the Trautman–Bondi mass, $M(u)$, for u approaching ∞, might have something to do with the area of the exceptional fiber, the black hole, $E(\underline{\lambda})$.

To understand the geometry of this space, including the notion of *gravity*, we shall reduce further to a plane in the light directions, i.e. we shall assume that $S^2 = E(\underline{\lambda})$ is reduced to a circle, with coordinate ϕ (say, putting $\theta = \frac{1}{2}\pi$). So, we consider the metric

$$g = \left(\frac{\rho - h}{\rho}\right)^2 d\rho^2 + (\rho - h)^2 d\phi^2 + d\lambda^2,$$

in the topological space

$$\mathbb{R} \times S^1 \times \mathbb{R}_+.$$

The corresponding equations for the geodesics in $\underline{\tilde{H}}$ are, see Laudal (2011),

$$\begin{aligned}
\frac{d^2\rho}{dt^2} &= -\left(\frac{h}{\rho(\rho - h)}\right)\left(\frac{d\rho}{dt}\right)^2 + \left(\frac{\rho^2}{(\rho - h)}\right)\left(\frac{d\phi}{dt}\right)^2, \\
\frac{d^2\phi}{dt^2} &= -2/(\rho - h)\frac{d\rho}{dt}\frac{d\phi}{dt}, \\
\frac{d^2\lambda}{dt^2} &= 0,
\end{aligned}$$

where t is time. But time is, by definition, the distance function in $\underline{\tilde{H}}$, so we must have

$$\left(\frac{\rho - h}{\rho}\right)^2 \left(\frac{d\rho}{dt}\right)^2 + (\rho - h)^2 \left(\frac{d\phi}{dt}\right)^2 + \left(\frac{d\lambda}{dt}\right)^2 = 1,$$

from which we find the time equation (TQ)

$$\left(\frac{d\rho}{dt}\right)^2 = \rho^2(\rho - h)^{-2}\left(1 - \left(\frac{d\lambda}{dt}\right)^2\right) - \rho^2\left(\frac{d\phi}{dt}\right)^2. \tag{5.2}$$

From the third equation, we find that $\frac{d\lambda}{dt}$, is constant, and $|\frac{d\lambda}{dt}|^2$ is the *rest-mass* of the system. Put $K^2 = (1 - |\frac{d\lambda}{dt}|^2)$, then K^2 is related to v^2 and will give us the *kinetic energy of the system.* The definition of time therefore gives us

$$\rho^{-2}\left(\frac{d\rho}{dt}\right)^2 = (\rho - h)^{-2}K^2 - \left(\frac{d\phi}{dt}\right)^2.$$

Put this into the first equation above, and obtain

$$\frac{d^2\rho}{dt^2} = -hK^2\left(\frac{\rho}{\rho - h}\right)\frac{1}{(\rho - h)^2} + \left(\frac{\rho + h}{\rho - h}\right)\rho\left(\frac{d\phi}{dt}\right)^2.$$

Assume now $r := \rho - h \approx \rho$, we find

$$\frac{d^2r}{dt^2} = -\frac{hK^2}{r^2} + r\left(\frac{d\phi}{dt}\right)^2, \tag{5.3}$$

i.e. Kepler's first law. The constant h, i.e. the radius of the exceptional fiber (here reduced to a circle), is thus also related to mass. Recall that the Schwarzschild radius, the Einstein equivalent to h, is assumed to be

$$r_s = 2GM/c^2,$$

where G = Newton's gravitational constant, M = mass, c = speed of light, which here, of course, is put equal to 1.

Then we find

$$hK^2 = GMv^2$$

close to the classical kinetic energy. As we have hinted at above, this suggests that *mass* is a property of the metric of the space $\underline{\tilde{H}}$.

In this case, it is a function of the surface of the exceptional fiber, i.e. the *black hole*, associated with the point $\underline{\lambda}$ in the ordinary 3-space $\underline{\Delta}$. Recall also Penrose's idea about mass in the Universe as the union of the areas of the black holes, and the Penrose inequality,

$$M^2 \leq \frac{A}{16\pi} = \frac{1}{4}h(\lambda)^2,$$

see Penrose (2010). One expects the density of gravitational "Mass," or "Energy," to be related to

$$\text{Area}(E(\underline{\lambda})) = 4\pi h(\underline{\lambda})^2.$$

This is now a consequence of the proposed constancy of mass in the Universe, i.e. that the mass of the Universe as it is at cosmological time λ is constant w.r.t. λ, see Chapter 7. Of course, we do not "see" the Universe "at cosmological time λ."

The second equation above gives us Kepler's second law,

$$r\left(\frac{d^2\phi}{dt^2}\right) + 2\left(\frac{dr}{dt}\right)\left(\frac{d\phi}{dt}\right) = 0. \tag{5.4}$$

Note that with the chosen metric, time, in light velocity direction, is *standing still* on the *horizon* $\rho = h$, of the *black hole* at $\underline{\lambda} \in \underline{\Delta}$. Therefore, no light can escape from the black hole. In fact, no geodesics can pass through $\rho = h$. Note also that, for a photon with light velocity, we have $K = 1$, so we may measure h, by measuring the trajectories of photons in the neighborhood of the *black hole.* Finally, see that if the distance between the two interacting points is close to constant, i.e. if we have a circular movement, the left side of the time-equation becomes zero, and we therefore have the following equation:

$$(\rho - h)d\phi = Kdt$$

which may be related to the perihelion precession, and also to the *Thomas Precession*, see Sachs and Wu (1977) and Weinberg (2005).

Now, suppose, we use the quantum-theoretical general force law of our metric, reduced to the commutative case, then, see the previous calculations, we obtain

$$d^2t_i = -\sum_{p,q}\Gamma^i_{p,q}dt_p dt_q.$$

We would have got the same Kepler's laws, of the form

$$\begin{aligned} d^2\rho &= -\frac{h}{\rho(\rho - h)}d\rho^2 + \frac{\rho^2}{(\rho - h)}d\phi^2, \\ d^2\phi &= -2\frac{1}{(\rho - h)}d\rho d\phi. \end{aligned}$$

Example 5.1. Let us compute the Killing vectors, i.e. the global gauge group $\mathfrak{g}_0$, for this metric. With the above indexes, one obtains

$$\gamma = \gamma_\rho \frac{\partial}{\partial\rho} + \gamma_\phi \frac{\partial}{\partial\phi} + \gamma_\lambda \frac{\partial}{\partial\lambda} \in \mathfrak{g}_0,$$

if and only if

$$\gamma(g_{i,j}) + \frac{\partial\gamma_p}{\partial t_i} g_{p,j} + \frac{\partial\gamma_p}{\partial t_j} g_{i,p} = 0.$$

There is just one solution, $\gamma_\rho = 0, \gamma_\phi = \text{const}, \ \gamma_\lambda = \text{const}$, so $\mathfrak{g}_0$ is just the rotations in ϕ and the translations in λ.

If we, in our metric, permit the radius of the black hole, h, to depend on $\lambda := |\underline{\lambda}|$, i.e. be given by a function of the form $h(\lambda)$, so that the metric looks like

$$\left(\frac{\rho - h(\lambda)}{\rho}\right)^2 d\rho^2 + (\rho - h(\lambda))^2 d\phi^2 + \kappa(\lambda) d\lambda^2,$$

then the force law formulas above become more involved,

$$\begin{aligned}
\frac{d^2\rho}{dt^2} &= -\left(\frac{h(\lambda)}{\rho(\rho - h(\lambda))}\right)\left(\frac{d\rho}{dt}\right)^2 + \left(\frac{2}{(\rho - h(\lambda))}\right)\left(\frac{dh}{d\lambda}\right)\left(\frac{d\rho}{dt}\right)\left(\frac{d\lambda}{dt}\right) \\
&\quad + \left(\frac{\rho^2}{(\rho - h(\lambda))}\right)\left(\frac{d\phi}{dt}\right)^2, \\
\frac{d^2\phi}{dt^2} &= -2/(\rho - h(\lambda))\frac{d\rho}{dt}\frac{d\phi}{dt} + 2/(\rho - h(\lambda))\left(\frac{dh}{d\lambda}\right)\left(\frac{d\phi}{dt}\right)\left(\frac{d\lambda}{dt}\right), \\
\frac{d^2\lambda}{dt^2} &= \left(\frac{(\rho - h(\lambda)}{\rho}\right)\left(\frac{1}{\kappa(\lambda)}\right)\left(\frac{dh}{d\lambda}\right)\left(\frac{d\rho}{dt}\right)^2 \\
&\quad + (\rho - h(\lambda))\left(\frac{1}{\kappa(\lambda)}\right)\left(\frac{dh}{d\lambda}\right)\left(\frac{d\phi}{dt}\right)^2 + 1/2\left(\frac{dln(\kappa)}{d\lambda}\right)\left(\frac{d\lambda}{dt}\right)^2,
\end{aligned}$$

where t, as above, is time.

We find that if $(\frac{dh}{d\lambda})$ is negative, then for $\rho \leq h(\lambda)$, the acceleration of ρ may be positive, and unlimited for ρ close to $h(\lambda)$, and the acceleration of λ is negative. We shall come back to this in relation to the Big Bang and inflation, where we also shall meet the Friedman–Robertson–Walker metric currently used in cosmology.

Since we then will have to compare our model with the standard models of General Relativity Theory (GRT), let us take a look at the classical Schwarzschild metric (for a black hole), which is given as

$$ds^2 = \left(1 - \frac{2GM}{rc^2}\right) dt^2 - \left(1 - \frac{2GM}{rc^2}\right)^{-1} dr^2 - r^2(d\phi^2 + \sin^2(\phi) d\theta^2).$$

Promoting ds to be our $d\lambda$, and solving for dt, we obtain

$$dt^2 = \left(\frac{r}{r-h}\right)^2 dr^2 + r^2 \left(\frac{r}{r-h}\right) (d\phi^2 + \sin^2(\phi) d\theta^2) + \left(\frac{r}{r-h}\right) d\lambda^2,$$

with $h = GMc^{-2}$. Consider the partition operator P, and see that r is considered to be the distance between the center of the "non-existing" black hole, neglecting the horizon, to the observer o, and ρ is the distance between the sphere $E(\underline{\lambda})$ and the observed p. Therefore, recalling the symmetry, $\rho \leftrightarrow (-\rho)$, we may put $r = -\rho + h$, then

$$\left(\frac{r}{r-h}\right) = \left(\frac{\rho - h}{\rho}\right), r^2 = (\rho - h)^2$$

and the time metric would look like,

$$\begin{aligned} dt^2 = & \left(\frac{\rho - h(\lambda)}{\rho}\right)^2 d\rho^2 + \left(\frac{\rho - h(\lambda)}{\rho}\right) (\rho - h(\lambda))^2 d\phi^2 \\ & + \left(\frac{\rho - h(\lambda)}{\rho}\right) d\lambda^2, \end{aligned}$$

and we are very close to our metric above.

Remark 5.2. Recall that we have formulated almost all our results about dynamical systems, see Chapter 4, in particular, the start of Section 4.3, in terms of affine spaces, or more precisely in affine open subsets. To be able to use them in our case here we have to find an atlas of charts of the form used in Section 4.3, of $\underline{\tilde{H}}$, and compute the metric in the corresponding coordinates. This is relatively easy. Obviously, $\underline{\Delta}$ is an affine space corresponding to the k-algebra $k[\underline{\lambda}]$, where we put $\underline{\lambda} = \{\lambda_1, \lambda_2, \lambda_3\}$, and for the purpose of having a manageable notation, we shall promote the $\underline{\lambda}$, to be t's, putting $t_{i+1} := \lambda_i$, $i = 1, 2, 3$.

Then, of course, we have, for the classical metric of the sphere, picking t_3 as one of the axes in the three-dimensional Euclidean space, in which we can consider the sphere,

$$\begin{aligned}
\rho^2 &= t_1^2 + t_2^2 + t_3^2, \\
t_3 &= \rho \cdot \cos(\theta), \\
t_1 &= \rho \cdot \sin(\theta)\cos(\phi), \\
t_2 &= \rho \cdot \sin(\theta)\sin(\phi), \\
\tan(\phi) &= t_2/t_1, \\
\cos(\theta) &= t_3/\rho, \\
d\phi &= 1/(t_1^2 + t_2^2)(t_2 dt_1 - t_1 dt_2),
\end{aligned}$$

and the metric of the unit sphere is given by

$$d\underline{\omega}^2 = d\theta^2 + \sin^2(\theta) d\phi^2.$$

Then our general metric of $\underline{\tilde{H}}$ above would look like

$$\begin{aligned}
g &= \left(\frac{\rho - h(\lambda)}{\rho}\right)^2 d\rho^2 + (\rho - h(\lambda))^2 d\omega^2 + \kappa(\lambda) d\underline{\lambda}^2 \\
&= \left(\frac{\rho - h(\lambda)}{\rho}\right)^2 dt_1^2 + \left(\frac{\rho - h(\lambda)}{\rho}\right)^2 dt_2^2 + \frac{(\rho - h(\lambda))^2}{(t_1^2 + t_2^2)} dt_3^2 \\
&\quad + \left(\frac{\rho - h(\lambda)}{\rho}\right)^2 \frac{t_1 t_3}{(t_1^2 + t_2^2)} dt_1 dt_3 + \left(\frac{\rho - h(\lambda)}{\rho}\right)^2 \frac{t_2 t_3}{(t_1^2 + t_2^2)} dt_2 dt_3 \\
&\quad + \kappa(t_4, t_5, t_6)(dt_4^2 + dt_5^2 + dt_6^2),
\end{aligned}$$

with

$$\kappa := \kappa(t_4, t_5, t_6)$$

some function of $\underline{\lambda}$.

This is the expression of the metric in the chart $D_3 := \rho_3 := (t_1^2 + t_2^2) \neq 0$. The obvious atlas $\{D_i\}$, $i = 3, 2, 1$, covers $\underline{\tilde{H}} - \underline{\tilde{\Delta}}$.

On the blow-up, $\underline{\tilde{\Delta}}$, the metric is given as

$$g|\underline{\tilde{\Delta}} := h(\lambda)^2(d\theta^2 + \sin^2(\theta)d\phi^2) + \kappa(t_4, t_5, t_6)(dt_4^2 + dt_5^2 + dt_6^2).$$

The determinant $\mathfrak{m} := \det(g_{i,j})$ of the metric is easily computed and turns out to be

$$\mathfrak{m} := \left(\frac{\rho - h(\lambda)}{\rho}\right)^6 \kappa^3,$$

vanishing in the subspace $\rho = h(\lambda)$. The inverse metric $(g^{i,j})$ is of course easy to compute, just a little more complicated.

Now, we have the material for applying the results of Chapter 4. In particular, we are free to apply Theorem 4.4, the generic time action to our situation, the sheaf of real functions $\tilde{H}$, and the metric g chosen above. Of course, here as there we have to be careful, restricting to affine opens, given by the *Charts*, chosen. We shall come back to this when we have introduced the universal local gauge group $\mathfrak{g}$ that drops out of the deformation theory of the associative four-dimensional k-algebra, U, our Big Bang singularity.

Example 5.2. But before this, let us go back to the mini-model above, the three-dimensional subspace of $\underline{\tilde{H}}$, with parameters $\{\rho, \phi, \lambda\}$. It corresponds to $\theta = \pi/2$, or $t_3 = 0$. The metric looks like

$$g = \left(\frac{\rho - h}{\rho}\right)^2 dt_1^2 + \left(\frac{\rho - h}{\rho}\right)^2 dt_2^2 + d\lambda^2$$

in the polynomial algebra $k[t_1, t_2, \lambda]$, localized in $\rho^2 = t_1^2 + t_2^2$. Computing, we find

$$\Gamma^1_{1,l} = \Gamma^2_{2,l} = t_l \left(\frac{\rho - h}{\rho}\right)^2 =: t_l G,$$

and putting $\dot{t_l} = v_l$,

$$\ddot{t_1} = -t_1 G {v_1}^2 - t_2 G v_1 v_2, \ \ddot{t_2} = -t_1 G v_1 v_2 - t_2 G {v_2}^2,$$

and finally

$$\ddot{\rho} = -\frac{h({v_1}^2 + {v_2}^2)}{\rho(\rho - h)} + \rho\dot{\phi}^2, \ (v_1^2 + v_2^2) = K^2,$$

i.e. the same Kepler's first law as we deduced from the Lagrangian, and the definition of time. The slight difference is due to our definition of time, the clock, that in particular implies that K^2 should be multiplied by $(\frac{\rho}{(\rho-h)})^2$.

5.2 Thermodynamics, the Heat Equation and Navier–Stokes

Let us now go back to Chapter 4, and consider the generic dynamical structure (σ_g), related to the metric on $\underline{\tilde{H}}$, of the form

$$g = h_1(\underline{\lambda}, \underline{\phi}, \rho)d\rho^2 + h_2(\underline{\lambda}, \underline{\phi}, \rho)d\underline{\omega}^2 + h_3(\underline{\lambda}, \underline{\omega}, \omega)d\underline{\lambda}^2,$$

where $d\underline{\omega}^2$ is the natural metric in $S^2 = E(\underline{\lambda})$.

Recall for $C = H = k[t_1, \ldots, t_6]$, and a non-singular Riemannian metric $g = 1/2\sum_{i,j=1,..,r} g_{i,j}dt_idt_j \in \mathrm{Ph}(C)$, the notations

$$\Gamma_p^{j,i} = \sum_k g^{j,k}\Gamma_{k,p}^i,\ R_{i,j} := [dt_i, dt_j], F_{i,j} := R_{i,j} - \sum_p(\Gamma_p^{j,i} - \Gamma_p^{i,j})dt_p,$$

and the general Force Law in $\mathrm{Ph}(C)$,

$$\begin{aligned} d^2t_i = &- \sum_{p,q}\Gamma_{p,q}^i dt_p dt_q - 1/2\sum_{p,q} g_{p,q}(F_{i,p}dt_q + dt_pF_{i,q}) \\ &+ 1/2\sum_{l,p,q} g_{p,q}[dt_p, (\Gamma_l^{i,q} - \Gamma_l^{q,i})]dt_l + [dt_i, T], \end{aligned}$$

generating the dynamical structure $\mathfrak{c} := \mathfrak{c}(g)$.

Remark 5.3. In principle, according to our philosophy, the natural common quantization of classical general relativity and Yang–Mills theory would be based on the dynamical properties of $\mathrm{Simp}_{\leq\infty}(\underline{\tilde{H}}(\mathfrak{c}))$, with respect to the versal family, $\tilde{\rho} : O_{\underline{\tilde{H}}(\mathfrak{c})} \to \mathrm{End}_{\underline{\tilde{H}}(\mathfrak{c})}(\tilde{V})$, where we have to consider $O_{\underline{\tilde{H}}(\mathfrak{c})}$ as a presheaf of associative k-algebras, defined in $\underline{\tilde{H}}$. As a first try, we shall concentrate on situations where the Dirac vectorfield $[\delta]$ vanishes at a chosen representation, a situation that we have termed *singular*. In particular, we shall treat the structure of the Levi-Civita representation.

It is reasonable to believe that the geometry of $(\underline{\tilde{H}}, g)$ might explain the notions like *energy*, mass, *charge*, etc. In fact, we tentatively propose that the source of mass and charge, etc. *is located* in the *black holes* $E(\underline{\lambda})$. This would imply that mass, charge, etc. are properties of the five-dimensional metric superstructure of our usual three-dimensional Euclidean space, essentially given by a *density* $h(\underline{\lambda}, \phi, \theta)$. This might bring to mind Kaluza–Klein theory. However, it seems to me that there are important differences, making comparison very difficult.

Recall that at a point $\underline{t} = (o, x) \in \underline{H} - \underline{\Delta}$, the tangent space, $\Theta_{\underline{\tilde{H}}}(\underline{t})$, is represented by the space of all pairs of three-vectors $\xi(\underline{t}) = (\xi_o, \xi_x)$, ξ_o fixed at o, and ξ_x fixed at the point x in $\mathbf{E}^3$. Moreover, any such tangent vector may, depending only upon the choice of metric, be decomposed into the sum $\xi = \xi_1 + \xi_2$, with $\xi_1 \in \tilde{\Delta}$, and $\xi_2 \in \tilde{c}$.

Recall also that the *center of gravity of the observer and the observed*, $c(o, x) \in \underline{\Delta}$, defined in terms of a Euclidean structure on our three-dimensional space, defines a unique point $\xi(\underline{t})$, of the blow-up of the diagonal, in the fiber of $\underline{\tilde{H}} \to \underline{H}$, above $c(o, x)$.

Now, consider the Levi-Civita connection,

$$\rho_\Theta : \tilde{H}(\sigma_g) \to \mathrm{End}_R(\Theta_{\underline{\tilde{H}}}),$$

together with the Hamiltonian, i.e. the Laplace–Beltrami operator, $Q \in \mathrm{End}_{\underline{\tilde{H}}}(\Theta_{\underline{\tilde{H}}})$.

Any *state* $\xi \in \Theta_{\underline{\tilde{H}}}$ may be interpreted as a (relative) momentum (ξ_o, ξ_x) of the pair of points $(o, x) \in \mathbf{E}^3$, defined for all pairs of points in the domain of definition for ξ. Write, as above,

$$\underline{\xi} = \underline{p} + \underline{m},\ \underline{p} = (p_1, p_2, p_3, 0, 0, 0) \in \tilde{c},\ \underline{m} = (0, 0, 0, m_1, m_2, m_3) \in \tilde{\Delta},$$

where we have introduced local coordinates, $(\lambda_1, \lambda_2, \lambda_3, x_1, x_2, x_3)$, such that

$$\tilde{\Delta} = \langle \delta_{\lambda_1}, \delta_{\lambda_2}, \delta_{\lambda_3} \rangle, \quad \text{and} \quad \tilde{c} = \langle \delta x_1, \delta x_2, \delta x_3 \rangle.$$

The norms

$$\mu := |\xi|,\ m := |\underline{m}|,\ \kappa := |\underline{p}|$$

defined by ξ are the *energy-density*, the *density of mass*, and the *density of kinetic momentum*, respectively. We find that (p_1, p_2, p_3) is a classical relative momentum-vector, and $v = (v_1, v_2, v_3) := \mu^{-1}(p_1, p_2, p_3)$ is a classical velocity vector.

Consider now the corresponding Schrödinger equation, our *Furniture equation*,

$$\frac{d}{dt}(\xi) = Q(\xi).$$

Time, here, is the notion used in quantum theory, and Q is the Laplace–Beltrami operator. We have, however, introduced another notion of time, the metric in general relativity theory, and they should be equal. Therefore, we must have

$$\frac{d}{dt} = \mu^{-1} D_{\xi},$$

as operators on $\Theta_{\underline{\tilde{H}}}$.

Let us compute the left-hand side of the Schrödinger equation. It is clear that

$$\begin{aligned}\frac{d}{dt}(\xi) =& \frac{d}{dt}(\underline{m}) + \frac{d}{dt}(\underline{p}) \\ =& \sum_{1}^{3} \mu^{-1} m_j D_{\lambda_j}(\underline{m}) + \sum_{1}^{3} \mu^{-1} m_j D_{\lambda_j}(\underline{p}) \\ &+ \sum_{1}^{3} v_j D_{x_j}(\underline{m}) + \sum_{1}^{3} v_j D_{x_j}(\underline{p}).\end{aligned}$$

Reduce to the space–time like sub-scheme $\underline{\tilde{M}}(l) \subset \underline{\tilde{H}}$, corresponding to a chosen line $l \subset \mathbb{A}^3$, see Section 5.1, and consult Laudal (2011). Then the three-dimensional vector $\underline{m} = \sum_{i=1}^{3} m_i \frac{\partial}{\partial \lambda_i}$ reduces to $m\frac{\partial}{\partial \lambda}$, and the term $\mu^{-1} m \frac{\partial}{\partial \lambda}$ may be compared to $\frac{\partial}{\partial \tau}$, where τ is the relativistic *proper time*.

The outcome of this is that, reduced to the sub-scheme $\underline{M}(l)$, the Schrödinger equation is, in a realistic classical Euclidean situation, where $D_{\lambda_j} = \frac{\partial}{\partial \lambda_j}$, the coupled equation, containing the general, relativistic Heat Equation (HE),

$$\frac{dm}{dt} = \frac{\partial m}{\partial \tau} + \sum_{j=1}^{3} v_j \frac{\partial m}{\partial x_j} = \Delta(m)$$

and the "relativistic" Navier–Stokes equation (NSE),

$$\frac{dp_i}{dt} = \frac{\partial p_i}{\partial \tau} + \sum_{j=1}^{3} v_j \frac{\partial p_i}{\partial x_j} = \Delta(p_i), \quad i = 1, 2, 3,$$

where $\Delta = Q = \frac{\partial^2}{\partial \lambda^2} + \sum_{j=1}^{3} \frac{\partial^2}{\partial x_j^2}$.

Computing, we find for NSE,

$$\frac{\partial p_i}{\partial \tau} = \mu \frac{\partial v_i}{\partial \tau} + \frac{\partial \mu}{\partial \tau} v_i, \quad \sum_{j=1}^{3} v_j \frac{\partial p_i}{\partial x_j} = \mu(v\nabla v_i) + (v\nabla\mu)v_i$$

where ∇ is the dimension 3 del-operator, with respect to the parameters $\underline{x} = (x_1, x_2, x_3)$. Let $\overline{\nabla}$ be the dimension 4 del-operator, with respect to the parameters (λ, x_1, x_2, x_3), put $\overline{v} = (\nu/\mu, v_1, v_2, v_3)$, and compute

$$\Delta(p_i) = \mu\Delta(v_i) + 2\overline{\nabla}\mu\overline{\nabla}v_i + \Delta(\mu)v_i,$$

from which we deduce an equation, close to the classical Navier–Stokes equation,

$$\mu\left(\frac{\partial v_i}{\partial \tau} + v\nabla v_i\right) = \mu\Delta(v_i) + 2\overline{\nabla}\mu\overline{\nabla}v_i + \Delta(\mu)v_i - (\overline{v}\overline{\nabla}\mu)v_i, \quad i = 1, 2, 3.$$

Remark 5.4. Any vector field $\xi \in \Theta_{\underline{\tilde{H}}}$ may be interpreted as a description of the *relative state* of the space, everywhere, a kind of mass–stress tensor, describing the situation of all pairs of points in our three-dimensional space, together with any corresponding pair of momenta. (The *Furniture* of our cosmos, referred to in the title of this chapter, will be states ξ of a much more complex representation, of the gauge groups, possibly an iterated extension of the simple "atoms" of $\Theta_{\underline{\tilde{H}}}$.)

In general, we might hope that knowing ξ, i.e. the six functions defined in $\underline{\tilde{H}}$, locally defining the vector ξ, the Schrödinger equation would determine the metric g, i.e. the six functions $h_\rho, h_\phi, h_{\underline{\lambda}}$. This would presumably lead to time-developments $\xi(T)$ and $g(T)$, determined by any given ground-state, ξ_*, and clocked by some parameter T. This would again have as a consequence that any cyclic behavior of the phenomenon modeled by $\xi(T)$ would lead to a gravitational wave defined by $g(T)$.

In particular, the collapsing of a star, or the Big Bang event, both usually modeled as a fluid depending on pressure, temperature, energy density, viscosity, entropy, would in the above scenario define a generalized gravitational wave, $g(T)$. See also Theorem 4.8 and the relationship between the changing metrics and the forces on the states of our representations.

We would therefore be tempted to consider the Schrödinger equation,

$$\frac{d}{dt}(\xi) = Q(\xi),$$

as another form of Field Equation, replacing the Einstein Field Equation. A solution would be a metric g determining the dynamics of the *past and the future* of our space. To make this reasonably understandable, we need a mathematical model of the beginning of it all, of the Big Bang. This is, however, the subject of Chapters 7 and 8.

As a first example, of the usefulness of the Schrödinger equation, in studying the geometry of our Universe, consider the very special case of the metric of the last section, defined by

$$h_\rho = \left(\frac{\rho - h}{\rho}\right)^2, \ h_\phi = (\rho - h)^2, \ h_3 = 1,$$

where h is a positive real number.

Put $\rho = t_1, \phi = t_2, \lambda = t_3$, then we find

$$\Gamma^1_{1,1} = h/\rho(\rho - h), \quad \Gamma^1_{2,2} = -\rho^2/(\rho - h),$$
$$\Gamma^2_{1,2} = 1/(\rho - h), \quad \Gamma^2_{2,1} = 1/(\rho - h),$$
$$\Gamma^3_{i,j} = 0.$$

All other components vanish. Since we now have the opportunity, let us also note the relations

$$\overline{\Gamma}^1_{2,2} = -\Gamma^1_{2,2}, \ \overline{\Gamma}^k_{i,j} = \Gamma^k_{i,j}, \quad \text{for} \quad (i,j) \neq (2,2).$$

Moreover, we find the following formulas:

$$D_\rho := \nabla_{\delta_1} = \frac{\partial}{\partial\rho} + \nabla_\rho,$$
$$D_\phi := \nabla_{\delta_2} = \frac{\partial}{\partial\phi} + \nabla_\phi,$$

$$D_\lambda := \nabla_{\delta_3} = \frac{\partial}{\partial \lambda} + \nabla_\lambda$$

$$Q = \sum_{i=1}^{3} 1/h_i \nabla_{\delta_i}^2$$

$$\rho(\delta^2(t_i)) = [Q, \rho(dt_i)] = 1/h_i[Q, \nabla_{\delta_i}].$$

Here, the h_i is the function defined above, i.e. $g_{i,i}$ in our metric, and

$$\nabla_\rho = \begin{pmatrix} h/\rho(\rho-h) & 0 & 0 \\ 0 & 1/(\rho-h) & 0 \\ 0 & 0 & 0 \end{pmatrix}, \quad D_\rho(\delta_\phi) = 1/(\rho-h),$$

$$\nabla_\phi = \begin{pmatrix} 0 & -\rho^2/(\rho-h) & 0 \\ 1/(\rho-h) & 0 & 0 \\ 0 & 0 & 0 \end{pmatrix},$$

$$\nabla_\lambda = 0,$$

$$[\nabla_\rho, \nabla_\phi] = \begin{pmatrix} 0 & -1/\rho(\rho-h) & 0 \\ -\rho/(\rho-h) & 0 & 0 \\ 0 & 0 & 0 \end{pmatrix},$$

$$\frac{\partial}{\partial \rho}\nabla_\rho = \rho^2(\rho-h)^{-2}$$

$$\times \begin{pmatrix} -h\rho^{-2}(\rho-h)^{-1} - h\rho^{-1}(\rho-h)^{-2} & 0 & 0 \\ 0 & -(\rho-h)^{-2} & 0 \\ 0 & 0 & 0 \end{pmatrix},$$

$$\frac{\partial}{\partial \rho}\nabla_\phi = (\rho-h)^{-2} \begin{pmatrix} 0 & 2h\rho & 0 \\ -1 & 0 & 0 \\ 0 & 0 & 0 \end{pmatrix}.$$

The Schrödinger equation looks like

$$D_\xi(\xi) = \mu\frac{\partial \xi}{\partial t} = \mu Q(\xi),$$

for a general vector field, $\xi = (f_1, f_2, f_3)$, we find, computer generated,

$$D_\xi(\xi) = \left(f_1\frac{\partial f_1}{\partial\rho}, f_1\frac{\partial f_2}{\partial\rho}, f_1\frac{\partial f_3}{\partial\rho}\right) + f_1(f_1 h(\rho-h))^{-1}, f_2 h(\rho-h)^{-1}, 0)$$
$$+ (f_2\frac{\partial f_1}{\partial\phi}, f_2\left(\frac{\partial f_2}{\partial\phi}, f_2\frac{\partial f_3}{\partial\phi}\right) + f_2(-f_2\rho^2(\rho-h)^{-1},$$
$$\times f_1 h(\rho-h)^{-1}, 0) + \left(f_3\frac{\partial f_1}{\partial\lambda}, f_3\frac{\partial f_2}{\partial\lambda}, f_3\frac{\partial f_3}{\partial\lambda}\right),$$

and, with the obvious simplified notations

$$Q(\xi) = -(\rho-h)^{-4}(\rho^4 f_{1:\rho\rho} + h^2 f_1 - \rho^2 f_1 + \rho^3 f_{2:\rho} + \rho^4 f_{1:\rho,\rho}$$
$$+ h^4 f_{1:\lambda,\lambda} - 2\rho^3 f_{2:\phi} + \rho^2 f_{1:\phi,\phi} + h^2 f1 : \phi,\phi - 4h\rho^3 f_{1:\lambda,\lambda}$$
$$+ 6h^2\rho^2 f_{1:\lambda,\lambda} - 2h\rho f_1 - h^2\rho f_{1:\rho} - 2h\rho^3 f_{1:\rho,\rho} + h^2\rho^2 f_{1:\rho,\rho}$$
$$- 4h^3\rho f_{1:\lambda,\lambda} + 2h\rho^2 f_{2:\phi} - 2h\rho f_{1:\phi,\phi})D_\rho - (\rho-h)^{-4}(3\rho^3 f_{2:\rho}$$
$$+ 2\rho f_{1:\phi} - 2h f_{1:\phi} + \rho^2 f_{2:\phi,\phi} + \rho^4 f_{2:\lambda,\lambda} + h^4 f_{2:\lambda,\lambda} + \rho^4 f_{2:\rho,\rho}$$
$$+ h^2 f_{2:\phi,\phi} - 4h\rho^2 f_{2:\rho} - 2h\rho f_{2:\phi,\phi} + 6h^2\rho^2 f_{2:\lambda,\lambda} - 2h\rho f_2$$
$$+ h^2\rho f_{2:\rho} + h^2\rho^2 f_{2:\rho,\rho} - 2h\rho^3 f_{2:\rho,\rho} - 4h^3\rho f_{2:\lambda,\lambda})D_\varphi$$
$$+ \rho_{-1}(\rho-h)^{-1}(\rho^2 f_{3:\rho} + \rho^3 f_{3:\rho,\rho} + \rho f_{3:\phi,\phi}$$
$$+ \rho^3 f_{3:\lambda,\lambda} - 2h\rho^2 f_{3:\lambda,\lambda} + h^2\rho f_{3:\lambda,\lambda})D_\lambda.$$

Put

$$\xi = (0, f_2(\rho), 0),$$

then $D_\xi(\xi) = (0,0,0)$, and the furniture equation reduces to the second-order differential equation, $Q(\xi) = 0$, that simplifies to

$$\rho^2(\rho-h)^2\frac{d^2 f_2}{d\rho^2} + \rho(\rho-h)(3\rho-h)\frac{df_2}{d\rho} - 2h\rho f_2 = 0,$$

with the easy solution

$$f_2 = (\rho-h)^{-2},$$

which means that the *fluid*, the content of the space, rotates about the Black Hole $\rho = 0$ with speed $(\rho-h)^{-1}$, so with infinite speed, close to the horizon, almost standing still at great distances, and therefore with lots of *shear*, or maybe some physicists would say, *tidal effects.* Note that this result is reasonable, and glues with the formula for $\Delta_\rho(\delta_\phi) = 1/(\rho - h)$, obtained above. Both may be considered as versions of Einstein's formula for light deflections, used in the works on Gravitational Lens Effect, see e.g. Refsdal (1964).

And note that, if we had used the general metric, we would have found something similar, but with an effect making the surface of the Horizon of the Black Hole shaded in a way that should make it visible as a dark disc.

Remark 5.5. Let us compute the curvature and the related "tensors," for our simple example,

$$g = 1/2\left(\left(\frac{\rho - h}{\rho}\right)^2 d\rho^2 + (\rho - h)^2\ d\phi^2 + d\lambda^2\right) = 1/2\sum_{i,j} g_{i,j}dt_i dt_j.$$

Using the general formula

$$R^l_{i,j,k} = \delta_i(\Gamma^l_{j,k}) - \delta_j(\Gamma^l_{i,k}) + \Gamma^l_{i,m}\Gamma^m_{j,k} - \Gamma^m_{i,k}\Gamma^l_{j,m},$$

we find, for $\rho = t_1, \phi = t_2, \lambda = t_3$,

$$R^3_{i,j,k} = R^l_{i,j,3} = R^l_{i,3,k} = R^l_{3,j,k} = 0,$$

$$R^2_{1,1,2} = -\frac{h\rho}{(\rho - h)^2},\ R^1_{2,1,2} = \frac{h\rho}{\rho(\rho - h)^2},$$

$$\mathrm{Ric}_{1,1} = \frac{h}{\rho(\rho - h)^2},\ \mathrm{Ric}_{2,2} = \frac{h\rho}{(\rho - h)^2},$$

$$R = 2\frac{h\rho}{(\rho - h)^4},$$

$$\mathfrak{G}_{1,1} = 0,\ \mathfrak{G}_{2,2} = 0,\ \mathfrak{G}_{3,3} = 1/2R.$$

Here, the components $\mathfrak{G}_{i,i}$ are the components of the Einstein tensor, $\mathfrak{G} := \mathrm{Ric} - 1/2R\overline{g}$. All other interesting tensors deduced from g vanish.

This implies that the Einstein field equation here takes the form

$$\mathfrak{G} = -\frac{h\rho}{(\rho - h)^4}d\lambda^2,$$

where the term $\frac{h\rho}{(\rho-h)^4}d\lambda^2 =: T^{0,0}$ is very close to the three-dimensional density of mass of the subspace of light-space, given by the cone less than ρ, and this is the time-component of the classical Energy–Stress–Mass tensor, as it should. Unluckily, in our book Eriksen *et al.* (2017), Example 5.1, the sum of the terms above came out wrong. It is, of course, reasonable that the Einstein tensor should, in this case, depend only on its term $d\lambda^2$, and so vanish when restricted to $\tilde{c}$.

Chapter 6

Entropy

Consider an algebraic geometric object X, and let $\mathfrak{aut}(X)$ be the Lie algebra of automorphisms of X. The sub-Lie algebra $\mathfrak{aut}_0(X)$ that lifts to automorphisms of the formal moduli of X is a Lie ideal. Put $\mathfrak{a}(X) := \mathfrak{aut}(X)/\mathfrak{aut}_0(X)$, then if $X(t)$ is a deformation of some X along a parameter t, we find $\dim_k \mathfrak{a}(X(t)) \leq \dim_k \mathfrak{a}(X)$. One may phrase this saying that an object X can never gain *information* when deformed. Moreover, deformation is, obviously, not a reversible process, so information can get lost. This measure of information losses, is related, as we shall see, to the notion of gain of entropy (energy and tropos=transform) coined by Clausius (1865) and generalized by Boltzmann and Shannon.

6.1 The Classical Commutative Case

In Laudal and Pfister (1988), studying moduli problems of singularities in (classical) algebraic geometry, we were led to consider the notion of *Modular Suite*. This is a canonical partition $\{\mathbf{M}_\alpha\}$, of the versal base space $\mathbf{M}$ of the deformation functor of an algebraic object, X. The different *rooms*, $\mathbf{M}_\alpha$, correspond to the subsets of equivalence classes of deformations in $\mathbf{M}$, along which the Lie algebra $\mathfrak{a} := \mathfrak{aut}/\mathfrak{aut}_0$ deforms as Lie algebras, and therefore conserves its dimension. Working with Thermodynamics, it occurred to me that the notion of entropy has an interesting parallel in deformation theory. In fact, I have proposed the following definition.

Definition 6.1. Fix an object X, and let $X(\underline{t})$ correspond to the point $\underline{t} \in \mathbf{M}_\alpha$, then we shall term *Entropy,* of the *state* $\underline{t}$, the integer,

$$S(\underline{t}) := \dim_k(\mathbf{M}_\alpha).$$

In this classical situation, assuming that the field is algebraically closed, and that $\mathbf{M}$ is of finite Krull dimension, the modular suite $\{\mathbf{M}_\alpha\}$ is finite, with an inner room, the *modular sub-stratum* and an *ambient* (open) *maximal entropy* stratum. But the structure of the modular suite may be very complex, even for simple singularities X, see the example of the quasi homogeneous plane curve singularity $x_1^5+x_2^{11}$, in Laudal and Pfister (1988). It is also clear that for any *algebraic dynamics* in $\mathbf{M}$, the entropy will always stay or grow, see again Laudal and Pfister (1988). To be able to construct situations where the entropy is lowered, or the information goes up, we must leave classical algebraic geometry, and venture into non-commutative algebraic geometry. Here is where non-commutative deformation theory comes into play.

6.2 The General Case

In the general situation, where our algebras of *observables* are associative, but not necessarily commutative, the first interesting cases are deformations of associative algebras A, or deformations of finite families of representations V_i of an associative algebra A. In Laudal (2013), we touched upon the first case, and in Chapter 3, we looked at the deformations of Swarms, i.e. families of representations V_i for which $\mathrm{Ext}^1(V_i, V_j)$ are of finite k-dimension. Here, we shall look at the general case, based upon the technique of the previous sections, in particular in Chapter 4.

There we worked with a polynomial algebra $C = k[t_1, \ldots, t_d]$ with a metric $g \in \mathrm{Ph}(C)$, and the Levi-Civita connection, considered as a representation

$$A := \mathrm{Ph}(C)/(\sigma_g) \to \mathrm{End}_k(\Theta_C).$$

The Dirac derivation $[\delta]$, in this case, vanished and the corresponding Hamiltonian turned out to be the Laplace–Beltrami operator, $Q \in \mathrm{End}_k(\Theta_C)$. We were then, in analogy with Quantum theory, led to consider, for every *state* $\xi \in \Theta_C$, the time development or

Schrödinger equation, and the corresponding *Furniture Equation*, see Laudal (2013),

$$\frac{d\xi}{dt} = Q(\xi).$$

We have, in Laudal (2013), shown that in the special case of our Toy Model, see Laudal (2011, 2013), this equation amounts to a combined heat and Navier–Stokes equation. The corresponding notion of entropy, in the above sense, might be defined by the modular suite of the versal base of the deformation functor of the corresponding representation,

$$\xi : \mathrm{Ph}(A) \to A = \mathrm{End}_A(A).$$

The formal moduli has a tangent space given by $\mathrm{Ext}^1_{\mathrm{Ph}(A)}(\xi, \xi)$, which is given by the classical long exact sequence

$$0 \to \mathrm{Hom}_{\mathrm{Ph}(C)}(\xi, \xi) \to \mathrm{Hom}_C(\xi, \xi)$$
$$\xrightarrow{\iota} \mathrm{Der}_A(\mathrm{Ph}(C), \mathrm{Hom}_C(\xi, \xi)) \xrightarrow{\kappa} \mathrm{Ext}^1_{\mathrm{Ph}(C)}(\xi, \xi) \to 0.$$

Here, $\iota = 0$, and $\mathrm{Hom}_C(\xi, \xi) = C$, so that

$$\mathrm{Ext}^1_{\mathrm{Ph}(C)}(\xi, \xi) = C^d \simeq \Theta_C,$$

as all relations $[dt_i, t_j] + [t_i, dt_j]$ in $\mathrm{Ph}(C)$ are mapped to 0 by any C-derivation into C.

The mini-versal base space of ξ is therefore of infinite dimension, and since the Lie algebra of the automorphism group of C is the Lie algebra $\mathrm{Der}_k(C) = \Theta_C$, the Lie algebra of the automorphism group of the representation ξ is

$$\mathfrak{aut}(\xi) := \{\eta \in \Theta_C | [\eta, \xi] = 0\}.$$

We should now define the entropy of the state ξ, as

$$S(\xi) := \dim\{\eta | \mathfrak{aut}(\xi) \vdash \mathfrak{aut}(\eta)\},$$

where $\mathfrak{aut}(\xi) \vdash \mathfrak{aut}(\eta)$ should mean that η is a deformation of ξ inducing a deformation of the Lie algebras of automorphisms. Obviously, this is unrealistic as most of the terms involved will be of infinite dimension.

We may try to overcome this difficulty by approximating the state ξ as a representation by a finite-dimensional representation, defined for every point-set object $\mathcal{P} = \{P_p\}$, of $\mathrm{Simp}_1(C)$, by

$$\xi_{\mathcal{P}} : \mathrm{Ph}(C) \to \mathrm{End}_k(\mathcal{P}),$$

where

$$\xi_{\mathcal{P}}(t_i) = \alpha_i^0(p),\ \xi_{\mathcal{P}}(dt_i) = \alpha_i^1(p)$$

and where we, in anticipation of the treatment via the technique of the next Section 6.3, have put

$$\alpha_i^0(p) = P_{p,i},\ \alpha_i^1(p) = \xi(P_p)_i.$$

We may look at the object $\xi_{\mathcal{P}}$ as a set of molecules in our observatory (or in the Universe), and ξ as the combined state of these, at the outset maybe considered independently. Extending the representation $\xi_{\mathcal{P}}$ to a representation

$$\xi_{\mathcal{P}} : \mathrm{Ph}^{\infty}(C) \to \mathrm{End}_k(\mathcal{P}),$$

or cutting it down to a representation of some dynamical system

$$\xi_{\mathcal{P},\sigma} : C(\sigma) := \mathrm{Ph}^{\infty}(C)/(\sigma) \to \mathrm{End}_k(\mathcal{P}),$$

for some reasonable dynamical structure (σ) of $\mathrm{Ph}^{\infty}(C)$, we can now use part of the technique of Chapter 4, and look at the versal family $\mathbf{M}$ of the representation $\xi_{\mathcal{P},\sigma}$ for some fundamental *state* ξ, of the *Furniture* of the Universe. There will be a canonically defined *Moduli Suite*, $\{\mathbf{M}_\alpha\}$, and for any deformation η of $\xi_{\mathcal{P},\sigma}$, there will be one α such that

$$\eta \in \mathbf{M}_\alpha$$

and the resulting definition of entropy would be

$$S(\eta) = \dim(\mathbf{M}_\alpha).$$

The goal is to show that for reasonable classical cases, this should come out close to Boltzmann's definition

$$S(\eta) := \log\ \mathrm{Vol}(M_\alpha), \eta \in M_\alpha,$$

where M_α is the sub-stratum of the corresponding *Coarse Graining* of the classical phase space of the situation, containing the *distribution* η, see again the very readable text of Penrose (2010).

This approximation also makes it possible to return to the general theory of Chapter 3, coupled with Section 4 of Laudal (2011), and find a very natural way of introducing analogues of Fock space and the Fock representation, and so including what physicists call the *second quantization*, in our picture. But first we have to take another look at the functor Ph*.

6.3 Representations of $\mathbf{Ph}^\infty$

Now let $A = k[t_1, \ldots, t_d]$, and consider a representation of $\mathrm{Ph}^\infty(A)$ as k-algebra,

$$\rho : \mathrm{Ph}^\infty(A) \to \mathrm{End}_k(V),$$

V a k-vector space. Put

$$D_i^0 := \rho(t_i), \quad D_i^p := \rho(d^p t_i), \quad p \geq 1.$$

The composition of $\exp(\tau\delta)$ and ρ is a homomorphism of k-algebras,

$$\mathrm{Ph}^\infty(A) \overset{\rho[\tau]}{\to} \mathrm{End}_k(V) \otimes_k k[[\tau]],$$

for which we have

$$X_i := \rho(\tau)(t_i) = \rho(\exp(\tau\delta)) = \sum_{p\geq 0} \tau^p/p! D_i^p.$$

Since $[t_i, t_j] = 0$, we must have $[X_i, X_j] = 0$, and, since the relations in $\mathrm{Ph}^\infty(A)$ are given by $\sum_{p+q=n\geq 0} 1/p!q!\ [d^p t_i, d^q t_j] = 0$, this is the condition,

$$\sum_{p+q=n\geq 0} 1/p!q!\ [D_i^p, D_j^q] = 0,$$

for the family of matrices $\{D_i^p,\ p \geq 0, \quad i = 1, \ldots, d\}$ to define a homomorphism ρ of k-algebras.

Clearly, if $\dim V = 1$, there are no conditions, and we may pick arbitrarily $D_i^p \in k$ and obtain formal power series

$$X_i = \sum_n D_i^n / n! \tau^n,$$

which, when convergent, give the dynamics of the point.

Example 6.1. Assume $\dim_k V = 2$ and put

$$\rho(t_i) = D_i^0 = \begin{pmatrix} x_i(1) & 0 \\ 0 & x_i(2) \end{pmatrix} =: \begin{pmatrix} \alpha_i^0(1) & 0 \\ 0 & \alpha_i^0(2) \end{pmatrix},$$

and $\alpha_i^0(r,s) := x_i(r) - x_i(s),\ r,s = 1,2$. Let, for $q \geq 0$,

$$D_i^q = \begin{pmatrix} \alpha_i^q(1) & r_i^q(1,2) \\ r_i^q(2,1) & \alpha_i^q(2) \end{pmatrix}.$$

Put

$$\alpha_i^l(r,s) : = \alpha_i^l(r) - \alpha_i^l(s), \quad r,s = 1,2,$$

$$r_i^k(r,s) = \sum_{l=0}^{k} \binom{k}{l} \sigma_{k-l}(r,s) \alpha_i^l(r,s), \quad r,s = 1,2,$$

where the sequence $\{\sigma_l(r,s)\},\ l = 0,1,\dots,$ is a sequence of arbitrary *coupling constants*, with $\sigma_0(r,s) = 0$. Then

$$\rho(d^n t_i) := D_i^n$$

defines a representation,

$$\rho : \mathrm{Ph}^\infty(A) \to \mathrm{End}_k(V),$$

if and only if

$$\sum_{p+q=n\geq 0} 1/p!q!\ [D_i^p, D_j^q] = 0,$$

which is exactly when

$$\sum_{p+q=n\geq 0} 1/p!q!(\alpha_i^p(r,s) r_j^q(s,r) - r_i^p(r,s) \alpha_j^q(s,r)) = 0.$$

Computing, we find the condition

$$\sum_{p+q=n\geq 0} 1/p!q!\sigma_l(r,s)(\alpha_i^p(r,s)\alpha_j^{q-l}(s,r) - \alpha_i^{q-l}(r,s)\alpha_j^p(s,r)) = 0$$

for $r, s = 1, 2,\ l \geq 1$. The situation above arises when we consider two (different) points

$$P_1 = (\alpha_1^0(1), \dots, \alpha_d^0(1)), \quad P_2 = (\alpha_1^0(2), \dots, \alpha_d^0(2)),$$

in space, with pre-described tangents, $\xi_1 = (\alpha_1^1(1), \dots, \alpha_d^1(1))$ and $\xi_2 = (\alpha_1^1(2), \dots, \alpha_d^1(2))$. For these two points, considered as dimension 1 representations of $\mathrm{Ph}(A)$, we saw that there is a one-dimensional space of *tangents* between the points, i.e. $\mathrm{Ext}^1_{\mathrm{Ph}(A)}(k(P_1), k(P_2)) = k$. This leads to possibly non-zero elements $r_i^1(1,2),\ r_i^1(2,1)$ in the matrix representation of the non-commutative deformation of the family $\{k(P_1), k(P_2)\}$ of $\mathrm{Ph}(A)$-modules.

We now have a much more complete picture of the situation. The dynamics of the pair of points is described by the Dirac derivation. Assuming that for time $\tau = 0$ we know the position $\alpha^0(1), \alpha^0(2)$ and the momenta $\alpha^1(1), \alpha^1(2)$ of the two points, then the dynamics is described in terms of the time τ by the matrices

$$X_i = \rho(\exp(\tau\delta))(t_i).$$

Putting

$$\alpha_i(r,s) = \sum_{n=0}^{\infty} \tau^n/n!\ \alpha_i^n(r,s),\ \sigma(r,s) = \sum_{n=0}^{\infty} \tau^n/n!\ \sigma_n(r,s),$$

we find the explicit formulas

$$X_i = \begin{pmatrix} \alpha_i(1) & \sigma(1,2)\alpha_i(1,2) \\ \sigma(2,1)\alpha_i(2,1) & \alpha_i(2) \end{pmatrix}, \quad i = 1, \dots, d.$$

The trace and determinant are

$$\mathrm{tr}(X_i) = (\alpha_i(1) + \alpha_i(2))$$
$$\det(X_i) = (\alpha_i(1)\alpha_i(2)) - \sigma(1,2))\alpha_i(1,2)\sigma(2,1))\alpha_i(2,1).$$

The *spectrum* of X_i, or the eigenvalues, are given as

$$X_i(r) = 1/2(\mathrm{tr}(X_i) \pm \sqrt{tr(X_i)^2 - 4\det(X_i)}), \quad r = 1, 2.$$

From this we see that if all coupling constants vanish, i.e. if $\sigma(r,s)=0$, $r,s=1,2$, then we have undisturbed independent motions of the two points, $X_i(r) = \alpha_i(r)$, $r = 1,2$. If the coupling constants are non-zero, the representation becomes simple, with trivial automorphism group, and so according to the definition of maximal entropy.

Moreover, assume $\sigma_l(1,2) = \sigma_l(2,1) = 0$ for $l = 0$, $l \geq 2$, then the conditions above become for $r,s = 1,2$, and for all $n \geq 0$,

$$\sigma_l(r,s) \cdot \sum_{p+q=n} 1/p!q!(\alpha_i^p(r,s)\alpha_j^{q-1}(s,r) - \alpha_i^{q-1}(r,s)\alpha_j^p(s,r)) = 0.$$

If $d = 3$, this simplifies to

$$\sigma_1(1,2) \cdot \sum_{p+q=n} 1/p!q!(\alpha^p(1,2) \times \alpha^{q-1}(2,1)) = 0,$$

where, of course, $\alpha^m(r,s) = -\alpha^m(s,r)$ for $r,s = 1,2,\ m \geq 0$.

In general, considering an *Object* $\mathcal{P}$ in space, consisting of r points $\{P_p\}_{p=1,\ldots,r}$, we find that the dynamics is closely related to the interaction process, involving non-commutative deformation of families of representations, described in Laudal (2011). In fact, we easily obtain the following theorem.

Theorem 6.1. *Given an* Object $\mathcal{P}$ *consisting of* r *points* $\{P_p\}_{p=1,\ldots,r}$ *in d-space,* $\mathrm{Simp}(A)$, *where* $A = k[t_1,\ldots,t_d]$. *With the notations above, in particular,* $\alpha_i^n(p,q) = \alpha_i^n(p) - \alpha_i^n(q)$, $i = 1,\ldots,d$, *consider the matrix*

$$D_i^n := \begin{pmatrix} \alpha_i^n(1) & r_i^n(1,2) & \ldots & r_i^n(1,r) \\ r_i^n(2,1) & \alpha_i^n(2) & \ldots & r_i^n(2,r) \\ . & . & \ldots & . \\ r_i^n(r,1) & r_i^n(r,2) & \ldots & \alpha_i^n(r)) \end{pmatrix}$$

with

$$r_i^0(p,q) = 0,\ r_i^n(p,q) = \sum_{l=0}^{n} \binom{n}{l} \alpha_i^l(p,q)\sigma_{n-l}(p,q),$$

where $\sigma_m(p,q) \in k$ *are arbitrary* coupling constants, *with* $\sigma_m = 0$, *for* $m \leq 0$. *Then these operators define a representation*

$$\rho : \mathrm{Ph}^\infty(A) \to M_r(k)$$

with

$$\rho(d^n t_i) = D_i^n$$

if and only if for all $n \geq 1$

$$\sum_h \binom{n}{h} \sigma_{n-h}(p,q)(\alpha_i^h(p,q)\alpha_j^0(p,q) - \alpha_i^0(p,q)\alpha_j^h(p,q))$$
$$= \sum_{k,l,m,s} \frac{n!\sigma_{n-k-m}(p,s)\sigma_{k-l}(s,q)}{l!m!(k-l)!(n-k-m)!}(\alpha_j^m(p,s)\alpha_i^l(s,q)$$
$$-\alpha_i^l(p,s)\alpha_j^m(s,q)).$$

Proof. Let us, as above, consider the matrix

$$X_i = \rho(\exp(\tau\delta))(t_i) = \sum_{n \geq 0} \tau^n/n! D_i^n.$$

Putting

$$\alpha_i(r) = \sum_{n=0}^{\infty} \tau^n/n! \ \alpha_i^n(r), \ \ \alpha_i(r,s)$$
$$= \sum_{n=0}^{\infty} \tau^n/n! \ \alpha_i^n(r,s), \ \ \sigma(r,s) = \sum_{n=0}^{\infty} \tau^n/n! \ \sigma_n(r,s),$$

we find the explicit formulas

$$X_i = \begin{pmatrix} \alpha_i(1) & \sigma(1,2)\alpha_i(1,2) & \dots & \sigma(1,r)\alpha_i(1,r) \\ \sigma(2,1)\alpha_i(2,1) & \alpha_i(2) & \dots & \sigma(2,r)\alpha_i(2,r) \\ \dots & \dots & \dots & \dots \\ \sigma(r,1)\alpha_i(r,1) & \sigma(r,2)\alpha_i(r,2) & \dots & \alpha_i(r) \end{pmatrix},$$

$i = 1, \dots, d.$

Now, compute

$$[X_i, X_j] = 0,$$

and see that the condition of the theorem emerges. □

Remark 6.1. We may consider the *space*

$$\mathbf{A}(r) = k[\alpha_i^n(p), \sigma_n(p,q)]/\mathfrak{a},$$

with coordinates

$$\{\alpha_i^n(p), \sigma_n(p,q), \quad i = 1,\dots,d,\ p,\ q = 1,\dots,r,\ n \geq 0,\ \sigma_0(p,q) = 0\},$$

and where the ideal $\mathfrak{a}$ is generated by the equations above as the versal base space for the versal family of the non-commutative deformation theory applied to the family of $\mathrm{Ph}^\infty(A)$ modules defined by the object $\mathcal{P}$, and with Dirac derivation δ acting as $[\delta](\alpha_i^n(p)) = \alpha_i^{n+1}(p)$.

Since $\mathrm{Ph}^\infty(A)$ is infinitely generated, there is, strictly speaking, no such thing, but we shall see that in special cases we can overcome this difficulty by finding clever dynamical structures.

In Section 3.2, we saw that to overcome another difficulty related to non-existing moduli spaces, we opted for approximating a state $\xi \in \Theta_A$ by a finite dimensional representation defined for every finite point set object $\mathcal{P} = \{P_p\}$ of $\mathrm{Simp}_1(A)$ by

$$\xi_{\mathcal{P}} : \mathrm{Ph}(C) \to \mathrm{End}_k(\mathcal{P}),$$

where

$$\xi_{\mathcal{P}}(t_i) = \alpha_i^0(p),\ \xi_{\mathcal{P}}(dt_i) = \alpha_i^1(p).$$

We wanted to look at the object $\mathcal{P}$ as the set of molecules in our observatory (or in the Universe), and at $\xi_{\mathcal{P}}$ as the combined state of these.

Extending the representation $\xi_{\mathcal{P}}$ to a representation

$$\xi_{\mathcal{P}} : \mathrm{Ph}^\infty(A) \to \mathrm{End}_k(\mathcal{P}),$$

or cutting it down to a representation of some dynamical system, $A(\sigma) := \mathrm{Ph}^\infty(A)/(\sigma)$,

$$\xi_{\mathcal{P},\sigma} : A(\sigma) \to \mathrm{End}_k(\mathcal{P}),$$

for some reasonable dynamical structure (σ), we might be able to compute the versal family $\mathbf{M}$ of the representation $\xi_{\mathcal{P},\sigma}$ for some fundamental *state* ξ of the *Furniture* of the Universe and also the *Moduli Suite* $\{\mathbf{M}_\alpha\}$.

For any deformation η of $\xi_{\mathcal{P},\sigma}$ represented by an equivalence class $\tilde{\eta}$ in the versal base space $\mathbf{M}$, there will be one α such that

$$\tilde{\eta} \subset \mathbf{M}_\alpha,$$

and the entropy of η would be

$$S(\eta) = \dim(\mathbf{M}_\alpha).$$

From the classification above, it follows that the minimal entropy would correspond to all $\sigma_l(p,q) = 0$, i.e. to some *dust-like* furniture of the space. And the maximal entropy would correspond to all $\sigma_l(p,q) \neq 0$, i.e. to some *black hole-like* object $\mathcal{P}$ containing a huge number of gravitational and/or other parameters.

The goal is still to show that, for reasonable classical cases, this should come out close to Boltzmann's definition

$$S(\eta) := \log \ \mathrm{Vol}(M_\alpha), \eta \in M_\alpha,$$

where M_α is the sub-stratum of the corresponding *Coarse Graining* of the classical phase space of the situation, containing the *distribution* η, see again Penrose (2010), Volume 2. In particular, consider Penrose's need (for) some clear-cut way of saying that "the gravitational degrees of freedom were not activated," and his need to identify the mathematical quantity that actually measures "gravitational degrees of freedom." I suggest that the above, coupled with my "Toy Model," see Laudal (2011), i.e. working on $\mathrm{Hilb}^{(2)}(\mathbf{A}^3)$ instead of the trivial affine space $\mathbf{A}^d = \mathrm{Spec}(k[t_1, \ldots, t_d])$, may be of interest for this quest.

We have already looked at the case $r = 2$, and seen that the result makes physical sense. For $r = 3, n = 2$, we find

$$\begin{aligned}&\sigma_1(p,q)(\alpha_i^1(p,q)\alpha_j^0(p,q) - \alpha_i^0(p,q)\alpha_j^1(p,q))\\&\quad = \sigma_1(p,s)\sigma_1(s,q)(\alpha_j^0(p,s)\alpha_i^0(s,q) - \alpha_i^0(p,s)\alpha_j^0(s,q)).\end{aligned}$$

In dimension $d = 3$, this has a particularly nice interpretation. Let $\alpha^0(i,j)$ be the vector starting at P_i and ending at P_j, and let ξ_i be a tangent vector at P_i for $i = 1, 2, 3$. Put $\alpha^1(i,j) = \xi_i - \xi_j$, then the condition above reads

$$\sigma_1(p,q)(\alpha^1(p,q) \times \alpha^0(p,q)) = -\sigma_1(p,s)\sigma_1(s,q)(\alpha^0(p,s) \times \alpha^0(s,q)),$$
$$\forall p, s, q = 1, 2, 3.$$

This says that for any two of the three points in space, the relative momentum must sit in the plane defined by the three points, the length being determined by the three coupling constants. Moreover, the sum of all three relative momenta must be 0. In fact, there are coefficients, $u, v \in k$, such that

$$\alpha^1(p,q) = u\alpha^0(p,s) + v\alpha^0(s,q).$$

Put this into the left-hand side of the formula above, and find

$$\sigma_1(p,q)(v-u) = \sigma_1(p,s)\sigma_1(s,q),$$

assuming that the three points are not co-linear.

Assume now that the coupling constants are given as

$$\sigma_1(p,q) := m(p,q)|\alpha^0(p,q)|^{-2},$$

then after some computation we find the following differential equations for the dynamics of the three points:

$$\begin{aligned}
\alpha^1(p,q) &= -m(q,p)|\alpha^0(p,q)|^{-1}\epsilon(p,q) - m(p,s)|\alpha^0(p,s)|^{-1}\epsilon(p,s) \\
&\quad + \rho\alpha^0(p,q), \\
\alpha^1(q,s) &= -m(s,q)|\alpha^0(s,q)|^{-1}\epsilon(q,s) - m(q,p)|\alpha^0(q,p)|^{-1}\epsilon(q,p) \\
&\quad + \rho\alpha^0(q,s), \\
\alpha^1(s,p) &= -m(p,s)|\alpha^0(p,s)|^{-1}\epsilon(s,p) - m(s,q)|\alpha^0(s,q)|^{-1}\epsilon(s,q) \\
&\quad + \rho\alpha^0(s,p).
\end{aligned}$$

Here, $\epsilon(i,j)$ is the unit vector from the point P_i to the point P_j.

Note that there is a different set-up, related to Grothendieck's generalized differential algebra. For A commutative, consider an A-module E, and an extension of this representation to

$$\rho : \mathrm{Ph}^\infty(A) \to \mathrm{End}_k(E).$$

We have seen that ρ must be given in terms of operators

$$D_i^p := \rho(d^p t_i) \in \mathrm{End}_k(E),$$

satisfying the conditions

$$\forall n \geq 0, \sum_{p+q=n} 1/p!q![D_i^p D_j^q] = 0.$$

There is an obvious family of solutions of these equations, given by any differential operator

$$Q \in \mathrm{Diff}(E),$$

with $D_i^0 = \rho(t_i) = t_i$, $D_i^p := \mathrm{ad}(Q)^p(t_i) \in \mathrm{Diff}(E)$, $p \geq 1$. Looking at the case of finite-dimensional representations treated above, one sees the difference between the two set-ups, and the much greater generality obtained by considering the representations of Ph^∞ the way we do.

Chapter 7

Cosmology, Cosmos, and Cosmological Time

7.1 Background, and Some Remarks on Philosophy of Science

> They say clearly that when the One had been constructed — whether *of planes or surface or seed or something they cannot express* — then immediately the nearest part of the Unlimited began "to be drawn and limited by the Limited" ...giving it (the Unlimited) *numerical structure* (Aristotle, on the creation of the world, as the Pythagoreans saw it).

In a first paper, Laudal (2005), touching upon this subject, we discussed the possibility of including a cosmological model in our Toy Model of time–space, and thereby giving some sense to this age-old struggle to cope with the notion of creation, and in particular, giving the Universe a *numerical structure.* See the very interesting book, Ferguson (2011), with the quotation above, on page 148.

The one-dimensional model we presented in Laudal (2005) was created by the deformations of the trivial singularity, $O := k[x]/(x)^2$. Using elementary deformation theory for algebras, we obtained amusing results, depending upon some rather bold mathematical interpretations of the, more or less accepted, cosmological vernacular.

Later, in the book, Laudal (2011), in many papers, and in particular in the book Eriksen *et al.* (2017), we have shown that our toy model, i.e. the moduli space, $\mathbf{H}$, of all pairs of two points in affine 3-space, or its étale covering, $\underline{\tilde{H}}$, is *created* by the (non-commutative) deformations of the obvious singularity in three-dimensions,

$$U := k\langle x_1, x_2, x_3\rangle/(x_1, x_2, x_3)^2.$$

Here, we shall show that there are reasons to believe that this model may be of some use, to fuse General Relativity (GR) and Quantum Field Theory (QFT), and thereby obtain a mathematical basis for the Standard Model (SM). It is, however, still just an example of the kind of mathematical models one would like to see in science, along the lines of Chapter 1, with no agreed-upon experimental confirmation.

The main axiom of the leading branches of cosmology seems to be that the space–time of the existing Universe can be described via a General Relativistic model, somehow given by Einstein's equation with respect to some mass–stress tensor, mass and energy being homogeneously and isotropically distributed in space. This leads to the assumption that the universe is a four-dimensional space–time of a form commonly called a Friedman–Robertson–Walker model, or sometimes, the FLRW-model (where one includes Lemaître). In particular, the space has an open-ended time-coordinate, leaving out the Big Bang, but still assuming that the point-like Big Bang is in the closure of even the shortest complete history of the universe.

There are a lot of assumptions here. One is that the space–time is capable of containing something, and that these things can be described as independent of the space, even though they curve space, and otherwise intervene in the dynamical process. For example, even at the very start of the universe, spin is assumed to be present. So, in mathematical terms, the space–time must be outfitted with a su(2) or an sl(2)-action of the tangent structure, obviously determined by the Big Bang event. Moreover, since the space–time of the model does not contain the prime event, and the jump between that, supposedly singular, point-like event and the mathematically well-defined space–time is not part of the model, we do not have a model of the Big Bang itself, but rather of what may have happened in our usual space, a long time ago, with respect to a rather artificially chosen time parameter.

In this chapter, I propose to show how the *time–space*, **H**, can be thought of as an immediate product of a mathematical scenario incorporating a Big Bang event, making this event mathematically sound. Starting with the pure notion of three-dimensionality, i.e. a k-scheme $\underline{U} = Spec(U)$ with only one point and a three-dimensional tangent space, in algebraical terms, the singularity,

$$U := k[x_1, x_2, x_3]/(x_1, x_2, x_3)^2,$$

we shall see, in the next sub-section, that we may construct, in a canonical way, a versal deformation base space, $\mathbf{M}$, and a corresponding versal family $\mathbf{U}_*$, containing all isomorphism classes of deformations of U, as associative algebra.

But first, note that any automorphism of U is reduced to a substitution,

$$y_i := \sum_{k=1}^{3} \alpha_{i,k} x_k, \alpha := (\alpha_{i,k}) \in \mathrm{GL}_k(3).$$

The technique for this general deformation theory can be found in Laudal (1979), see also Laudal and Pfister (1988). In the last book, we introduced the notion of *Moduli suite.* This corresponds here to a partition of the space $\mathbf{M}$, in a series of *rooms*, containing an inner room, composed of just one point $*$, corresponding to the singularity we start with, U, and a very special component that turns out to be $\mathbf{H} := \mathrm{Hilb}^2(\mathbf{A}^3)$, and where the family $\mathbf{U}$, the restriction of $\mathbf{U}_*$ to $\mathbf{H}$, has corresponding to a point $(o, p) \in \underline{H} - \underline{\Delta}$ the fiber

$$U(o,p) := k\langle x_1, x_2, x_3\rangle/(x_i x_j - o_i x_j - p_j x_i + o_i p_j),$$

where we have used the coordinates x_1, x_2, x_3 to express the two points o and p in 3-space $\mathbf{A}^3$, by coordinates $\{o_i\}, \{p_j\}, i, j = 1, 2, 3$.

If we change the coordinates by the automorphism above, then with obvious indexes,

$$U_x(o,p) \simeq U_y(\alpha(o), \alpha(p)).$$

Moreover, if $o = p$, $U(o, p)$ is isomorphic to U. But $\mathbf{U}$ has, nevertheless, a unique extension to all of $\underline{\tilde{H}}$, and the Z_2-action also extends.

There is also a special tangent vector, defined by the derivation that maps $x_i x_j$ to $\epsilon_{i,j,k}$. This leads to a room consisting of one single point, the Quaternions, $\mathbf{Q}$. In fact, $\mathbf{Q}$ is a deformation of U, given by

$$\mathbf{Q} = k\langle x_1, x_2, x_3\rangle/(x_i x_j - \epsilon_{i,j,k} x_k + \delta_{i,j}),$$

where $\epsilon_{i,j,k}$ and $\delta_{i,j}$ are the usual notations for $0, 1$. Obviously, $\mathbf{Q}$ has no non-trivial deformations, so it is isolated in the moduli suite of U.

Note, for later use that the discoverer of the Quaternions, Hamilton, wrote about his algebra as the *science of pure time*, see Øhrstrøm (1985).

Recall that we have, in Chapter 6, seen that the notion of moduli suite, in deformation theory may serve as a model for both *Entropy* and *Information*, needed to connect to the present day Cosmology. There is a readable text by Penrose (2010).

7.2 Deformations of Associative Algebras

Given an associative k-algebra A, the tangent space of the formal moduli of A, as an associative k-algebra is, by deformation theory, see Laudal (1979) and Laudal and Pfister (1988),

$$T_* := A^1(k, A; A) = \mathrm{Hom}_F(\ker \rho, A)/\,\mathrm{Der},$$

where $\rho : F \to A$ is any surjective homomorphism of a free k-algebra F, onto A, Hom_F means the F-bilinear maps, and Der denotes the subset of the restrictions to $I := \ker \rho$ of the derivations from F to A.

As an example, let $A = k[x_1, \ldots, x_d]$ be the polynomial algebra, then we find

$$A^1(k, A; A) = \mathrm{Hom}_F(\ker \rho, A)/\,\mathrm{Der},$$

where $F = k\langle x_1, \ldots, x_d\rangle$, and $\ker \rho = \langle [x_i, x_j]\rangle$, and any element in $A^1(k, A; A)$ is a generalized Poisson structure. Note also that there is a universal action of $\mathrm{Der}_k(A)$ on $A^1(k, A; A)$, see, e.g. Eriksen *et al.* (2017), that we shall come back to.

The tangent space of the formal moduli of the singularity

$$U := k[x_1, x_2, x_3]/(x_1, x_2, x_3)^2,$$

as an associative k-algebra is now

$$T_* := A^1(k, U; U) = \mathrm{Hom}_F(\ker \rho, U)/\,\mathrm{Der},$$

where $\rho : F \to U$ is the obvious surjection of the free k-algebra $F = k\langle x_1, x_2, x_3\rangle$, with $\ker \rho = (\underline{x})^2$, generated as F bi-module by the family $\{x_{i,j} := x_i x_j\}$.

Any F-bilinear morphism $\phi : (\underline{x})^2 \to U$ must be of the form

$$\phi(x_{i,j}) = a^0_{i,j} + \sum_{l=1}^{3} a^l_{i,j} x_l,$$

and the bi-linearity is seen to imply that $a^0_{i,j} = 0$. Thus, the dimension of $\mathrm{Hom}_F(I, U)$ is 27.

Any derivation $\delta \in \mathrm{Der}$ must be given by

$$\delta(x_i) = b^0_i + \sum_{l=1}^{3} b^l_i x_l$$

and the restriction of this map, to the generators of $I = (\underline{x})^2$, must have the form

$$\delta(x_{i,j}) = b^0_j x_i + b^0_i x_j,$$

therefore determined by the $b^0_i s$. It follows that the tangent space T_* is of dimension $27 - 3 = 24$.

Let $o, p \in \mathbb{A}^3$ be two points, $o = (o_1, o_2, o_3), p = (p_1, p_2, p_3)$, with respect to the coordinate system, $\underline{x}$, and put

$$\phi_{o,p}(x_{i,j}) = p_j x_i + o_i x_j,$$

then it is easy to see that the maps $\{\phi_{o,p}\}$ generate a six-dimensional sub-vector subspace T_0 of T_*. Note that, if $o = p$, then $\phi_{o,p}$, is a derivation, thus 0 in T_*.

Now, the rather unexpected happens. We may integrate the tangent subspace T_0, and obtain a family of flat deformations of U. In fact, it is easy to see that

$$U(o, p) := k\langle x_1, x_2, x_3\rangle/(x_i x_j - o_i x_j - p_j x_i + o_i p_j),$$

is an associative k-algebra of dimension 4, and a deformation of U, in a direction of $\underline{H}$. This defines a family of associative $H := k\,[o, p]$ algebras,

$$\mathbf{U} := H\langle x_1, x_2, x_3\rangle/(x_i x_j - o_i x_j - p_j x_i + o_i p_j),$$

where normally $k = \mathbb{R}$, the real numbers. Let us put

$$\mathbf{x}_{i,j} := (x_i - o_i)(x_j - p_j) = x_i x_j - o_i x_j - p_j x_i + o_i p_j,$$

$$o := (o_1, o_2, o_3), p = (p_1, p_2, p_3) \in H^3.$$

Note that if $o = p$, then $U(o, p)$ is isomorphic to U, as it should, and that, $U(o, p) \simeq U(-o, -p)$. Moreover, for any non-zero element

$\kappa \in k$, and any three-vector $c \in \mathbb{A}^3$, we have

$$U(o,p) \simeq U(\kappa o, \kappa p),\ U(o,p) \simeq U(o-c, p-c).$$

Choosing $c = 1/2(p+o)$, we find $o' := o - c = -(p-c) =: -p'$, and it is easy to see that if $o' \neq 0$ is the sub-Lie algebra above generated by $\{x_1, x_2, x_3\}$ in $U(o', p')$, is isomorphic to the standard three-dimensional Lie algebra with relations, $[y_1, y_2] = y_2, [y_1, y_3] = y_3, [y_2, y_3] = 0$. Moreover, choosing $c = (p+o)$, we find an isomorphism

$$U(o,p) \simeq U(-p,-o) \simeq U(p,o),$$

which should be related to the action of Z_2 on $\underline{H}$, and thus, according to our philosophy, to the mathematical reason for the CPT-equivalence, see Laudal (2011, (4.9)).

Remark 7.1 (The modular suite of U). We have found a subspace $\underline{H}$ of the versal base space of U, of dimension 6, corresponding to the deformations $U(o,p)$. Note also that the algebra,

$$\mathbf{Q} := k\langle x_1, x_2, x_3\rangle/(x_i x_j - \epsilon_{i,j,k} x_k + \delta_{i,j}),$$

where $\epsilon_{i,j,k}$ and $\delta_{i,j}$ are the usual indices, the first one non-zero only for $\{i,j,k\} = \{1,2,3\}$, and the last one the usual delta function, is isomorphic to the quaternions, which therefore is another non-trivial deformation of U.

Thanks to Gunnar Fløystad, *I* know that the two-dimensional matrix algebra M_2 and the external algebra $\wedge^* k^2$ are also non-commutative deformations of U. However, I have not been able to compute the complete modular suite for U. We shall, nevertheless, be able to guess a maximal modular room, taking care of the maximal entropy in this case. This is the object of Remark 7.2.

Consider now the restriction to the sub-scheme $\underline{H} - \underline{\Delta}$, of the family $\mathbf{U}$, denoted by

$$\nu' : \mathbf{U}' \to \underline{H} - \underline{\Delta}.$$

Since for all non-zero $\kappa \in k$, we have $U(\lambda + \kappa u, -\kappa u + \lambda) \simeq U(\kappa u, -\kappa u) \simeq U(u, -u)$, this family extends uniquely to a family

$$\nu : \tilde{\mathbf{U}} \to \underline{\tilde{H}}.$$

Let us compute the algebras $U(o,p)$, and their Lie algebras of derivations, $\mathfrak{g}(\underline{t}) := \mathrm{Der}_k(U(\underline{t}))$.

First, the four-dimensional k-algebras $U(o,p)$, with relation

$$x_{i,j} = (x_i - o_i)(x_j - p_j),$$

with $o \neq p$ are all isomorphic, since in this case there is an element $\alpha \in \mathrm{GL}_k(3)$ sending (o,p) onto any other pair (o',p'), with $o' \neq p'$. Let us see this, using the generalized Burnside theorem, see Laudal (2011), (3.2)). Obviously, $U(o,p)$ has only two simple representations, of dimension 1, call them k_o and k_p. By the O-construction, there is an isomorphism

$$\eta : U(o,p) \to \begin{pmatrix} H_{1,1} \otimes \mathrm{End}(k_o) & H_{1,2} \otimes \mathrm{Hom}_k(k_o, k_p) \\ H_{2,1} \otimes \mathrm{Hom}_k(k_p, k_o) & H_{2,2} \otimes \mathrm{End}_k(k_p) \end{pmatrix},$$

where $H_{1,1}$ is a formal algebra with tangent space $\mathrm{Ext}^1_{U(o,p)}(k_o,k_o)$, $H_{2,2}$ is a formal algebra with tangent space $\mathrm{Ext}^1_{U(o,p)}(k_p,k_p)$, and $H_{1,2}$, is a bi-module generated by $\mathrm{Ext}^1_{U(o,p)}(k_o,k_p)^*$, respectively. There are no problems computing the Ext-groups. Recall that

$$\mathrm{Ext}^1_{U(o,p)}(V_1,V_2) = \mathrm{Der}_k(U(o,p), \mathrm{Hom}_k(V_1,V_2))/\,\mathrm{Triv},$$

and that $u \in U(o,p)$ operates on $\phi \in \mathrm{Hom}_k(V_1,V_2)$, as

$$(u\phi)(v_1) = u\phi(v_1), (\phi u)(v_1) = \phi(uv_1).$$

In the general case (one may test it in the interesting case, $o = (1,0,0), p = (0,0,0)$ above), we obtain

$$\mathrm{Ext}^1_{U(o,p)}(k_o,k_o) = \mathrm{Ext}^1_{U(o,p)}(k_p,k_p) = \mathrm{Ext}^1_{U(o,p)}(k_o,k_p) = 0,$$
$$\mathrm{Ext}^1_{U(o,p)}(k_p,k_o) = k^2.$$

Therefore,

$$\eta : U(o,p) \to \begin{pmatrix} k & 0 \\ \langle \xi_1, \xi_2 \rangle & k \end{pmatrix}$$

is an isomorphism. Here, $\xi_i \cdot 1 = \xi_i$. We may pick generators of this algebra,

$$x_1 := \begin{pmatrix} 0 & 0 \\ 0 & 1 \end{pmatrix}, x_2 := \begin{pmatrix} 0 & 0 \\ \xi_1 & 0 \end{pmatrix}, x_3 := \begin{pmatrix} 0 & 0 \\ \xi_2 & 0 \end{pmatrix},$$

and obtain the relations corresponding to the choice of $o = (1,0,0), p = (0,0,0)$, or we may pick

$$x_1 := \begin{pmatrix} 0 & 0 \\ \xi_1 & 0 \end{pmatrix}, x_2 := \begin{pmatrix} 0 & 0 \\ \xi_2 & 0 \end{pmatrix}, x_3 := \begin{pmatrix} 1 & 0 \\ 0 & -1 \end{pmatrix},$$

and obtain the relations corresponding to the choice of $o = (0,0,-1)$, $p = (0,0,1)$. We have therefore obtained an algebraic subspace $\underline{\tilde{H}}$, of the mini-versal base space $\mathbf{M}$ of the algebra U, corresponding to the algebras $U(o,p)$ that are all isomorphic. This subspace is therefore a *trivializing section* of this mini-versal base space.

Remark 7.2 (Deformations of $U(o,p)$, maximal entropy). Using the same technique as above, computing the deformations of one of these isomorphic algebras, we may show that the tangent space of $\mathsf{Def}_{U(o,p)}$ is trivial. In fact, as above, this tangent space is given by

$$A^1(U(o,p), U(o,p)) = \mathrm{Hom}_F(J, U(o,p))/\,\mathrm{Der},$$

where $J = \ker(\pi)$ and $\pi : F \to U(o,p)$ is a surjective homomorphism of the free k-algebra $F = k\langle x_1, x_2, x_3\rangle$ onto $U(o,p)$. Obviously, $J = \ker(\pi)$ is generated by the elements $\{x_{i,j} := x_i x_j - o_i x_j - p_j x_i + o_i p_j\}$, and we have, in J, the relations

$$x_{i,j}x_k + o_i x_{j,k} + p_j x_{i,k} = x_i x_{j,k} + o_j x_{j,k} + p_k x_{i,j}.$$

Let $o = (1,0,0), p = (0,1,0)$, then an easy, but quite lengthy, computation shows that these relations imply that any bilinear homomorphism $c \in \mathrm{Hom}_F(J, U(o,p))$ is the restriction of a derivation $\beta \in \mathrm{Der}_k(F, U(o,p))$, proving that

$$A^1(U(o,p), U(o,p)) = 0.$$

To see this, compute first the Der-part of A^1. Let $\beta \in \mathrm{Der}_k(F, U(o,p))$, and put $\beta(x_i) = \beta_i^p x_p = \beta_i$, and compute

$$\beta(x_{i,j}) =: c_{i,j}^s x_s + c_{i,j}, \quad i,j,s = 1,2,3.$$

The result is

$$c^1 := \begin{pmatrix} (2\beta_1 + \beta_1^1 + \beta_1^2) & (\beta_2 + \beta_2^2) & (\beta_3 + \beta_3^2) \\ (\beta_2 + \beta_2^1) & 0 & 0 \\ (\beta_3 + \beta_3^1) & 0 & 0 \end{pmatrix},$$

$$c^2 := \begin{pmatrix} 0 & (\beta_1 + \beta_1^1) & 0 \\ (\beta_1 + \beta_1^2) & (2\beta_2 + \beta_2^1 + \beta_2^2) & (\beta_3 + \beta_3^2) \\ 0 & (\beta_3 + \beta_3^1) & 0 \end{pmatrix},$$

$$c^3 := \begin{pmatrix} 0 & 0 & (\beta_1 + \beta_1^1) \\ 0 & 0 & (\beta_2 + \beta_2^1) \\ (\beta_1 + \beta_1^2) & (\beta_2 + \beta_2^2) & (2\beta_3 + \beta_3^1 + \beta_3^2) \end{pmatrix},$$

$$c := \begin{pmatrix} (-\beta_1 - \beta_1^2) & (-\beta_1 - \beta_1^1 - \beta_2 - \beta_2^2) & (-\beta_3 - \beta_3^2) \\ 0 & (-\beta_2 - \beta_2^1) & 0 \\ 0 & (-\beta_3 - \beta_3^1) & 0 \end{pmatrix}.$$

Now, any bi-module homomorphism $c \in \mathrm{Hom}_F(J, U(o,p))$ must be of the same form,

$$c(x_{i,j}) =: c_{i,j}^s x_s + c_{i,j}, \quad i, j, s = 1, 2, 3,$$

and the relations in J give us 27 equations, resulting in the following equations:

$$c := \begin{pmatrix} c_{1,1} & c_{1,2} & c_{1,3} \\ 0 & c_{2,2} & 0 \\ 0 & c_{3,2} & 0 \end{pmatrix},$$

$$c^1 := \begin{pmatrix} c_{1,1}^1 & (-c_{1,1} - c_{1,2} - c_{1,1}^1) & -c_{1,3} \\ -c_{2,2} & 0 & 0 \\ -c_{3,2} & 0 & 0 \end{pmatrix},$$

$$c^2 \; : = \begin{pmatrix} 0 & c_{1,1}^1 & 0 \\ -c_{1,1} & -c_{3,2} & -c_{1,3} \\ 0 & -c_{3,2} & 0 \end{pmatrix},$$

$$c^3 := \begin{pmatrix} 0 & 0 & (c_{1,1} + c_{1,1}^1) \\ 0 & 0 & -c_{2,2} \\ c_{1,1} & (c_{2,2} - c_{3,2}) & (-c_{1,3} - c_{3,2}) \end{pmatrix}.$$

Counting dimensions now gives the result, $\dim_k \mathrm{Hom}_F (J, U(o,p)) = \dim_k \mathrm{Der}$, implying that $A^1(U(o,p), U(o,p)) = 0$. So, it is not unreasonable to consider the subset $\underline{H}$ as the maximal entropic one in the modular suite of U.

7.3 The Universal Gauge Groups and SUSY

Consider the Lie algebra

$$\mathfrak{g} := \mathrm{Der}_H(\mathbf{U})$$

as a principal Lie algebra bundle on the space $\underline{\tilde{H}}$. Any element $\delta \in \mathrm{Der}_H(\mathbf{U})$ must be given by its values on the coordinates

$$\delta(x_i) = \delta_i^0 + \delta_i^1 x_1 + \delta_i^2 x_2 + \delta_i^3 x_3,\ \ \delta_i^j \in H.$$

Now, let us define

$$\tilde{\Theta}_{\tilde{H}} := \{\kappa \in \mathrm{End}_{\tilde{H}}(\mathbf{U}), \delta(\mathbf{1}) = 0\}.$$

Obviously,

$$\mathfrak{g} \subset \tilde{\Theta}.$$

Any $\kappa \in \tilde{\Theta}_{\tilde{H}}$ will correspond to $\kappa_i := \kappa(x_i) \in U(o,p), i = 1,2,3$, i.e. to a matrix of the type

$$M := \begin{pmatrix} 0 & 0 & 0 & 0 \\ \kappa_1^0 & \kappa_1^1 & \kappa_1^2 & \kappa_1^3 \\ \kappa_2^0 & \kappa_2^1 & \kappa_2^2 & \kappa_2^3 \\ \kappa_3^0 & \kappa_3^1 & \kappa_3^2 & \kappa_3^3 \end{pmatrix},$$

where $\kappa_i := (\kappa_i^0, \kappa_i^1, \kappa_i^2, \kappa_i^3)$. Moreover, it is clear that $\tilde{\Theta}$ is a Lie algebra, and that $\mathfrak{g}$ is a natural sub-Lie algebra, of this matrix algebra.

Put

$$\bar{o} = (1, o_1, o_2, o_3), \quad \bar{p} = (1, p_1, p_2, p_3),$$

and consider now the four-vectors,

$$\delta_i = (\delta_i^0, \delta_i^1, \delta_i^2, \delta_i^3), \quad i = 1,2,3.$$

Suppose $\delta \in \mathfrak{g}$, then we find the formula, in $\mathbf{U}$,

$$\delta(x_i x_j - o_i x_j - p_j x_i + o_i p_j) = (\delta_i \cdot \bar{o})x_j - (\delta_i \cdot \bar{o})p_j + x_i(\delta_j \cdot \bar{p}) - o_i(\delta_j \cdot \bar{p})$$

which leads to

$$\delta \in \mathrm{Der}_H(\mathbf{U})$$

if and only if

$$\delta_i \cdot \bar{o} = \delta_i \cdot \bar{p} = 0, \quad i = 1, 2, 3.$$

Consider, from now on,

$$\mathbf{x}_{i,j} = x_i x_j - o_i x_j - p_j x_i + o_i p_j \in H\langle x_1, x_2, x_3 \rangle =: \mathbf{F}.$$

We find that if $\delta \in \mathrm{Der}_H(\mathbf{U})$, then in the free algebra $\mathbf{F}$ we have

$$\delta(\mathbf{x}_{i,j}) = \sum_{p=1}^{3} \delta_i^p(\mathbf{x}_{p,j}) + \sum_{p=1}^{3} \delta_j^p(\mathbf{x}_{i,p}).$$

Let $\xi \in A^1(H, \mathbf{U}; \mathbf{U})$ be represented by $\xi(\mathbf{x}_{i,j}) = \xi_{i,j}^0 + \sum_{p=1}^{3} \xi_{i,j}^p x_p$. Recall that ξ is zero if it is of the form

$$\xi(\mathbf{x}_{i,j}) = -(\eta_i \cdot \bar{o})p_j - o_i(\eta_j \cdot \bar{p}) + (\eta_i \cdot \bar{o})x_j + (\eta_j \cdot \bar{p})x_i$$

for some derivation $\eta \in \mathrm{Der}_H(\tilde{F}, \mathbf{U})$.

Choose $\underline{t} = (o, p) \in \underline{H} - \underline{\Delta}$, and consider the relation of $U(\underline{t})$,

$$x_{i,j} := x_i x_j - o_i x_j - p_j x_i + o_i p_j.$$

Let us compute the Lie algebra $\mathfrak{g}(\underline{t}) := \mathrm{Der}_k(U(\underline{t}))$. Any element $\delta \in \mathrm{Der}_k(U(\underline{t}))$ must have the form

$$\delta(x_i) = \delta_i^0 + \delta_i^1 x_1 + \delta_i^2 x_2 + \delta_i^3 x_3, \quad \delta_i^p \in k.$$

Put, as above, $\bar{o} = (1, o_1, o_2, o_3), \bar{p} = (1, p_1, p_2, p_3)$, and consider the four-vectors $\delta_i = (\delta_i^0, \delta_i^1, \delta_i^2, \delta_i^3), \quad i = 1, 2, 3.$

As above, we find that

$$\delta \in \mathrm{Der}_k(U(\underline{t}))$$

if and only if

$$\delta_i \cdot \bar{o} = \delta_i \cdot \bar{p} = 0, \quad i = 1, 2, 3.$$

Given a point $\underline{t} = (o, p) \in \underline{H}$, let us compute the Lie algebra $\mathfrak{g}(\underline{t}) := \mathrm{Der}_k(U(\underline{t}))$. Any element $\delta \in \mathrm{Der}_k(U(\underline{t}))$ must have the form

$$\delta(x_i) = \delta_i^0 + \delta_i^1 x_1 + \delta_i^2 x_2 + \delta_i^3 x_3, \ \delta_i^p \in k.$$

Put, as above, $\bar{o} = (1, o_1, o_2, o_3), \bar{p} = (1, p_1, p_2, p_3)$, and consider the four-vectors $\delta_i = (\delta_i^0, \delta_i^1, \delta_i^2, \delta_i^3), i = 1, 2, 3$. As above, we find that

$$\delta \in \mathrm{Der}_k(U(\underline{t}))$$

if and only if

$$\delta_i \cdot \bar{o} = \delta_i \cdot \bar{p} = 0, \quad i = 1, 2, 3.$$

The tangent space $\Theta_{H,\underline{t}}$ of $\underline{H}$, at $\underline{t}$, is represented by the space of all pairs of three-vectors, (ξ, ν), and we are interested in the action of $\mathfrak{g}(\underline{t})$ on this tangent space. Since all $U(o,p)$ are isomorphic, there must, for any tangent, (ξ, ν), exist an isomorphism of $k[\epsilon]$-algebras,

$$\eta : U(o,p) \otimes k[\epsilon] \to U(o + \xi\epsilon, p + \nu\epsilon),$$

commuting with the projection onto $U(o,p)$. It must be given by formulas

$$\eta(x_i) = x_i + \kappa(x_i)\epsilon, \;\; \kappa(x_i) = \kappa_i^0 + \kappa_i^1 x_1 + \kappa_i^2 x_2 + \kappa_i^3 x_3 \in U(o,p),$$
$$i = 1,2,3.$$

Put $\kappa_i := (\kappa_i^0, \kappa_i^1, \kappa_i^2, \kappa_i^3)$, then

$$\kappa \in \tilde{\Theta}_k.$$

A little computation now shows that we must have

$$\xi_i = \kappa_i \cdot \overline{o}, \;\; \nu_i = \kappa_i \cdot \overline{p}, \quad i = 1,2,3.$$

Therefore, given a point $\underline{t} = (o,p)$, and the corresponding generators $\{x_i, i = 1,2,3\}$ of $U(o,p)$, any $\kappa \in \tilde{\Theta}_k$ will correspond to $\kappa_i := \kappa(x_i) \in U(o,p), i = 1,2,3$, and therefore to a tangent of $\underline{H}$ at the point $\underline{t} = (o,p)$,

$$(\xi = \overline{\kappa} \cdot \overline{o}, \;\; \nu = \overline{\kappa} \cdot \overline{p}) \in \Theta_{\underline{H},\underline{t}}.$$

We therefore find an exact sequence of bundles on $\underline{\tilde{H}}$,

$$0 \to \mathfrak{g} \to \tilde{\Theta}_{\tilde{H}} \to \Theta_{\tilde{H}} \to 0.$$

The Lie algebra $\mathfrak{g}$ is now seen to operate naturally on $\tilde{\Theta}_{\tilde{H}}$. Any $\delta \in \mathfrak{g}$ operates on $\kappa \in \tilde{\Theta}_{\tilde{H}}$ as $\delta(\kappa) = \delta \cdot \kappa - \kappa \cdot \delta$. Given

$$(\xi = \overline{\kappa} \cdot \overline{o}, \;\; \nu = \overline{\kappa} \cdot \overline{p}) \in \Theta_{\underline{H},\underline{t}},$$

we may therefore define the operation of δ on $\Theta_{\underline{H},\underline{t}}$ by

$$\delta(\xi, \nu) := (\delta(\kappa)\overline{o}, \delta(\kappa)\overline{p}).$$

Since $\delta \cdot \overline{o} = \delta \cdot \overline{p} = 0$, we find

$$\delta(\xi, \nu) := (\delta \cdot \kappa \cdot \overline{o}, \delta \cdot \kappa \cdot \overline{p}) = (\delta(\xi), \delta(\nu)).$$

Observe that, since $(\overline{o} - \overline{p}) = (o - p)$, the Lie algebra representation of $\mathfrak{g}$ on the tangent space $\Theta_{\tilde{H},\underline{t}}$, at the point $\underline{t} = (o,p)$, kills the subspace generated by the vectors $\{(\xi, 0), (0, \nu)\}, \xi = (o - p), \nu = (p - o)$.

If $o \neq p$, it follows that $\bar{o}$ and $\bar{p}$ are linearly independent, in a four-dimensional vector space, therefore, each vector $\delta_i, i = 1, 2, 3$ is confined to a two-dimensional vector space. Consequently, $\mathfrak{g}(\underline{t}) := \mathrm{Der}_k(U(\underline{t}))$ is of dimension six. Using the isomorphism $U(o, p) \simeq U(o - c, p - c)$, mentioned above, we may choose coordinates such that $o = (0, 0, 0), p = (1, 0, 0)$. In fact, we may first put $c = o$, and reduce to the situation where $o = 0$, and p is a non-zero three-vector. Any $\delta \in \mathrm{Der}_k(U(o, p))$ will then be represented by a matrix of the form

$$M := \begin{pmatrix} \delta_1^1 & \delta_1^2 & \delta_1^3 \\ \delta_2^1 & \delta_2^2 & \delta_2^3 \\ \delta_3^1 & \delta_3^2 & \delta_3^3 \end{pmatrix},$$

where $M(p) = 0$, and we know that the Lie structure is the ordinary matrix Lie-products. Now, clearly we may find a non-singular matrix N such that $N(p) = (1, 0, 0)$, and the Lie algebra of matrices M will be isomorphic to the Lie algebra of the matrices, NMN^{-1}, which are those corresponding to $p = e_1 := (1, 0, 0)$, and we are working with $U(0, e_1)$. Note that in this picture, the fundamental vector $\overline{op} = (1, 0, 0)$. With this it is seen that $\delta \in \mathfrak{g}(\underline{t})$ imply

$$\delta_i^0 = \delta_i^1 = 0, \quad i = 1, 2, 3.$$

The following result is now easily proved.

Theorem 7.1. *The Lie algebra $\mathfrak{g}(\underline{t})$ is isomorphic to the Lie algebra of matrices of the form*

$$\begin{pmatrix} 0 & \delta_1^2 & \delta_1^3 \\ 0 & \delta_2^2 & \delta_2^3 \\ 0 & \delta_3^2 & \delta_3^3 \end{pmatrix}.$$

The radical $\mathfrak{r}$ is generated by three elements, $\{u, r_1, r_2\}$, with

$$u = \begin{pmatrix} 0 & 0 & 0 \\ 0 & 1 & 0 \\ 0 & 0 & 1 \end{pmatrix}, r_1 = \begin{pmatrix} 0 & 1 & 0 \\ 0 & 0 & 0 \\ 0 & 0 & 0 \end{pmatrix}, r_2 = \begin{pmatrix} 0 & 0 & 1 \\ 0 & 0 & 0 \\ 0 & 0 & 0 \end{pmatrix},$$

where $u \notin [\mathfrak{g}, \mathfrak{g}]$, $[u, r_i] = -r_i, [r_1, r_2] = 0$, *and the quotient*

$$\mathfrak{g}(\underline{t})/\mathfrak{r} = \mathfrak{sl}(2),$$

with the usual generators $h := u_0, e := u_1, f := -u_2$,

$$u_0 = \begin{pmatrix} 0 & 0 & 0 \\ 0 & 1 & 0 \\ 0 & 0 & -1 \end{pmatrix}, u_1 = \begin{pmatrix} 0 & 0 & 0 \\ 0 & 0 & 1 \\ 0 & 0 & 0 \end{pmatrix}, u_2 = \begin{pmatrix} 0 & 0 & 0 \\ 0 & 0 & 0 \\ 0 & -1 & 0 \end{pmatrix}.$$

In particular, we find that $\mathfrak{sl}(2) \subset \mathfrak{g}(\underline{t})$.

Remark 7.3. At this point, it may be useful to refer to the examples and the explicit calculations, coming up in Sections 9.3 and 9.4. Given a point $\underline{t} \in \underline{H}$, the tangent space at this point is, of course, nicely represented by the space of all pairs of three-vectors, (ξ, μ), and, as we have seen, it is easy to compute the action of $\mathfrak{g}(\underline{t})$ on this six-dimensional vector space. Just as in the case of the action of $\mathrm{Der}_k(U)$ on $\underline{H}$, any $\delta \in \mathfrak{g}(\underline{t})$ with $\delta_i^0 = 0$, $i = 1, 2, 3$, acts as $\delta(\xi, \mu) = (\delta(\xi), \delta(\mu))$, and in each coordinate, the action is that of the matrix algebra above.

The Lie algebra $\mathfrak{sl}_2(\underline{t})$ acts as follows. There are natural three-dimensional sub-bundles Θ_o, Θ_p of the tangent bundle of $\underline{H}' := \underline{H} - \underline{\Delta}$, such that $\Theta_{\underline{H}'} = \Theta_o \oplus \Theta_p$. We may find a natural basis for both components, for Θ_o as well as for Θ_p, $\{\mathfrak{l}, \nu_1, \nu_2\}$, where $\mathfrak{l}$ is the special tangent vector given by $\overline{po}$, i.e. the tangent direction in our Euclidean three-space, *in which we are looking.* It is obvious from the above matrix-bases of $\mathfrak{g}(\underline{t})$ that $\mathfrak{g}(\underline{t})$ kills $\mathfrak{l}$. Therefore, assuming some metric given, on this basis, $\mathfrak{sl}(2)$ acts on the planes normal to $\mathfrak{l} = \overline{po}$. As a consequence, if we pick any line $\mathbf{l} \subset \mathbb{A}^3$, then the tangent space of $\underline{H}(\mathbf{l})$ is killed by $\mathfrak{g}(\underline{t})$, for all $\underline{t} \in \underline{H}(\mathbf{l})$. We have therefore seen that for any point $\underline{t} \in \underline{\tilde{H}}$ the $\mathfrak{sl}(2)$ component of the Lie algebra of infinitesimal automorphisms of the universal algebra $U(\underline{t})$ act on $\Theta_{\underline{\tilde{H}}}$ in a particularly nice way. The generators u_0, u_1, u_2 (the generators of $\mathfrak{sl}(2)$, normally denoted h, e, f) act on sections of the sub-bundle $B_o \oplus B_p$ of the tangent bundle $\Theta_{\underline{\tilde{H}}}$, just as we described, geometrically, in Chapter 1 and in Laudal (2005, 2011). This $\mathfrak{sl}(2)$-action takes, as we shall see later on, care of the *Isospin* and the *Electromagnetic Forces*, and the action of $\mathfrak{r}$ corresponds to the *Electroweak* sector, the generators $\{u, r_1, r_2\}$ corresponding to the Bosons

$\{Z, W_+, W_-\}$, responsible for the *Electroweak Forces*, acting on the quarks, see Chapter 12.

Now, as in Laudal (2005), consider the Kodaira–Spencer map of the family

$$\nu : \mathbf{U} \longrightarrow \underline{\tilde{H}}.$$

It is the linear map

$$\gamma : \Theta_H = \mathrm{Der}_k(\tilde{H}, \tilde{H}) \to A^1(H, \mathbf{U}, \mathbf{U}),$$

defined by

$$\begin{aligned}\gamma\left(\frac{\partial}{\partial o_i}\right) &= \{\mathbf{x}_{i,j} = (x_i x_j - o_i x_j - p_j x_i + o_i p_j) \mapsto \frac{\partial}{\partial o_i}(r_{i,j}) \\ &= -x_j + p_j\}, \\ \gamma\left(\frac{\partial}{\partial p_j}\right) &= \{\mathbf{x}_{i,j} = (x_i x_j - o_i x_j - p_j x_i + o_i p_j) \mapsto \frac{\partial}{\partial p_i}(r_{i,j}) \\ &= -x_i + o_i\}.\end{aligned}$$

Recall that,

$$A^1(H, \mathbf{U}, \mathbf{U}) = \mathrm{Hom}_H(\tilde{I}, \mathbf{U}) / \mathrm{Der},$$

where we have picked as in Section 7.2, a surjection ρ of a free H-algebra $\tilde{F}$ onto $\mathbf{U}$ and $\tilde{I} = \ker(\rho)$. Der is then the sub-module of $\mathrm{Hom}_{H,H}(\tilde{I}, \mathbf{U})$ generated by the elements of $\mathrm{Der}_H(\tilde{F}, \mathbf{U})$, restricted to $\tilde{I}$. In our case, $\tilde{I}$ is generated by $(\mathbf{x}_{i,j} := x_i x_j - o_i x_j - p_j x_i + o_i p_j), i, j = 1, 2, 3$. Therefore, since any derivation $\eta \in \mathrm{Der}_H(\tilde{F}, \mathbf{U})$ has the form

$$\eta(x_i) = \eta_i^0 + \eta_i^1 x_1 + \eta_i^2 x_2 + \eta_i^3 x_3, \; \eta_i^l \in H, \quad i = 1, 2, 3, \quad l = 0, 1, 2, 3,$$

we find that

$$\begin{aligned}\eta(r_{i,j}) = \eta(x_i)x_j + x_i\eta(x_j) - o_i\eta(x_j) - \eta(x_i)p_j = (\eta_j \cdot \bar{o})(x_i - o_i) \\ + (\eta_i \cdot \bar{o})(x_j - p_j).\end{aligned}$$

Since we have seen that all $\mathfrak{g}(\underline{t})$, $\underline{t} \in \underline{\tilde{H}}$ are isomorphic, it is reasonable that the Kodaira–Spencer map η vanishes on $\underline{\tilde{H}}$, implying the following.

Lemma 7.1. *The Kernel of the Kodaira–Spencer map, also called the* Gauss–Manin–Kernel *of the family* ν, *is*

$$\tilde{\mathfrak{g}} := \ker(\gamma) = \left\{ \xi = \sum_{i=1}^{3} (\eta_i \cdot \bar{o}) \frac{\partial}{\partial o_i} + \sum_{j=1}^{3} (\eta_i \cdot \bar{p}) \frac{\partial}{\partial p_j} \right.$$

$$\left. \times \mid \eta \in \mathrm{Der}_H(\tilde{F}, \mathbf{U} \right\} \simeq \Theta_{\underline{\tilde{H}}}.$$

This is equivalent to $\mathfrak{g}_{\mathfrak{g}} = \Theta_H$.

Note that, so far, we have not been forced to choose an origin in our affine space $\mathbb{A}^3$. All results regarding the vectors $o, p, \overline{o}, \overline{p}$ have been independent of this choice. However, for the next theorem we shall have to fix an origin. This may seem strange, but as it will be explained in Section 7.4, this is a natural consequence of the introduction of a Big Bang event, the singularity U, and its versal base space. Note that this choice of origin in the affine space $\mathbb{A}^3$ has a consequence for our definition of the cosmological zero-velocity subbundle, $\tilde{\Delta}$, that we shall have to come back to. Compare also the next result with the discussion of Bosonic and Fermionic representations.

Theorem 7.2. *There is a well-defined* $\tilde{H}$*-linear map*

$$\kappa : \tilde{\mathfrak{g}} \to \tilde{\Delta} \subset \Theta_{\underline{\tilde{H}}},$$

defined by

$$\kappa(\delta) = \sum_{i=1}^{3} \left((\delta_i \cdot o) \frac{\partial}{\partial o_i} + (\delta_i \cdot p) \frac{\partial}{\partial p_i} \right) = -\sum_{i=1}^{3} \delta_i^0 \left(\frac{\partial}{\partial o_i} + \frac{\partial}{\partial p_i} \right),$$

where $\delta_i \cdot o = -\delta_i^0 = \delta_i \cdot p$.

Assume the vectors $o := (o_1, o_2, o_3), p := (p_1, p_2, p_3)$ *are linearly independent, and let* $\alpha := (\alpha_1, \alpha_2, \alpha_3) = o \times p$. *Then,*

$$\tilde{\mathfrak{k}} := \ker(\kappa).$$

is a rank 3 $\tilde{H}$*-sub-Lie algebra of* $\tilde{\mathfrak{g}}$ *generated by elements* $\{u, v, w\}$ *with Lie structure*

$$[u, v] = -\alpha_2 u + \alpha_1 v, \ [u, w] = -\alpha_3 u + \alpha_1 w, \ [v, w] = -\alpha_3 v + \alpha_2 w,$$

which at each point $\underline{t} \in \underline{\tilde{H}}$ *is isomorphic to* $rad(\mathfrak{g}(\underline{t})) \subset \mathfrak{g}(\underline{t})$.

If for any $\delta \in \tilde{\mathfrak{g}}$ *we define*

$$\tilde{\delta} := \delta + \kappa(\delta),$$

then, applying the adjoint (regular) representation of $\mathfrak{g}$*, we have the following:*

$$\tilde{\delta} \in \mathrm{End}_k(\tilde{\mathfrak{g}}),$$
$$\kappa(\tilde{\delta}) = \kappa(\delta),$$
$$\kappa([\tilde{\delta}, \tilde{\eta}] - \widetilde{[\delta, \eta]}) = [\kappa(\delta), \kappa(\eta)].$$

Finally, κ *defines an isomorphism of* $\tilde{H}$*-modules*

$$\tilde{\kappa} : \tilde{\mathfrak{g}}/\tilde{\mathfrak{k}} \to \tilde{\Delta},$$

where $\mathfrak{g}/\mathfrak{k} \simeq \mathfrak{sl}(2)$*, and we may identify the base elements*

$$\tilde{\kappa}(h) = d_3, \ \ \tilde{\kappa}(e) = d_1, \ \ \tilde{\kappa}(f) = d_2.$$

Proof. Put

$$\bar{\delta}_i : = (\delta_i^0, \delta_i^1, \delta_i^2, \delta_i^3),$$
$$\bar{o} : = (1, o_1, o_2, o_3),$$
$$\bar{p} : = (1, p_1, p_2, p_3).$$

Let $\delta \in \mathrm{Der}$, and compute, as above,

$$\begin{aligned}\delta(r_{i,j}) &= \delta(x_i)x_j + x_i\delta(x_j) - o_i\delta(x_j) - \delta(x_i)p_j \\ &= (\bar{o} \cdot \bar{\delta}_i)x_j + (\bar{p} \cdot \bar{\delta}_j)x_i - \bar{(}o \cdot \bar{\delta}_i)p_j - (\bar{p} \cdot \bar{\delta}_j)o_i.\end{aligned}$$

From this we learn two things. First, we see that $\delta \in \mathrm{Der}_H(\mathbf{U}, \mathbf{U})$, if and only if $\bar{o} \cdot \bar{\delta}_i = 0$, $\bar{p} \cdot \bar{\delta}_i = 0$, for all $i = 1, 2, 3$, which we already knew, second, we see that κ is well-defined. The computation of $[\tilde{\delta}, \tilde{\eta}]$, and its image by κ, is left as an exercise.

But then we see that $\tilde{\mathfrak{k}} := \ker(\kappa) = \{\delta \in \mathrm{Der}_H(\mathbf{U}, \mathbf{U}) | \delta_i^0 = 0, i = 1, 2, 3\}$, is a sub-Lie algebra, though not necessarily a Lie ideal. By definition of $\tilde{\mathfrak{k}}$, if $\delta \in \tilde{\mathfrak{k}}$, we must have $\delta_i^0 = 0$, for all $i = 1, 2, 3$, and so also

$$o \cdot \delta_i = 0, \ \ p \cdot \delta_i = 0, \quad \forall i = 1, 2, 3.$$

It follows that

$$(\delta_i^1, \delta_i^2, \delta_i^3) = c_i(\alpha_1, \alpha_2, \alpha_3), \quad c_i \in H.$$

If δ, δ' corresponds to the two vectors $c := (c_1, c_2, c_3)$ and $c' := (c_1', c_2', c_3')$, then $[\delta, \delta']$ is seen to correspond to the vector $c \times c' \times \alpha$, from where the structural constants may be read off. Moreover, since the determinant of the corresponding matrix

$$\begin{pmatrix} -\alpha_2 & \alpha_1 & 0 \\ -\alpha_3 & 0 & \alpha_1 \\ 0 & -\alpha_3 & \alpha_2 \end{pmatrix}$$

is zero, it is clear that $[\mathfrak{k}, \mathfrak{k}]$ is of dimension two, and abelian. Since $\mathfrak{k}$ is solvable, and obviously maximal, the contention follows.

Note that the map κ, obviously, is not a Lie algebra morphism. Nevertheless, the kernel is an $\tilde{H}$-sub–Lie algebra, identified, at every point $\underline{t} \in \underline{\tilde{H}}$, with the radical of $\mathfrak{g}(\underline{t})$. Therefore, we may, for every $\delta \in \mathfrak{g}$, talk about its class, modulo $\mathfrak{k}$, i.e. in $\mathfrak{sl}(2)$. The last contention therefore follows from the surjectivity of κ. □

Now, recall that $\tilde{\Delta} \subset \Theta_{\tilde{H}}$ is the sub-bundle defined, at the point $\underline{t} = (o, p)$, as the set of tangents of the form (ξ, ξ). Given a metric on $\underline{\tilde{H}}$, then $\mathfrak{sgl}(3)$ acts on $\tilde{\Delta}$, as a gauge group; the observer, o, observing an observed, p, cannot observe a translation of the affine three-space!

We might also look at the action of $\mathfrak{su}(3)$ on the complexified tangent fields, $\tilde{\Delta} \otimes \mathcal{C}$. It acts in the obvious way on the lower right corner of $\Theta = \tilde{c} \oplus \tilde{\Delta}$, like

$$\gamma = \begin{pmatrix} 0 & 0 & 0 & 0 & 0 & 0 \\ 0 & 0 & 0 & 0 & 0 & 0 \\ 0 & 0 & 0 & 0 & 0 & 0 \\ 0 & 0 & 0 & * & * & * \\ 0 & 0 & 0 & * & * & * \\ 0 & 0 & 0 & * & * & * \end{pmatrix}.$$

Since the zero-velocity direction defined at (o, p) is $(o - p, o - p)$, which here is d_3, we may, in an essentially unique way, decompose the Cartan sub-algebra $\mathfrak{h} \subset \mathfrak{su}(3)$, into the Cartan sub-algebra $\mathfrak{h}_1$, for the $\mathfrak{su}(2)$-component leaving δ_3 invariant, and the part $\mathfrak{h}_2 \subset \mathfrak{h}$

perpendicular, in the Killing metric, to $\mathfrak{h}_1$. They are generated by the operators

$$\mathfrak{h}_1 = \begin{pmatrix} 0 & 0 & 0 & 0 & 0 & 0 \\ 0 & 0 & 0 & 0 & 0 & 0 \\ 0 & 0 & 0 & 0 & 0 & 0 \\ 0 & 0 & 0 & 1/2 & 0 & 0 \\ 0 & 0 & 0 & 0 & -1/2 & 0 \\ 0 & 0 & 0 & 0 & 0 & 0 \end{pmatrix},$$

$$\mathfrak{h}_2 = \begin{pmatrix} 0 & 0 & 0 & 0 & 0 & 0 \\ 0 & 0 & 0 & 0 & 0 & 0 \\ 0 & 0 & 0 & 0 & 0 & 0 \\ 0 & 0 & 0 & 1/3 & 0 & 0 \\ 0 & 0 & 0 & 0 & 1/3 & 0 \\ 0 & 0 & 0 & 0 & 0 & -2/3 \end{pmatrix},$$

with respect to the basis $\{c_1, c_2, c_3, d_1, d_2, d_3\}$.

See Laudal (2011), and Section 10.3, where $\mathfrak{h}_1$ is going to be associated to the Spin, and $\mathfrak{h}_2$ defines the notion of Charge.

It is clear that this, together with the formulas mentioned above, give good reasons to believe that there is a relation to the Standard Model. Moreover, here all ingredients are universally given by the information contained in the singularity U, the Big Bang, in my tapping. The choice of metric, i.e. time and so gravitation, will have to be made on the basis of the nature of what I have called the Furniture of the model, see above, at the end of Chapter 5, where the simple metric proposed for the purpose of comparing our gravitation laws with the theory of Kepler and Newton seems to fit with the present theory of black holes.

7.4 The Singular Sub-Scheme of SUSY

These results are going to have several important consequence, in particular, for the derivation of a general analogy of the Dirac equation, and for the construction, in our case, of a kind of supersymmetry. But first, let us consider the *singular sub-scheme* of the morphism

κ above, see Theorem 7.2, i.e. the subspace $\underline{M}(B)$ of $\underline{\tilde{H}}$, where o and p are not linearly independent. It turns out to be a four-dimensional subspace of $\underline{\tilde{H}}$, fibered over the exceptional fiber $E(\underline{0})$, associated to the Big Bang, $\underline{0} \in \underline{\Delta}$. We have the following diagram:

$$\begin{array}{ccccc} \underline{M}(B) \xrightarrow{\mu} E(\underline{0}) \hookrightarrow & & \underline{\tilde{\Delta}} \hookrightarrow & & \underline{\tilde{H}} \\ \downarrow & & \downarrow & & \downarrow \\ \{\underline{0}\} \hookrightarrow & & \underline{\Delta} \hookrightarrow & & \underline{H}, \end{array}$$

where μ is the map that for all directed lines $l \subset \mathbb{A}^3$ through the Big Bang, i.e. such that $\underline{0} \in l$ maps the subspace $\underline{\tilde{H}}(l) := \{(o,p)|o, p \in l\} \subset \underline{\tilde{H}}$ into the point of $E(\underline{0})$ corresponding to l.

Remark 7.4. $\underline{M}(B)$ looks like the product $\underline{c}(\underline{0}) \times \mathbb{R}$, where $\underline{c}(\underline{0})$ is the light-space of the Big Bang, and where the $\mathbb{R}$-coordinate is the unique zero-velocity coordinate of $\underline{\tilde{H}}(l)$, which we have called λ_0. If we make a picture of this, with λ pointing upwards, we have, as we shall see better at the end of Chapter 8, the usual cosmological four-dimensional space–time picture, with the not so unimportant difference that the usual time coordinate, pointing upwards, is replaced by *proper time*.

Recall now, for later use in Chapter 8, that if we start with $U = U(0,0)$, the Big Bang, we observe that the *Dirac derivation* $\delta = id$ acts on the subspace T_0 of its tangent space, $T_* = A^1(k, U, U)$, corresponding to the creation of $\underline{\tilde{H}}$, like the vector field

$$[\delta_0] := \sum_{i=1,2,3} o_i \frac{\partial}{\partial o_i} + \sum_{j=1,2,3} p_j \frac{\partial}{\partial p_j}.$$

Evaluated at a point $\underline{t} = (o,p)$, we see that

$$\delta(o,p) = (-1/2(\overline{op}), 1/2(\overline{op})) + (1/2(o+p), 1/2(o+p)).$$

The last summand is in $\tilde{\Delta}$, i.e. it is a zero-velocity, and the only one contained in $M(B)$. This statement is, however, not entirely kosher, since the notion of $\tilde{\Delta}$ in this case where we have fixed a point in $\underline{\Delta}$ is not well-defined. One might say that *standing still* in $M(B)$ means to be carried along by the *cosmic stream* defined by the Dirac derivation δ of U. The first summand of the vector δ is a classical light-direction, and represents the extension of the visible Universe. Moreover, we shall see that this means that at any point $(o,p) \in M(B)$, *an observer* o *observing an observed* p does so in the direction of the Big Bang event. This is a consequence of equating local time, i.e. the metric at a point $(o,p) \in \underline{\tilde{H}}$, in the space direction, $\tilde{c}(\underline{\lambda})$, with the cosmological time, λ.

In fact, this implies immediately that the tangent space of the past visible space, at any point, is normal (in the metric g) to the vector field $[\delta]$.

Having shown that the gauge Lie algebra of the Standard Model (SM), operates naturally on the non-commutative space $\mathrm{Ph}(\underline{\tilde{H}})$, and that there is defined on $\underline{\tilde{H}}$ a canonical principal bundle

$$\mathfrak{g} := \mathrm{Der}_H(\mathbf{U}),$$

we have now a good measure of the ingredients of the (SM). In fact, we see that the choice of a metric g defines a complex structure on $\Theta_{\underline{\tilde{H}}}$. Moreover, as we have seen that $\mathfrak{su}(2)$, and also complexified $\mathfrak{sl}(2)$, acts naturally on complexified $B_o \oplus B_p$, and that $\mathfrak{su}(3)$ acts on complexified $\tilde{\Delta}$. We therefore see that we have available most of the ingredients of a *canonical Yang-Mills Theory*, defined on $\mathbf{H}$.

It is therefore tempting to propose that the (SM), itself, is concerned with the geometry of the non-commutative quotient scheme

$$\tilde{H}(\sigma_g)/\mathfrak{g}_0 \oplus \mathfrak{g}_1, \;\; \mathfrak{g}_1 = \mathfrak{g} \oplus \mathfrak{sgl}(3),$$

see Chapter 9 for the relation between $\mathfrak{s}gl(3)$ and $\mathfrak{su}(3)$ considered above. Put

$$\mathbf{GQR} := \mathrm{Simp}(\tilde{H}(\sigma_g), (\mathfrak{g}_o \oplus \mathfrak{g}_1)).$$

Taking this as a model for a combined GR and SM, let us make sure that the main ingredients of SM are available here.

First of all, we have the necessary *Gauge Fields*, since we have the principal bundles,

$$\mathfrak{u}(1) \simeq B_o \simeq B_p, \tag{7.1}$$

$$\mathfrak{g} \simeq \mathrm{Der}_{\tilde{H}}(\mathbf{U}), \tag{7.2}$$

$$\mathfrak{s}gl(3), or\ \mathfrak{su}(3) := \mathfrak{su}_{\tilde{H}}(\tilde{\Delta}_{\mathbf{C}}), \tag{7.3}$$

where $\tilde{\Delta}_{\mathbf{C}} := \tilde{\Delta} \otimes \mathbf{C}$. Since we have shown that we may operate with any field k, and certainly go from the reals $\mathbb{R}$ to the complex $\mathbb{C}$, without any mathematical problems, we shall normally omit the $\mathbb{C}$ in $\tilde{\Delta}_{\mathbb{C}}$.

The canonical action of these principal bundles on the (complexified) tangent bundle of $\underline{\tilde{H}}$ give us a lot of possible *force and matter fields.* Moreover, we have the $\tilde{H}$-linear isomorphism, defined in Theorem 7.2

$$\kappa : \tilde{\mathfrak{sl}}(2) \simeq \mathfrak{g}/\mathfrak{k} \simeq \tilde{\Delta}, \tag{7.4}$$

which defines the two obvious $\tilde{H}$-endomorphisms,

$$Q_i \in \mathrm{End}_{\tilde{H}}(\tilde{\mathfrak{sl}}(2) \oplus \tilde{\Delta}), \quad i = 1, 2, \tag{7.5}$$

corresponding to $(\kappa, 0)$, respectively, $(0, \kappa^{-1})$, such that

$$\{Q_1, Q_2\} = id, \ [Q_i, P_\mu] = 0, \tag{7.6}$$

where P_μ is *the infinitesimal translation operator.* The last equation following from the fact that P_μ obviously commutes with any $\kappa(\delta)$, see Theorem 7.2.

We have got a graded Lie algebra,

$$\tilde{\mathfrak{g}} \subset \mathrm{End}_{\tilde{H}}(\tilde{\mathfrak{sl}}(2) \oplus \tilde{\Delta})$$

generated by the complexified adjoint operations of $\mathfrak{sl}(2)$ and $\mathfrak{su}(3)$, together with Q_i, $i = 1, 2$. An element is *even*, or *odd*, according to whether it contains an even or odd number of factors of the type Q_i. Even operators, take "Bosonic states," $\tilde{\mathfrak{sl}}(2)$, into bosonic states, and also "Fermionic states," $\tilde{\Delta}$, into fermionic states. And, obviously, odd operators take bosons into fermions, and vice versa. With this interpretation, our model has acquired an $N = 1$, SUSY-like structure, defined in $\underline{\tilde{H}}$ outside of $\underline{M}(B)$. On this singular subset, the symmetry is "broken."

But, of course, in quantum field theory, Bosons and Fermions are observables of type a^+ or a, having the correct "statistics," i.e. being eigen-operators of $\mathrm{ad}(Q)$, where Q is our Hamiltonian, the Laplace–Beltrami operator on $\Theta_{\underline{\tilde{H}}}$, the least eigenvalue of which is the Planck's constant. See Section 10.4, and also Laudal (2011, (4.6), p. 70).

From this point of view, Bosons are observables corresponding to even elements in $\mathfrak{g}*$, and Fermions are observables corresponding to odd elements.

We have, in Laudal (2011), discussed the notions of Chirality, the PST invariance, stemming from the Hilbert scheme structure of $\mathbf{H} = \underline{\tilde{H}}/Z_2$, and Spinors, with the action on the tangent bundle, of two copies of $\mathfrak{sl}(2)$, together with $\mathfrak{su}(3)$. We saw how the charges of the up and down quarks where defined by the split form of the Cartan sub-algebras $\mathfrak{h}_1 \times \mathfrak{h}_2$ of the Lie algebras $\mathfrak{su}(2) \subset \mathfrak{su}(3)$, respectively, canonically defined at any point in $\underline{\tilde{H}}$, see Section 7.3, and further remarks in Chapter 14.

Here, we have made all this a unique consequence of the Big Bang event, mathematically played by the versal family of associative non-commutative four-dimensional k-algebras, deforming the algebra $U := k[x_1, x_2, x_3]/(x_1, x_2, x_3)^2$. As we have seen, U (and thus also its versal base space) contains a lot of information, in its nine-dimensional Lie algebra of derivations, although it is just modeling a single point, together with a three-dimensional tangent space.

We are tempted to express the content of this sub-section, by saying that a substantial part of (SM), including its spin structure and a canonical SUSY structure, turns out to be an immediate consequence of a Big Bang scenario.

Nothing less.

Chapter 8

The Universe as a Versal Base Space

So, where was the Big Bang (BB), in relation to our time–space, and what on Earth is the meaning of the terms: cosmological time, expansion of the universe, red-shift? How can one fill into this geometric picture the more down to Earth notions like: matter, stress, pressure, charge, and forces, like: gravitation, electromagnetism, weak and strong forces, acting on: elementary particles, quarks, and their multiple combinations?

We should not have to goose-feed the Big Bang-created geometric picture with this additional structure. It should all be part of Creation! Otherwise, it would be difficult to believe in the existence of this prime event.

Going back to the constructed family, the universal family of the Hilbert scheme of sub-schemes of length 2 in $\mathbf{A}^3$,

$$\pi : \mathbf{E} \longrightarrow \mathbf{H},$$

we have just proved that this family may be complemented with another family, no longer a universal one, but just part of a versal family

$$\nu : \mathbf{U} \longrightarrow \mathbf{H},$$

of four-dimensional associative algebras. The three-dimensional space $\underline{\Delta}$ is not a subspace of $\mathbf{H}$, in fact, any point of this ghost space corresponds to the same four-dimensional algebra, namely to U, the BB itself. A metric defined on $\underline{\Delta}$, therefore, measures time at BB, before the creation of the Universe, when "God did nothing," see St. Augustin (1861), and see Chapter 14, where we shall comment on several proposals by Hawking (1988), related to the notion of time.

8.1 First Properties

So, let us fix a point $* \in \underline{\Delta}$, the origin of the coordinate system (x_1, x_2, x_3), used to define U, thereby fixing the whereabouts of BB, clearly outside of our Universe, even though time is already there, as the metric in $\underline{\Delta}$, measuring zero-velocities of U.

We have already observed that the component $\tilde{\Delta} \subset \Theta_{\tilde{H}}$, of the canonical zero-velocity momenta, is no longer uniquely defined as above, since the action of the additive group k^3 on $\tilde{\underline{H}}$ creating $\tilde{\Delta}$ does not keep the $* \in \underline{\Delta}$ fixed. However, we shall see that the Dirac derivation, of U, again comes into play, and recovers the structure of the zero-velocities, $\tilde{\Delta} \subset \Theta_{\tilde{H}}$.

Recall from Section 7.4, the diagram

$$\begin{array}{ccccc} \underline{M}(B) \xrightarrow{\ \mu\ } E(\underline{0}) \hookrightarrow & & \tilde{\underline{\Delta}} \hookrightarrow & & \tilde{\underline{H}} \\ \downarrow & & \downarrow & & \downarrow \\ * = \{\underline{0}\} \hookrightarrow & & \underline{\Delta} \hookrightarrow & & \underline{H}, \end{array}$$

where μ is a map that for all directed lines $l \subset \mathbb{A}^3$ through the BB, i.e. such that $\underline{0} \in l$ maps the subspace $\tilde{\underline{H}}(l) \subset \underline{M}(B) \subset \tilde{\underline{H}}$ into the point of $E(\underline{0})$ corresponding to l. Recall also that, $\mathfrak{g}(*) = \mathrm{Der}_k(U) = \mathfrak{gl}_3(k)$ acts on the tangent space of the versal base space, T_*, and in fact on the subspace T_0 identified with $\underline{H}$, creating a very special derivation, again the *Dirac Derivation*, (maybe a model for Black Energy) in this situation, denoted,

$$\delta_0 \in \mathfrak{g}(*),\ \delta_0 = \begin{pmatrix} 1 & 0 & 0 \\ 0 & 1 & 0 \\ 0 & 0 & 1 \end{pmatrix},$$

i.e. the unit element. We have seen earlier, in Remark 7.2, that δ_0 acts on $\underline{H}$ as

$$[\delta_0] := \sum_{i=1,2,3} o_i \frac{\partial}{\partial o_i} + \sum_{j=1,2,3} p_j \frac{\partial}{\partial p_j},$$

so this operation will stretch all vectors in $\underline{\Delta}$ proportional to their length (in the Euclidean metric). The value of $[\delta_0]$ at a point (o,p) is the tangent,

$$[\delta_0]((o,p)) = 1/2(\overline{op}),\ -1/2(\overline{op})) + (1/2((o+p),\ 1/2((o+p)),$$

where $(\overline{op}) = o - p$. The first summand,

$$(1/2(\overline{op}), -1/2(\overline{op})) = \rho\frac{\partial}{\partial\rho}$$

sits in $\tilde{c}$ and the last one,

$$(1/2((o+p), 1/2((o+p)) = \lambda\frac{\partial}{\partial\lambda},$$

is in the subspace we have called $\tilde{\Delta}$, i.e. it is a relative "zero-velocity," and the only one in $\underline{M}(B)$. Moreover, with the notations above, we find that

$$[\delta_0] = \rho\frac{\partial}{\partial\rho} + \lambda\frac{\partial}{\partial\lambda}$$

is a vector field of $M(B)$.

The subset, $\tilde{\Delta}_* := \langle[\delta_0], \underline{\omega}_\lambda\rangle \subset \Theta_{\tilde{H}}$, where $\underline{\omega}_\lambda$ are the coordinates of the unit sphere in $\underline{\Delta}$, with center in the BB, are universally defined by the structure of the BB, i.e. by the Dirac derivation of the primal object, $\underline{U}$.

This $\tilde{\Delta}_*$, for λ constant, is now going to be our general zero-velocity tangent space, in cosmology, replacing $\tilde{\Delta}$. Clearly, $[\delta_0]$ is stable with respect to the action of $\mathfrak{g}$, since $\frac{\partial}{\partial\rho} = c_3$ and $\frac{\partial}{\partial\lambda} = d_3$ both are killed by $\mathfrak{g}$.

Now, as a start, assume we concentrate on the first question of this sub-section, and assume the metric is the trivial one, so that mass, stress, and charge, etc. can be neglected.

Then, we find that the velocity associated to the direction of the tangent vector $(\rho o, \rho p)$ at $\underline{t} := (o,p)$ is given as $v = \sin(\theta)$, where

$$tg(\theta) = |1/2(\rho\cdot\overline{op}),\ -1/2(\rho\cdot\overline{op}))|/|1/2(\rho\cdot(o+p),\ 1/2(\rho\cdot(o+p))|.$$

From this we deduce two versions of the Hubble formula

$$v = |\overline{op}|/\sqrt{|\overline{op}|^2 + |(o+p)|^2} = r/t,$$

and,

$$v/\sqrt{1-v^2} = |\overline{op}|/|(o+p)| = r/T,$$

where T is cosmological time, and t is "real time" since the BB. In fact, $r = \rho = 1/2|\overline{op}|$, $T = \lambda = 1/2|(o+p)|$, and $t = 1/2\sqrt{|\overline{op}|^2 + |(o+p)|^2}$ is the distance in $\underline{H}$ from $*$ to (o,p) in the Euclidean metric. Of course, this is purely formal, since t is the distance covered by a point in our space left alone, carried away by the intrinsic expansion of space.

The term r/T in the last formula is, in an obvious sense, the speed of the expansion of the Universe, with respect to cosmological time. It is seen to approach infinity when the real speed of the expansion v comes close to maximum, 1. This lead us to think about the inflation-scenario, more or less accepted in cosmology, and we shall have reasons to return to the problem.

But there is a problem here, the expansion of the Universe should not be measurable, in the way we assumed above, within the Universe, and talking about outside the Universe seems to be non-sense. Measuring the cosmological time seems also to be delicate. We might solve these fundamental problems by looking at the world purely locally. At the point $(o,p) \in \tilde{\underline{H}}$, we have a space around us (see Section 8.4), in which we have projective coordinates of the tangent space, given by the action of $\mathfrak{g}$, called $\langle c_1, c_2, c_3, d_1, d_2, d_3 \rangle$. Let ϕ be the angle between $[\delta_0]$, and $\tilde{\Delta} = \langle d_1, d_2, d_3 \rangle$. then the light velocity vector decomposes into a non-observable vector parallel with $[\delta_0]$, and a light velocity vector normal to $[\delta_0]$. The value of this last velocity, w.r.t. this coordinate system, where the velocity of light of course is set to be 1, would be $\cos(\phi)$, explaining the "red-shift," as an observable closely related to the distance between o and p and, thus, given the cosmic microwave background, to the age of the Universe.

Now, we need to introduce a metric that would fit with the one we introduced in Chapter 5, defining the gravitation of Cosmos. Of course, the metric must be invariant under the gauge group $\mathfrak{g}$, and we should be able to deduce the metric from the equations of Chapters 4 and 5. One should also be able to show how to construct the content of the Universe, i.e. the *furniture*, call it $\mathfrak{F}$, and solve equations of the form

$$\mathfrak{G} = \delta(\mathfrak{F}),$$

the Furniture equation, where $\mathfrak{G}$ is our version of the Einstein tensor.

This seems to be what cosmologists are trying out, and we shall, return to the question in Chapter 10. There are several non-trivial problems to ponder. The light-component of the furniture propagates, of course, in time but also in cosmological time. It looks like the propagation of a light wave, ψ, which, due to the structure imposed on our "Toy Model," by the choice of a BB event, is now perpendicular to $\tilde{\Delta}_*$, and so perpendicular to $[\delta_0]$.

This will have an important consequence on the metric, and therefore on the history of our "Present Universe."

Consider again the metric proposed in Chapter 5, of the form

$$g = h_\rho(\underline{\lambda}, \underline{\omega}, \rho)d\rho^2 + h_{\underline{\omega}}(\underline{\lambda}, \underline{\omega}, \rho)d\underline{\omega}^2 + h_{\underline{\lambda}}(\underline{\lambda}, \underline{\omega}, \rho)d\underline{\lambda}^2,$$

where $d\underline{\omega}^2 = d\theta^2 + \sin^2\theta d\phi^2$ is the natural metric in $S^2 = E(\underline{\lambda})$, and $h_{\underline{\lambda}}(\underline{\lambda}, \underline{\omega}, \rho)d\underline{\lambda}^2$ is the metric of $\tilde{\Delta}$. We shall assume the metric invariant under the gauge group $\mathfrak{g}$, which imposes strong conditions on the components; $h_{\underline{\omega}}(\underline{\lambda}, \underline{\omega}, \rho)$ should not depend on θ. In fact, it seems that all $h_\rho, h_{\underline{\omega}}, h_{\underline{\lambda}}$ should depend just on ρ and λ, with $d\rho$ and $d\lambda$ duals of c_3, respectively, d_3.

Recall also the classical metrics of Schwarzschild treated in Section 5.2, and that of Friedman, Robertson, and Walker,

$$ds^2 = -dt^2 + a(t)^2(dr^2 + r^2 d\underline{\omega}^2).$$

Since, as we have seen above in Chapter 5, the natural transformation from the classical language, to the Toy Model, is to assume ds "imaginary" and promote it to be our $d\lambda$, to obtain the metric defined on $M(B)$,

$$dt^2 := g = a(t)^2(dr^2 + r^2 d\underline{\omega}^2) + d\lambda^2.$$

Then we would be tempted to put

$$a(t) = \left(\frac{\rho - h(\lambda)}{\rho}\right), r = \rho$$

obtaining the metric used in Chapter 5, on the Kepler movements,

$$g = \left(\frac{\rho - h(\lambda)}{\rho}\right)^2 d\rho^2 + (\rho - h(\lambda))^2 d\underline{\omega}^2 + d\lambda^2,$$

At this point we shall, as an example, try out the metric

$$g = \overline{g} = \left(\frac{\rho - h(\lambda)}{\rho}\right)^2 d\rho^2 + (\rho - h(\lambda))^2 d\phi^2 + \kappa(\lambda)d\lambda^2,$$

introduced in Chapter 5, for the simplified space, in which $E(\underline{\lambda})$ is reduced to a circle, $\underline{\omega}$ is the corresponding angle ϕ, and the coordinates $\underline{\lambda}$ are reduced to one parameter $\lambda = |\underline{\lambda}|$. This corresponds to reducing our Universe to the sub-universe $M(b) \subset M(B)$, parametrized by (λ, ϕ, ρ), putting $\theta = \pi/2$. The decomposition $\Theta = \tilde{c} \oplus \tilde{\Delta}$ is now replaced by

$$\Theta_{M(b)} = \tilde{c}_* \oplus \tilde{\Delta}_*,$$

where $\tilde{\Delta}_* = \langle \delta_0 \rangle$, and where the summands should be orthogonal in the above metric.

Recall that time is the metric, so we must have the same relations here as in Chapter 5

$$\left(\frac{\rho - h(\lambda)}{\rho}\right)^2 \left(\frac{d\rho}{dt}\right)^2 + (\rho - h(\lambda))^2 \left(\frac{d\phi}{dt}\right)^2 + \kappa(\lambda) \left(\frac{d\lambda}{dt}\right)^2 = 1.$$

Now, choose $h(\lambda) = h/\lambda$, and $\kappa(\lambda) = 1$. An easy calculation shows that the vector field

$$\xi = \frac{\partial}{\partial \rho} - (\rho/\lambda)(1 - (h/\rho\lambda))^2 \frac{\partial}{\partial \lambda}$$

is orthogonal to $[\delta_0]$, and we may choose

$$\tilde{c}_* = \left\langle \frac{\partial}{\partial \rho} - (\rho/\lambda)(1 - (h/\rho\lambda))^2 \frac{\partial}{\partial \lambda}, \frac{\partial}{\partial \phi} \right\rangle.$$

At a point on the horizon, where $1 - (h/\rho\lambda) = 0$, we observe that the light directions are given by

$$\tilde{c}_* = \left\langle \frac{\partial}{\partial \rho}, \frac{\partial}{\partial \phi} \right\rangle,$$

just as in the Kepler–Newton case. Moreover, since for $\lambda > 0$, $|\xi|^2$ vanish at the horizon, an integral curve of ξ will never cross the horizon. As a consequence, we might say that an observer observing an observed not too far away would do it in the same way as within the Kepler–Newton space. However, he/she will receive light signals from all the way back to the infinite horizon of the BB, in line with the Cosmic Microwave Radiation (CMR), that reaches us from "everywhere."

8.2 Density of Mass, Inflation, and Cyclical Cosmology

In the general $\underline{\tilde{H}}$-case, with a general metric g, we have reasons to believe that invariance of g with respect to the universal gauge group, $\mathfrak{g}*$, fixes the form of the metric g,

$$g = h_\rho(\underline{\lambda}, \underline{\omega}, \rho) d\rho^2 + h_{\underline{\omega}_c}(\underline{\lambda}, \underline{\omega}, \rho) d\underline{\omega}_c^2 + h_{\underline{\lambda}}(\underline{\lambda}, \underline{\omega}, \rho) d\underline{\lambda}^2 + h_{\underline{\omega}_\lambda}(\underline{\lambda}, \underline{\omega}, \rho) d\underline{\omega}_\lambda^2,$$

where

$$\underline{\omega} := \underline{\omega}_c \times \underline{\omega}_\lambda$$

are the coordinates of $S^2 \times S^2$, $(\rho, \underline{\omega}_c)$ being the cylindrical coordinates of the light-space, and $(\lambda, \underline{\omega}_\lambda)$ the spherical coordinates of $\underline{\Delta}$, cantered at BB.

This should be a consequence of the fact that $\mathfrak{g}$ keeps the spheres S_c^2 and S_λ^2 fixed, therefore, also $d\underline{\omega}^2$. Moreover, $\mathfrak{g}$ keeps c_3 and d_3 fixed, and in the space of isomorphism classes of metrics,

$$\mathbf{M} := \mathfrak{M}/\mathfrak{g}$$

the tangents of a class of metrics, would have the same form.

We have already computed the Force Laws of the simplified "Universe" of Section 5.1, and they look like,

$$\frac{d^2\rho}{dt^2} = -\left(\frac{h(\lambda)}{\rho(\rho - h(\lambda))}\right)\left(\frac{d\rho}{dt}\right)^2 + \left(\frac{2}{(\rho - h(\lambda))}\right)\left(\frac{dh}{d\lambda}\right)\left(\frac{d\rho}{dt}\right)\left(\frac{d\lambda}{dt}\right) + \left(\frac{\rho^2}{(\rho - h(\lambda))}\right)\left(\frac{d\phi}{dt}\right)^2,$$

$$\frac{d^2\phi}{dt^2} = -2/(\rho - h(\lambda))\frac{d\rho}{dt}\frac{d\phi}{dt} + 2/(\rho - h(\lambda))\left(\frac{dh}{d\lambda}\right)\left(\frac{d\phi}{dt}\right)\left(\frac{d\lambda}{dt}\right),$$

$$\frac{d^2\lambda}{dt^2} = -\left(\frac{(\rho - h(\lambda)}{\rho}\right)\left(\frac{1}{\kappa(\lambda)}\right)\left(\frac{dh}{d\lambda}\right)\left(\frac{d\rho}{dt}\right)^2 - (\rho - h(\lambda))\left(\frac{1}{\kappa(\lambda)}\right)\left(\frac{dh}{d\lambda}\right)\left(\frac{d\phi}{dt}\right)^2 + 1/2\left(\frac{dln(\kappa)}{d\lambda}\right)\left(\frac{d\lambda}{dt}\right)^2,$$

where t, as above, is time. The corresponding constant mass is, of course, $2\pi h_0$.

We observe that $(\frac{dh}{d\lambda}) = -h_0\lambda^{-2}$ is always negative. This means that for $\rho \leq h(\lambda)$, the acceleration of ρ may be positive, and unlimited close to the Horizon, i.e. for ρ close to $h(\lambda)$. In the same region, assuming that $\kappa(\lambda)$ is constant, the acceleration of λ is negative, vanishing on the Horizon.

In particular, we see that *gravitation* may be an expanding *force* inside the Horizon, and a contracting one outside, giving ideas about *inflation* in cosmology.

Moreover, $h(0)$ is infinite, implying that $\underline{M}(B)$ looks like the product, $\mathbb{P}^2_{\mathbb{R}} \times H(l)$, which is the usual cosmological four-dimensional space–time picture, the L–F–R–W model, with the not so unimportant difference that the usual time coordinate, pointing upwards, is replaced by *proper time.*

Moreover, in general, for λ approaching infinity, $h(\lambda)$ vanish, and the restriction of the metric g to $\underline{c}(\underline{\lambda})$ becomes Euclidean. Dividing g with $(\frac{\rho-h(\lambda)}{\rho})^2$, we see that the restriction of this conformally equivalent metric, to $\underline{c}(0)$, the "space" of the BB, is also Euclidean. This seems to relate to the "cyclical universe" proposed by Penrose, see Manin and Marcolli (2014) and Section 8.3.

We have also found a startling analogy to the present day assumption of Inflation. Time is 0 on the Horizon, so creation of photons at a Horizon close to the BB, the assumed origin of the CMR, may seem to "happen at the same time." In $M(B)$, the origin of any creation should be BB, or really $E(*)$, which is given as $\lambda = 0$. This might be the reason why CMR seems to be in coherent phase, an argument for Inflation brought forward by, among others, Dodelson and Silverstein (2017).

Note that in the general case, $\tilde{c}_*$ and $\tilde{\Delta}_*$ are orthogonal, by definition. We would, however, in relation to the general Dirac equation, have liked to find coordinates t_i such that $\tilde{c} = \left\langle \frac{\partial}{\partial t_1}, \frac{\partial}{\partial t_2}, \frac{\partial}{\partial t_3} \right\rangle$, and $\tilde{\Delta} = \left\langle \frac{\partial}{\partial t_4}, \frac{\partial}{\partial t_5}, \frac{\partial}{\partial t_6} \right\rangle$, for which the corresponding metric, $g = \sum g_{p,q} dt_p dt_q$, satisfy,

$$\frac{\partial}{\partial t_i}(g^{j,q}) = \frac{\partial}{\partial t_j}(g^{i,p}) = 0, \quad \forall p, q, 1 \leq i \leq 3, 4 \leq j \leq 6.$$

Then we would have called $\tilde{c}$ and $\tilde{\Delta}$ *normal*, or normally orthogonal, noted $\tilde{c} \perp \tilde{\Delta}$, a notion we shall come back to in Chapter 10.

Consider again the coordinates of the sphere $E(\underline{\lambda})$, for $\underline{\lambda} = c(o,p)$, where $c(o,p)$ is the center of $\overline{op}$. At the point (o,p), we have three light-velocity tangents c_1, c_2, c_3, forming two planes through $c(o,p)$, $\langle c_1, c_3 \rangle$, and $\langle c_2, c_3 \rangle$, defining two angles ϕ_1 and ϕ_2, which turn out to be uniquely defined. Clearly, one may promote one of them to become ϕ in the classical coordinate system of the sphere, used above.

8.3 A Conformally Trivial Cosmological Model

Consider the following equalities

$$\left(\frac{\rho - h(\lambda)}{\rho}\right)^2 = \left(\frac{\lambda - h(\rho)}{\lambda}\right)^2, (\rho - h(\lambda))^2 = \rho^2\lambda^{-2}(\lambda - h(\rho))^2,$$

where we have put $h(x) = h_0 x^{-1}$. We might then try out the metric

$$g = \left(\frac{\rho - h(\lambda)}{\rho}\right)^2 d\rho^2 + \left(\frac{\lambda - h(\rho)}{\lambda}\right)^2 d\lambda^2 + (\rho - h(\lambda))^2 d\underline{\omega}_c^2$$
$$+ (\lambda - h(\rho))^2 d\underline{\omega}_\lambda{}^2,$$

where we identify $\underline{\tilde{H}}$ with $\mathbb{R}^3 \times S^2 \times \mathbb{R}_+$, and where $\underline{\omega}_c$ and $\underline{\omega}_\lambda$ are the coordinates of the unit two-spheres, such that the coordinates of $E(\underline{\lambda})$ become $h(\lambda)\underline{\omega}_c$.

Since $(\frac{\rho - h(\lambda)}{\rho}) = (1 - h(\rho\lambda))$, we find

$$g = (1 - h(\rho\lambda))^2(d\rho^2 + d\lambda^2 + \rho^2 d\underline{\omega}_c^2 + \lambda^2 d\underline{\omega}_\lambda^2),$$

where now $d\rho^2 + \rho^2 d\underline{\omega}_c^2$ is the "Euclidean metric" of $\underline{c}_\lambda$, or rather of $\underline{c}_\lambda$ with $E(\underline{\lambda})$ collapsed, and $d\lambda^2 + \lambda^2 d\underline{\omega}_\lambda^2$ is the Euclidean metric of $\underline{\Delta}$, in spherical coordinates. Moreover, put $\Omega = (1 - h(\rho\lambda))$, and let

$$M := \underline{\tilde{H}}, \overline{M} = \{(o,p) \in M \mid \Omega(o,p) > 0\},$$

then $\overline{M}$ is a smooth sub-space of M with Euclidean metric $\overline{g}$ and such that,

$$g = \Omega^2 \overline{g}$$

is the metric of M. In particular, we find that g is, outside of the horizon, conformally equivalent to the Euclidean metric, so it fits with the proposals of Penrose *et al.* on cyclicality and duality, $\rho \leftrightarrow \lambda$, see Manin and Marcolli (2014), see also the chapter by Hugget *et al.* (1988).

In the Euclidean metric, the normals to $[\delta]$ are spheres in $\underline{H} = \underline{\Delta} \times \underline{C}$, where $\underline{C}$ is the normal subspace to $\underline{\Delta}$ in $\underline{H}$ (see Section 7.1) with center in $(0,0) \in \underline{\Delta}$, and radius the cosmological time measured in $\underline{\Delta}$. The length of a normal curve in the metric g, i.e. the time a light signal (a photon), emitted close to the BB, at cosmological time λ_0, has used to reach the observer, sitting close to the horizon, at cosmological time λ_1, can be given by the integral

$$\rho(\lambda_0, \lambda) = \int_{\phi_0}^{\phi_1} (1 - h(\lambda\rho))\kappa d\phi,$$

where $\phi \in [0, \pi/2], \lambda = \kappa \cdot \sin(\phi), \rho = \kappa \cdot \cos(\phi), \lambda_i = \kappa \cdot \sin(\phi_i)$, $i = 0, 1$ and

$$(1 - h(\lambda\rho)) = (1 - h_0(\kappa^2 \cos(\phi)\sin(\phi))^{-1}).$$

Now, since $\int (\cos(\phi)\sin(\phi))^{-1} d\phi = \ln(tg(\phi))$, we find

$$\int_{\phi_0}^{\phi_1} (1 - h(\lambda\rho))\kappa d\phi = \lambda_1(\phi_1 - \phi_0) + (h_0/\kappa)\ln(tg(\phi_0)/tg(\phi_1)).$$

Since for any point (o, p) outside the Horizon, $\rho > h(\lambda)$, so that $1 > h(\rho\lambda)$, we obtain $0 < (1 - h(\lambda\rho))\kappa < \kappa$. When both of the points corresponding to ϕ_0 and ϕ_1 are outside the horizon, this shows that

$$\rho(\lambda_0, \lambda_1) < (\pi/2)\kappa.$$

Maximal length will occur when the photon is emitted "at the horizon" and received at the horizon. This may be interpreted as follows: In any direction, an observer observes light emanating from "the horizon of the BB," or just outside.

The age of these photons are all less than $\pi/2$ multiplied with the cosmological time of the event, and the speed of the photons measured at the reception is given by $\cos(\phi)$, and provokes a red-shift. We shall come back to this after some words on the notions of observer and observed. See also Chapter 10 for more, related to red-shift and distances in Space.

This looks nice, but we have neglected a big problem. What is the structure of $E(\underline{\lambda})$? And what is the mass density of Space,

$\eta(\underline{\lambda}) := \text{Area}(E(\underline{\lambda}))$? We should have an evolution equation for $\eta(\underline{\lambda})$. Now, $h(\lambda) = h_0/\lambda$ is just the mean density of the gravitational mass, computed so that the gravitational mass rests constant, with respect to cosmological time. Obviously, looking at the night sky, there must be huge volumes in $\underline{\Delta}$ where $\eta(\underline{\lambda})$ is negligible.

Dirac and Feynman introduced the notion of positive and negative energy content. Together, the integral of $4\pi h(\lambda)^2$ or η on the sphere with radius λ should be the "negative" energy of the Universe, the positive should be the energy stored in the objects, the representations constructed as iterated extensions of the elementary particles, using gravitational "waves," as energy.

We therefore need a theory that takes care of the "time"-development of time itself, i.e. a way of measuring the changes of the function $h(\underline{\lambda})$, or maybe the structure of $E(\underline{\lambda})$, as function of the cosmological time, i.e. as function of the spherical coordinates of $\underline{\Delta}$, based at the BB.

Moreover, we should have a theory that makes it possible for any observer to observe the neighborhood of the horizon of any Black Hole $E(\underline{\lambda})$. This is the subject considered in Section 8.4.

Note, in relation to the cyclicality of the Universe, that this follows from the duality of ρ and λ in the metric g above. Note also the obvious Arrow of Time that follows from this Cosmological Model. Any observer observing anything obtains information from the past. There is in fact no Present, and we obtain no information from the Future, a possibility we sometimes are made to believe. This will be commented upon in Chapter 14, in relation to some of the more known texts on this subject.

Remark 8.1 (Coordinate relations). We will, from time to time, have to compare the expressions of the metric, in the different coordinate systems, the original one for $\underline{H} = \text{Spec}(H)$, $H = k[o,p] = k[t_1,t_2,t_3,t_4,t_5,t_6]$, where $\underline{c} \cap \underline{H} \subset \text{Spec}(k[t_1,t_2,t_3]), \underline{\Delta} = \text{Spec}(k[t_4,t_5,t_6])$ and the one used above, $\{\rho, \lambda, \underline{\omega}_c, \underline{\omega}_\lambda\}$, see Chapter 5.

The following formulas may be useful to remember:

$$\rho^2 = t_1^2 + t_2^2 + t_3^2,$$

$$t_3 = \rho \cdot \cos(\theta_c),$$

$$
\begin{aligned}
t_1 &= \rho \cdot \sin(\theta_c)\cos(\phi_c),\\
t_2 &= \rho \cdot \sin(\theta_c)\sin(\phi_c),\\
tan(\phi_c) &= t_2/t_1,\\
\cos(\theta_c) &= t_3/\rho,\\
d\phi_c &= 1/(t_1^2+t_2^2)(t_2dt_1 - t_1dt_2),\\
d\theta_c &= -(t_1^2+t_2^2)^{-1}(dt_3 + t_3\rho^{-1}d\rho),\\
\lambda^2 &= t_4^2+t_5^2+t_6^2,\\
t_6 &= \lambda \cdot \cos(\theta_\lambda),\\
t_4 &= \lambda \cdot \sin(\theta_\lambda)\cos(\phi_\lambda),\\
t_5 &= \lambda \cdot \sin(\theta_\lambda)\sin(\phi_\lambda),\\
tan(\phi_\lambda) &= t_5/t_4,\\
\cos(\theta_\lambda) &= t_6/\lambda,\\
d\phi_\lambda &= 1/(t_4^2+t_5^2)(t_5dt_4 - t_4dt_5),\\
d\theta_\lambda &= -(t_4^2+t_5^2)^{-1}(dt_6 + t_6\lambda^{-1}d\lambda,\\
d\underline{\omega_c}^2 &= d\theta_c^2 + \sin^2(\theta_c)d\phi_c^2,\\
d\underline{\omega_\lambda}^2 &= d\theta_\lambda^2 + \sin^2(\theta_\lambda)d\phi_\lambda^2.
\end{aligned}
$$

Then our conformally trivial metric of $\underline{\tilde{H}}$, see Section 8.3,

$$g = (1-h(\rho\lambda))^2(d\rho^2 + d\lambda^2 + \rho^2 d\underline{\omega}_c^2 + \lambda^2 d\underline{\omega}_\lambda^2)$$

may be written down in terms of the parameters $\{t_1, t_2, t_3, t_4, t_5, t_6\hat{E}, \phi_c, \theta_c, \phi_\lambda, \theta_\lambda\}$!

8.4 Where Are We, the Observers, in This Universe?

The model, worked out so far is perhaps interesting as a global picture of a Universe, filled with time, gravitation, and objects, the furniture. This furniture, that we shall come back to, is modeled as representations of $\tilde{H}(\sigma)$, invariant with respect to the actions of the different gauge groups, collected into the Lie algebra $\mathfrak{g}*$.

But, where is the night-sky, the stars, and the earthly objects that we see and live with? For the Model to be credible, these objects must be "localizable" and give us a picture of our daily world as a "human" observer would recognize. So, how does the local filtration of the visible Universe into a union of three-dimensional light-spaces, each associated to a black hole, relate to our immediate "vision" of our world around us?

Clearly, the global model makes us believe that the structure, having gravitational influence, is given by a metric, defined on the topological space $\underline{\tilde{H}}$, with a singular Horizon. Every point in our "real world" should be related to a Black Hole, i.e. a two-dimensional extra-spatial topological sphere of some sort. In fact, consider the obvious isomorphism

$$\mathbb{A}^3 \simeq \underline{\Delta},$$

where we treat the affine space as a vector space of dimension three, with the BB as origin. Each point, or vector $o \in \mathbb{A}^3$, corresponds to the point $\underline{\lambda} := (o, o) \in \underline{\Delta}$. The sphere bundle

$$\tilde{e} : \tilde{E} = \underline{\tilde{\Delta}} \to \underline{\Delta} = \mathbb{A}^3,$$

then relates every point in the "real world," $o \in \mathbb{A}^3$ to the fiber $E(\underline{\lambda})$ of $\tilde{e}$, therefore to a light-space, $\underline{c}(\underline{\lambda})$, a section of the time–space $\underline{\tilde{H}}$.

From now on, we shall use the notation $\underline{c}(\underline{o}) := \underline{c}(\underline{\lambda})$.

This light-space $\underline{c}(\underline{o})$, with respect to a metric, like the one treated above,

$$g = \left(\frac{\rho - h(\lambda)}{\rho}\right)^2 d\rho^2 + \left(\frac{\lambda - h(\rho)}{\lambda}\right)^2 d\lambda^2 + (\rho - h(\lambda))^2 d\underline{\omega}_c^2$$
$$+ (\lambda - h(\rho))^2 d\underline{\omega}_\lambda{}^2,$$

where $\underline{\lambda} \in \underline{\Delta}$, $\underline{\omega} \in E(\underline{\lambda})$ is now my (or "the observers") eye looking at the world, of course from slightly outside the horizon!

In the language we started with, (o, p) was a pair of an "observer," o observing an "observed" p, or inversely, so that the light-space, of light waves, relating these objects, would be $\underline{c}(\underline{\lambda})$, where $\underline{\lambda} := c(o, p) = (1/2(o+p), 1/2(o+p))$. The corresponding cylindrical coordinates of (o, p) would then be $(\underline{\lambda}, \omega, \rho)$, where $\omega \in E(\underline{\lambda})$ is defined by the direction from o to p, i.e. by $(p - o)$, and $\rho = 1/2|(p - o)|$.

From now on, we shall identify "the observer" of the point (o, p) with the light-cylinder $\underline{c}(o, p)$, of the point $c(o, p)$, as defined in our Cosmological Model. The Night Sky of an observer, is therefore the light of all "objects" younger than $c(o, p)$. Note that time vanishes along the Horizon of $\underline{c}(o, p)$, so one may compare the observer with "the centre of the retina of an eye looking out."

With a given subspace $Y \subset \underline{\Delta} \simeq \mathbb{A}^3$, we may consider the corresponding union of the light-spaces, $\underline{c}(\underline{\lambda})$ for all $\underline{\lambda} := (y, y) \in Y$, call it

$$\underline{c}(Y) \subset \underline{\tilde{H}}.$$

It is the union of all light-lines observed reaching (not emanating from) Y. Given a special "observer o," let

$$\underline{c}(Y : o) \subset \underline{\tilde{H}}$$

be the set of points $\{(y : o) := (y - o, y + o) | y \in Y\} \subset \underline{c}(Y)$, i.e. the set of "observers" in Y that can see o.

In Chapter 10, we shall see that these preparations lead, in some special cases, to interesting results in Geometric Optics, including the problems of çaustics in singularity theory.

The gravitational properties of these light-subspaces will obviously depend entirely on the metric g, but in general, to treat the dynamics of the Furniture of the Universe, we must use the quantum method explained in Section 4.8, to which we shall come back. Obviously, the complete theory will have to be concerned with the totality of "gravitational" objects, thus with a multi-body problem, and much more subtle changes of the metric, so with gravitational waves, and in a complicated relation with the changes of the Furniture of the Cosmos.

We shall have to generalize the notion of density of mass in the Universe, by putting

$$\rho(\underline{\lambda}) = \text{Area}(E(\underline{\lambda})),$$

and imposing the condition that this still keeps the mass of the Universe constant, in some sense.

As we have referred to above, the cosmologists like to model "us" observing something "far away," that being at some "future null infinity," relative to the "objects" we observe close to us, see the work of

Penrose, and also Bondi and Trautman on the notion of "null infinity," referred to in the "non-technical" paper Sormani (2017), or in Newman (2017).

Here, we have seen that it is reasonable to consider that $\underline{\lambda} = c(o, p)$ is an "observer," "observing" the "photons" emanating from the object p, or o, some time in the "past." We have, in examples, given equations of motion for (o, p) or (p, o) inside the light-space, with topology

$$\underline{c}(\underline{\lambda}) \simeq \mathbf{S}^2 \times \mathbb{R}^+,$$

and with parameters $\underline{\omega} \in E(\underline{\lambda})$ and $\rho \in \mathbb{R}$. But we also have equations for how these observers, and therefore these light-spaces, move relatively to each other. The solution of the combined equations seems to be difficult; maybe they are related to the so-called 5th force, or the Nordtvedt effect, see Wikipedia.

There is, of course, a massive literature here, but look at Newman (2017). I propose that the coordinates of $\mathfrak{I}^+$ in the last paper are the same as ours, with $(u, \zeta, \overline{\zeta})$ corresponding to $(\rho, \underline{\omega})$ in $\underline{c}(\underline{\lambda})$. But, of course, the reasoning behind the choice of coordinates is very different.

Given the Conformally Trivial Cosmological Model, with metric g, of Section 8.3, we may consider the $\tilde{E} := E(\underline{\lambda})$ as a representation of the algebra $\tilde{H}(\sigma_g)$, and $h(\lambda)$ as a section of $\tilde{E}$. The equation of evolution might tell us something about the possible configurations of gravitational objects in the Universe. We might use the tools of Section 6.3, to treat discrete objects, like a finite subset of points $\{P_i\}$ in $\underline{\Delta}$.

Accepting this would, of course, imply a whole reformulation of most of the existing quantum-gravity theory.

8.5 The Speed of Photons, and the Red-Shift

Above we found that the age of the photons had to be less than $\pi/2$ multiplied with the cosmological time of the event, and the red-shift should be given by $\cos(\phi)$.

This last statement, in the context of Section 8.3, comes out as follows. At any point (o, p), the tangent space is seen from the local point of view isomorphic to $\tilde{\Delta}_* \times \tilde{c}_*$, where $\tilde{\Delta}_*$ are the null-velocities, and $\tilde{c}_*$ are the light-velocities, and therefore normal to $[\delta_0]$. But at a point (o, p) corresponding to an angle ϕ between d_3 and $[\delta_0]$, the light velocities as "measured" by an observer close to the horizon of $E(\underline{\lambda})$ are $\sin(\pi/2 - \phi) = \cos(\phi)$, provoking the "red-shift."

There are several interesting points pertaining to this model. One is that, our model of the BB would have been isomorphic to the versal deformation space of any one of the $U(\underline{\lambda}), \underline{\lambda} \in \underline{\Lambda}$. Therefore, one might say that the Universe is continuously creating itself. See Chapter 10, for more philosophy in this direction.

Chapter 9

Worked Out Formulas

9.1 Some Examples

As we have seen, $\mathrm{Ext}^1_A(V,V)$ is the tangent space of the mini-versal deformation space of V as an A-module, so that the non-commutative space $\mathrm{Ph}(A)$ also parametrizes the set of *generalized momenta*, i.e. the set of pairs of an A-module V, and a tangent vector of the formal moduli of V, at that "point." In particular, any rank 1 representation of $\mathrm{Ph}(A)$ is represented by a pair, (q,p), of a closed point, q of $\mathrm{Spec}(A)$, and a tangent, p at that point. For $A = k[x_1,\ldots,x_d]$, and two such points, (q_i,p_i), $i = 1,2$, we proved in Theorem 2.1,

$$\begin{aligned}
\dim_k \mathrm{Ext}^1_{\mathrm{Ph}(A)}(k(q_1,p_1),\, k(q_2,p_2)) &= 2n, \text{ for } \ (q_1,p_1) = (q_2,p_2),\\
\dim_k \mathrm{Ext}^1_{\mathrm{Ph}(A)}(k(q_1,p_1),\, k(q_2,p_2)) &= \ n, \ \text{ for } \ q_1 = q_2, p_1 \neq p_2,\\
\dim_k \mathrm{Ext}^1_{\mathrm{Ph}(A)}(k(q_1,p_1),\, k(q_2,p_2)) &= \ 1, \ \text{ for } \ q_1 \neq q_2.
\end{aligned}$$

Moreover, there is a generator of

$$\begin{aligned}
\mathrm{Ext}^1_{\mathrm{Ph}(A)}(k(q_1,p_1),k(q_2,p_2)) = \mathrm{Der}_k(&\mathrm{Ph}(A),\mathrm{Hom}_k(k(q_1,p_1)),\\
&\times k(q_2,p_2)))/\,\mathrm{Triv},
\end{aligned}$$

uniquely characterized by the tangent line defined by the vector $\overline{q_1q_2}$.
Consider the following example.

Example 9.1. Let $A = k[x_1,x_2,x_3]$, and consider now the space of two-dimensional representation of $\mathrm{Ph}(A)$. It is an easy computation that any such is given by the actions

$$x_1 = \begin{pmatrix} a_1 & 0 \\ 0 & a_2 \end{pmatrix}, \quad x_2 = \begin{pmatrix} b_1 & 0 \\ 0 & b_2 \end{pmatrix}, \quad x_3 = \begin{pmatrix} c_1 & 0 \\ 0 & c_2 \end{pmatrix},$$

and

$$dx_1 = \begin{pmatrix} \alpha_1 & \sigma(a_1 - a_2) \\ \sigma(a_2 - a_1) & \alpha_2 \end{pmatrix},$$

$$dx_2 = \begin{pmatrix} \beta_1 & \sigma(b_1 - b_2) \\ \sigma(b_2 - b_1) & \beta_2 \end{pmatrix},$$

$$dx_3 = \begin{pmatrix} \gamma_1 & \sigma(c_1 - c_2) \\ \sigma(c_2 - c_1) & \gamma_2 \end{pmatrix}.$$

The *angular momentum* is now given by,

$$L_{1,2} := x_1 dx_2 - x_2 dx_1 = \begin{pmatrix} (a_1\beta_1 - b_1\alpha_1) & \sigma(a_2 b_1 - a_1 b_2) \\ \sigma(a_1 b_2 - a_2 b_1) & (a_2\beta_2 - b_2\alpha_2) \end{pmatrix}.$$

And *the isospin* has the form

$$I_1 := [x_1, dx_1] = \begin{pmatrix} 0 & \sigma(a_1 - a_2)^2 \\ \sigma(a_2 - a_1)^2 & 0 \end{pmatrix}.$$

Use the result, Theorem 4.1, and obtain for the k-algebra, $\mathrm{Ph}(A)$, the existence of a k-algebra $C(2)$, an open sub-scheme $\underline{U}(2) \subset Spec(C(2))$, an étale morphism

$$\pi : \underline{U}(2) \rightarrow \mathrm{Simp}_2(\mathrm{Ph}(A))$$

and a versal family

$$\rho : \mathrm{Ph}(A) \rightarrow M_2(C(2)).$$

Any section ψ, of the $\underline{C}(2)$-bundle $M_2(C(2))$ defined in $\underline{U}(2)$, determines a vector field $\xi \in \Theta_{\underline{\tilde{H}}}$. In fact, ψ defines for every point in $\underline{C}(2)$, the points

$$p_i := (a_i, b_i, c_i) \in \mathbf{A}^3, \quad i = 1, 2$$

together with the vectors

$$\xi_i := (\alpha_i, \beta_i, \gamma_i), \quad i = 1, 2,$$

which together define a vector field in $\underline{H}$. The extension to $\underline{\tilde{H}}$ is left as an exercise!

The appearance of the *coupling constant* σ in the formulas for ρ, which is just dependent upon the pair of points p_1, p_2, shows that a quantum field theory, in the language of Chapter 1, defined by $\mathrm{Simp}_2(\mathrm{Ph}(A))$, and a Dirac derivation defined in $\mathrm{Ph}(A)$, tell us much more about the dynamics of the vector fields than the covariant derivations of a connection. The next results will show that this model contains an infinitely more complex extension, that someone should take seriously.

Example 9.2. Let $A = M_2(k)$, and let us compute $\mathrm{Ph}(A)$. Clearly, the existence of the canonical homomorphism, $i : M_2(k) \to \mathrm{Ph}(M_2(k))$, shows that $\mathrm{Ph}(M_2(k))$ must be a matrix ring, generated, as an algebra, over $M_2(k)$ by $d\epsilon_{i,j}$, $i, j = 1, 2$, where $\epsilon_{i,j}$ is the elementary matrix. A little computation shows that we have the following relations:

$$d\epsilon_{1,1} = \begin{pmatrix} 0 & (d\epsilon_{1,1})_{1,2} = -(d\epsilon_{2,2})_{1,2} \\ (d\epsilon_{1,1})_{2,1} = -(d\epsilon_{2,2})_{2,1} & 0 \end{pmatrix},$$

$$d\epsilon_{2,2} = \begin{pmatrix} 0 & (d\epsilon_{2,2})_{1,2} = -(d\epsilon_{1,1})_{1,2} \\ (d\epsilon_{2,2})_{2,1} = -(d\epsilon_{1,1})_{2,1} & 0 \end{pmatrix},$$

$$d\epsilon_{1,2} = \begin{pmatrix} \epsilon_{1,2}(d\epsilon_{2,2})_{2,1} & (d\epsilon_{1,2})_{1,2} = -(d\epsilon_{2,1})_{2,1} \\ 0 & -(d\epsilon_{2,2})_{2,1}\epsilon_{1,2} \end{pmatrix},$$

$$d\epsilon_{2,1} = \begin{pmatrix} (d\epsilon_{2,2})_{1,2}\epsilon_{2,1} & 0 \\ (d\epsilon_{2,1})_{2,1} = -(d\epsilon_{1,2})_{1,2} & \epsilon_{2,1}(d\epsilon_{1,1})_{1,2} \end{pmatrix}.$$

From this follows that any co-section, $\rho : \mathrm{Ph}(M_2(k)) \to M_2(k)$, of $i : M_2(k) \to \mathrm{Ph}(M_2(k))$ is given in terms of an element $\phi \in M_2(k)$ such that $\rho(da) = [\phi, a]$.

9.2 Action of $\mathfrak{g}$, and a Canonical Basis for Vector Fields

Let us go back to the Section 7.3, Theorem 7.1.

If $o = (0,0,0)$, $p = (1,0,0)$, we have seen that the Lie algebra $\mathfrak{g}(\underline{t})$ comes out isomorphic to the Lie algebra of matrices of the form

$$\begin{pmatrix} 0 & \delta_1^2 & \delta_1^3 \\ 0 & \delta_2^2 & \delta_2^3 \\ 0 & \delta_3^2 & \delta_3^3 \end{pmatrix}.$$

The radical $\mathfrak{r}$, is generated by three elements, $\{u, r_1, r_2\}$, with

$$u = \begin{pmatrix} 0 & 0 & 0 \\ 0 & 1 & 0 \\ 0 & 0 & 1 \end{pmatrix}, \quad r_1 = \begin{pmatrix} 0 & 1 & 0 \\ 0 & 0 & 0 \\ 0 & 0 & 0 \end{pmatrix}, \quad r_2 = \begin{pmatrix} 0 & 0 & 1 \\ 0 & 0 & 0 \\ 0 & 0 & 0 \end{pmatrix},$$

where $u \notin [\mathfrak{g}, \mathfrak{g}]$, and

$$[u, r_i] = -r_i, [r_1, r_2] = 0,$$

and the quotient

$$\mathfrak{g}(\underline{t})/\mathfrak{r} = \mathfrak{sl}(2),$$

with the usual generators h, e, f,

$$h = u_0 = \begin{pmatrix} 0 & 0 & 0 \\ 0 & 1 & 0 \\ 0 & 0 & -1 \end{pmatrix}, \quad e = u_1 = \begin{pmatrix} 0 & 0 & 0 \\ 0 & 0 & 1 \\ 0 & 0 & 0 \end{pmatrix},$$

$$f = u_2 = \begin{pmatrix} 0 & 0 & 0 \\ 0 & 0 & 0 \\ 0 & 1 & 0 \end{pmatrix}.$$

In particular, we find that $\mathfrak{sl}(2) \subset \mathfrak{g}(\underline{t})$.

Note also that in this case, the unique zero-tangent line at the point $\underline{t}_0 = (o, p)$, $o = (0,0,0), p = (1,0,0)$, killed by $\mathfrak{g}$, is represented by the pair $d_3 := ((1,0,0),(1,0,0))$, and the unique light-velocity line is represented by $c_3 := ((1,0,0),(-1,0,0))$. Let $d_1 := ((0,1,0),(0,1,0))$, $d_2 := ((0,0,1),(0,0,1))$ and let $c_1 := ((0,1,0),(0,-1,0))$, $c_2 := ((0,0,1),(0,0,-1))$. Then $\{c_1, c_2, c_3, d_1, d_2, d_3\}$ is a basis for the tangent space $\Theta_{\underline{t}_0}$, and $\{d_1, d_2, d_3\}$ is a basis for $\tilde{\Delta}_{\underline{t}_0}$.

We observe that the generator h of the Cartan sub-algebra $\mathfrak{h} \subset \mathfrak{g}$ acts in this basis as

$$h = \begin{pmatrix} 1 & 0 & 0 & 0 & 0 & 0 \\ 0 & -1 & 0 & 0 & 0 & 0 \\ 0 & 0 & 0 & 0 & 0 & 0 \\ 0 & 0 & 0 & 1 & 0 & 0 \\ 0 & 0 & 0 & 0 & -1 & 0 \\ 0 & 0 & 0 & 0 & 0 & 0 \end{pmatrix},$$

which makes the choice of basis above canonical, i.e. determines $\{c_1, c_2, d_1, d_2\}$ as (± 1) eigenvectors of h, in $\tilde{c}$, respectively, in $\tilde{\Delta}$. The actions of the gauge fields $\tilde{\mathfrak{g}}$ can then be given canonically: The generators, $h, e, f \in \mathfrak{sl}(2) \subset \mathfrak{g}$ act, in the above basis, like

$$h = \begin{pmatrix} 1 & 0 & 0 & 0 & 0 & 0 \\ 0 & -1 & 0 & 0 & 0 & 0 \\ 0 & 0 & 0 & 0 & 0 & 0 \\ 0 & 0 & 0 & 1 & 0 & 0 \\ 0 & 0 & 0 & 0 & -1 & 0 \\ 0 & 0 & 0 & 0 & 0 & 0 \end{pmatrix},$$

$$e = \begin{pmatrix} 0 & 1 & 0 & 0 & 0 & 0 \\ 0 & 0 & 0 & 0 & 0 & 0 \\ 0 & 0 & 0 & 0 & 0 & 0 \\ 0 & 0 & 0 & 0 & 1 & 0 \\ 0 & 0 & 0 & 0 & 0 & 0 \\ 0 & 0 & 0 & 0 & 0 & 0 \end{pmatrix},$$

$$f = \begin{pmatrix} 0 & 0 & 0 & 0 & 0 & 0 \\ 1 & 0 & 0 & 0 & 0 & 0 \\ 0 & 0 & 0 & 0 & 0 & 0 \\ 0 & 0 & 0 & 0 & 0 & 0 \\ 0 & 0 & 0 & 1 & 0 & 0 \\ 0 & 0 & 0 & 0 & 0 & 0 \end{pmatrix}.$$

Note that $e-f$ acts on $\{c_1, c_2\}$ and $\{d_1, d_2\}$ as the imaginary unit $\imath$, that is $(e-f)^2 = -id$. Since we want to stay "real," we shall therefore when needed, use the notations

$$c*_i := (e-f)c_i, \; d_j^* := (e-f)d_j.$$

The generators $u, r_1, r_2 \in rad(\mathfrak{g})$ act, in the above basis, like

$$u = \begin{pmatrix} 1 & 0 & 0 & 0 & 0 & 0 \\ 0 & 1 & 0 & 0 & 0 & 0 \\ 0 & 0 & 0 & 0 & 0 & 0 \\ 0 & 0 & 0 & 1 & 0 & 0 \\ 0 & 0 & 0 & 0 & 1 & 0 \\ 0 & 0 & 0 & 0 & 0 & 0 \end{pmatrix},$$

$$r_1 = \begin{pmatrix} 0 & 0 & 0 & 0 & 0 & 0 \\ 0 & 0 & 0 & 0 & 0 & 0 \\ 1 & 0 & 0 & 0 & 0 & 0 \\ 0 & 0 & 0 & 0 & 0 & 0 \\ 0 & 0 & 0 & 0 & 0 & 0 \\ 0 & 0 & 0 & 1 & 0 & 0 \end{pmatrix},$$

$$r_2 = \begin{pmatrix} 0 & 0 & 0 & 0 & 0 & 0 \\ 0 & 0 & 0 & 0 & 0 & 0 \\ 0 & 1 & 0 & 0 & 0 & 0 \\ 0 & 0 & 0 & 0 & 0 & 0 \\ 0 & 0 & 0 & 0 & 0 & 0 \\ 0 & 0 & 0 & 0 & 1 & 0 \end{pmatrix}.$$

9.3 The 8-Fold Way of Gell-Mann: The "Real" Story

Now, go back to Theorem 7.2, recall that $\tilde{\Delta} \subset \Theta_{\tilde{H}}$ is the sub-bundle defined at the point $\underline{t}$ as the space of tangents of the form (ξ, ξ). Given a metric on $\underline{\tilde{H}}$, we may look at the action of $\mathfrak{su}(3)$ on $\tilde{\Delta} \otimes \mathbb{C}$. Knowing that $\Theta = \tilde{c} \oplus \tilde{\Delta}$, it acts in the obvious way on the lower right corner, like

$$\gamma = \begin{pmatrix} 0 & 0 & 0 & 0 & 0 & 0 \\ 0 & 0 & 0 & 0 & 0 & 0 \\ 0 & 0 & 0 & 0 & 0 & 0 \\ 0 & 0 & 0 & * & * & * \\ 0 & 0 & 0 & * & * & * \\ 0 & 0 & 0 & * & * & * \end{pmatrix}.$$

Note that the zero-velocity direction defined at (o, p), by the affine line $\overline{op}$ which here is $d_3 = (p - o, p - o)$, is unique, see Section 6.1. We have also seen in Chapter 4 that we may, in an essentially unique way, decompose the Cartan sub-algebra $\mathfrak{h} \subset \mathfrak{su}(3)$, into the Cartan sub-algebra $\langle \mathfrak{h}_1 \rangle$, for the sub-Lie algebra $\mathfrak{su}(2) \subset \mathfrak{su}(3)$, leaving δ_3 invariant, and the part $\langle \mathfrak{h}_2 \rangle \subset \mathfrak{h}$ perpendicular, in the Killing metric, to $\mathfrak{h}_1$. This part is generated by the two elements

$$\mathfrak{h}_1 = \pm \begin{pmatrix} 0 & 0 & 0 & 0 & 0 & 0 \\ 0 & 0 & 0 & 0 & 0 & 0 \\ 0 & 0 & 0 & 0 & 0 & 0 \\ 0 & 0 & 0 & -1/2 & 0 & 0 \\ 0 & 0 & 0 & 0 & 1/2 & 0 \\ 0 & 0 & 0 & 0 & 0 & 0 \end{pmatrix},$$

$$\mathfrak{h}_2^{\pm} = \pm \begin{pmatrix} 0 & 0 & 0 & 0 & 0 & 0 \\ 0 & 0 & 0 & 0 & 0 & 0 \\ 0 & 0 & 0 & 0 & 0 & 0 \\ 0 & 0 & 0 & -1/3 & 0 & 0 \\ 0 & 0 & 0 & 0 & -1/3 & 0 \\ 0 & 0 & 0 & 0 & 0 & 2/3 \end{pmatrix}.$$

Classically, one picks eight base elements of $\mathfrak{su}(3)$, as λ_i, $i = 1, \ldots, 8$, operating on $\tilde{\Delta} \otimes \mathbb{C}$ in our basis, as

$$\lambda_1 = \begin{pmatrix} 0 & 1 & 0 \\ 1 & 0 & 0 \\ 0 & 0 & 0 \end{pmatrix}, \lambda_2 = \begin{pmatrix} 0 & -i & 0 \\ i & 0 & 1 \\ 0 & 0 & 0 \end{pmatrix}, \lambda_3 = \begin{pmatrix} 1 & 0 & 0 \\ 0 & -1 & 0 \\ 0 & 0 & 0 \end{pmatrix},$$

$$\lambda_4 = \begin{pmatrix} 0 & 0 & 1 \\ 0 & 0 & 0 \\ 1 & 0 & 0 \end{pmatrix}, \lambda_5 = \begin{pmatrix} 0 & 0 & -i \\ 0 & 0 & 0 \\ i & 0 & 0 \end{pmatrix},$$

$$\lambda_6 = \begin{pmatrix} 0 & 0 & 0 \\ 0 & 0 & 1 \\ 0 & 1 & 0 \end{pmatrix}, \lambda_7 = \begin{pmatrix} 0 & 0 & 0 \\ 0 & 0 & -i \\ 0 & i & 0 \end{pmatrix},$$

$$\lambda_8 = 1/\sqrt{3} \begin{pmatrix} 1 & 0 & 0 \\ 0 & 1 & 0 \\ 0 & 0 & -2 \end{pmatrix}.$$

Treating the theory of quarks, it is usual to consider the following real matrices, generating (upon intelligent multiplication with the imaginary unit i) the algebra $\mathfrak{L}(\mathfrak{su}(3))$,

$$h_1 = \mathfrak{h}_1, \quad h_2 = -(\sqrt{3}/2)\mathfrak{h}_2,$$

and the restrictions to $\tilde{\Delta}$ of the operator $\mathfrak{e}^i_\pm$, $i = 1, 2, 3$. These operators, in the basis $\{d_1, d_2, d_3\}$, are given by

$$\mathfrak{e}^1_+ = \begin{pmatrix} 0 & 1 & 0 \\ 0 & 0 & 0 \\ 0 & 0 & 0 \end{pmatrix}, \ \mathfrak{e}^2_+ = \begin{pmatrix} 0 & 0 & 0 \\ 0 & 0 & 1 \\ 0 & 0 & 0 \end{pmatrix}, \ \mathfrak{e}^3_+ = \begin{pmatrix} 0 & 0 & 1 \\ 0 & 0 & 0 \\ 0 & 0 & 0 \end{pmatrix},$$

and their "duals,"

$$\mathfrak{e}^1_- = \begin{pmatrix} 0 & 0 & 0 \\ 1 & 0 & 0 \\ 0 & 0 & 0 \end{pmatrix}, \mathfrak{e}^2_- = \begin{pmatrix} 0 & 0 & 0 \\ 0 & 0 & 0 \\ 0 & 1 & 0 \end{pmatrix}, \mathfrak{e}^3_- = \begin{pmatrix} 0 & 0 & 0 \\ 0 & 0 & 0 \\ 1 & 0 & 0 \end{pmatrix}.$$

We observe that the action of $\mathfrak{e}^1_+$ and $\mathfrak{e}^1_-$ on $\tilde{\Delta}$ coincide with that of e and f of $\mathfrak{g}$. Moreover, $\mathfrak{e}^2_- = r_2$ and $\mathfrak{e}^3_- = r_1$, and

$$[u, \mathfrak{e}^i_\pm] = \pm\mathfrak{e}^i_\pm,$$

$\mathfrak{e}^2_-, \mathfrak{e}^3_-$ are the "anti-operators" of $\mathfrak{e}^2_+$, respectively, $\mathfrak{e}^3_+$.

Let us also compute the commutators,

$$[\mathfrak{h}_1, \mathfrak{h}_2] = 0, [\mathfrak{h}_1, \mathfrak{e}^1_\pm] = \pm\mathfrak{e}^1_\pm, [\mathfrak{h}_1, \mathfrak{e}^2_\pm] = \mp 1/2\mathfrak{e}^2_\pm, [\mathfrak{h}_1, \mathfrak{e}^3_\pm] = \pm 1/2\mathfrak{e}^3_\pm,$$

together with the following ones:

$$[\mathfrak{h}_2, \mathfrak{e}^1_\pm] = 0, [\mathfrak{h}_2, \mathfrak{e}^2_\pm] = \pm\mathfrak{e}^2_\pm, [\mathfrak{h}_2, \mathfrak{e}^3_\pm] = \pm\mathfrak{e}^3_\pm.$$

Note, for later use, the Casimir element

$$C = \mathfrak{e}^1_+ \cdot \mathfrak{e}^1_- + \mathfrak{e}^2_+, \mathfrak{e}^2_- + \mathfrak{e}^3_- \mathfrak{e}^3_+ = 1.$$

This completes the computation of the action of $\mathfrak{g}$ on $\Theta_{\tilde{H}}$. Since the tangent space of $\underline{H}(\overline{op})$, generated by c_3, d_3, is uniquely determined, say by (6.1), the effective gauge group of our system turns out to be the *Real Lie algebra*

$$\mathfrak{g}* := \mathfrak{g} \oplus \langle \mathfrak{h}_2 \rangle,$$

inducing the gauge group of Gell-Mann generated by $\{\mathfrak{h}_1, \mathfrak{h}_2, \mathfrak{e}^i_\pm, i = 1, 2, 3\}$, and therefore also the theory of quarks. The color charges, $\langle green, blue, red \rangle$ being simply a choice of basis, among $\langle d_1, d_2, d_3 \rangle$. For the use of this in physics, see the physics literature, an easy to understand, and easy to find text, on the net, is "Notes on SU(3) and the Quark Model," on Workspace.

Theorem 9.1 (Scholie). *Together, these formulas show that the quotients of the* $\mathfrak{g}*$*-representation* $\Theta_{\tilde{H}}$ *are the following:*

- $\tilde{c}$ *and therefore also the* photon $\{c_1, c_2\}$ *and a singleton,* $\{c_3\}$, *both simple.*
- $\tilde{\Delta}$ *and therefore the* "tenebron" $\{d_1, d_2\}$ (*see Chapter* 10), *and a singleton,* $\{d_3\}$, *both simple.*
- *Weyl spinors,* B_o, B_p, *and Dirac spinors,* $B_o \oplus B_p$.

The non-trivial simple quotients of the $\mathfrak{su}(3)$*-representation* $\Theta_{\tilde{H}}$ *are reduced to*

- *The quarks* $\tilde{\Delta}$.

It is now easy to see that the Pauli matrices are found as follows:

$$\sigma^1 = e + f = \begin{pmatrix} 0 & 1 \\ 1 & 0 \end{pmatrix},$$

$$\sigma^2 = i(e - f) = \begin{pmatrix} 0 & \imath \\ -\imath & 0 \end{pmatrix},$$

$$\sigma^3 = h = \begin{pmatrix} 1 & 0 \\ 0 & -1 \end{pmatrix}.$$

Moreover, the parity operator P, *the generator* γ *of the symmetry group* $\mathbb{Z}_2$, *operating on* $\underline{\tilde{H}}$, *acts on* $\tilde{c}$, *and* $\tilde{\Delta}$, *as multiplication by* (-1), *respectively* $(+1)$, *see* Laudal (2011). *Consequently, we have a "partition" isomorphism*

$$P : B_o \to B_p,$$

and a corresponding Parity operator

$$P : B_o \times B_p \to B_o \times B_p,$$

mapping the basis of B_o, $(c_1 + d_1), (c_2 + d_2)$ *to the basis of* B_p, $(-c_1 + d_1), (-c_2 + d_2)$.

In the basis $\{c_1, c_2, c_3, d_1, d_2, d_3\}$, *of* $\Theta_{\tilde{H}} = \tilde{c} \oplus \tilde{\Delta}$, *the morphism* P *is given by the matrix*

$$P := \gamma = \begin{pmatrix} -id & 0 \\ 0 & id \end{pmatrix},$$

which turns left handedness to right handedness, with respect to the direction (o, p), *resp.* (p, o).

This shows that it is meaningful to consider the representations given by the Dirac matrices

$$\gamma^0 = \begin{pmatrix} -1 & 0 \\ 0 & 1 \end{pmatrix}, \ \gamma^k = \begin{pmatrix} 0 & \sigma^k \\ \pm\sigma^k & 0 \end{pmatrix}, \ k = 1, 2, 3,$$

as well as the new operators

$$\gamma^{k+3} = \begin{pmatrix} \sigma^k & 0 \\ 0 & -\sigma^k \end{pmatrix}, \quad k = 1, 2, 3,$$

acting on $B_o \oplus B_p$ *such that*

$$\forall p \neq q, \quad \gamma^p \gamma^q = -\gamma^q \gamma^p, \quad \gamma^p \gamma^p = \pm 1, \quad p, q = 1, 2, 3, 4, 5, 6.$$

Proof. Scholie; German word for: Scholium: note which amplifies a proof. □

9.3.1 *Charge, and the charge conjugation operator C*

The charge of the quarks are defined as the eigenvalues of the Charge Energy Operator, $\mathfrak{h}_2^+$, usually called Q, on $\Theta_{\tilde{H}}$,

$$Q := \mathfrak{h}_2^+ = \begin{pmatrix} 0 & 0 & 0 & 0 & 0 & 0 \\ 0 & 0 & 0 & 0 & 0 & 0 \\ 0 & 0 & 0 & 0 & 0 & 0 \\ 0 & 0 & 0 & -1/3 & 0 & 0 \\ 0 & 0 & 0 & 0 & -1/3 & 0 \\ 0 & 0 & 0 & 0 & 0 & 2/3 \end{pmatrix},$$

and the Charge Conjugation Operator, C, is defined in the above basis of $\Theta_{\tilde{H}} = \tilde{c} \oplus \tilde{\Delta}$, as

$$C = \begin{pmatrix} \mathrm{id} & 0 \\ 0 & -\mathrm{id} \end{pmatrix}.$$

Obviously, $\mathfrak{h}_2^+ = \mathfrak{h}_2^- C$, so C turns a particle into its antiparticle. We shall, in Chapter 13, come back to the relation of this.

Note also that the "potentials," $\psi \in \mathfrak{P}$, such as $\psi_1 = \mathfrak{h}_1$, $\psi_2 = \mathfrak{h}_2^+ C$, and $\psi_3 = \mathfrak{h}_2^- C$, define elements in Ext^1, so "forces," changing the given representation $\rho : \tilde{H}(\sigma_g) \to \mathrm{End}_k(\tilde{\Delta})$, a momentum of ρ_0, into another representation on to the sum $\tilde{\Delta} \oplus \tilde{\Delta}$, thereby acting as forces in $\tilde{\Delta}$, between particles.

Moreover, since the Weyl as well as the Dirac operators on $\Theta_{\tilde{H}}$, in terms of the $\tilde{H}$-basis $\{c_1, c_2, c_3, d_1, d_2, d_3\}$, are all in $\mathfrak{g}$, and constant, their action on the Potentials, $\mathfrak{P}$, are trivial, see Section 4.5.1. In particular, any representation (or object, see Chapter 10) $\rho : \tilde{H}(\sigma_g) \to \mathrm{End}_k(\mathfrak{B})$, defined by $\rho(dt_i) = \xi + \psi_i$, is isomorphic to the one defined by, $\rho(dt_i) = \xi + \psi_i + [\gamma, \psi_i]$, for any $\gamma \in \mathfrak{g}$.

Remark 9.1. There are faithful representations of $\mathfrak{g}$ in the bundle $B_o \oplus B_p$, and $\mathfrak{g}$ kills $A_{o,p}$. The representations, B_o and B_p, the two-component *Weyl-Spinors* of physicists, are not P-invariant, but the space $B_o \oplus B_p$ the *Dirac–Spinors* are P-invariant points of the non-commutative quotient of the moduli space $\tilde{H}$, by the gauge group $\mathfrak{g}$.

Thus, we see that the universal action by Z_2 takes care of all the CPT-equivalences. Let

$$T = -\,\mathrm{id},$$

be the time inversion operator, turning time around, then we find

$$CPT = \mathrm{id}.$$

We shall come back to the classical CP-problems, in the context of the weak interaction, in Chapter 12.

It is clear that this, together with the formulas above, give good reasons to believe that there is a relation between this model and the Standard Model, and so also to the eight-fold way of Gell-Mann. The *color charges*, red, blue, and green, seem to be related to the fact that the quark-space, $\tilde{\Delta}$, as defined above, is of dimension three. The *gluons* being identified with a basis of $\mathfrak{g}*$, like the $\{\mathfrak{e}^i_{\pm}\}$, above.

The three flavors distinguishing the three possible masses of each quark, should also be due to this dimension, see Chapter 10. Moreover, here, all ingredients are universally given by the information contained in the singularity U, the BB, in my tapping. The choice of metric, i.e. time and so gravitation, will have to be made on the basis of the nature of what I have called the Furniture of the model, see Section 1.4.

Note that topologists, in studying the spin structures related to the monopole equations and invariants for three- and four-dimensional Riemannian varieties, are using the skew-adjoint form of the Pauli matrices, i.e.

$$\sigma_1 = \imath\sigma^3 = \begin{pmatrix} \imath & 0 \\ 0 & -\imath \end{pmatrix},$$

$$\sigma_2 = \imath\sigma^2 = \begin{pmatrix} 0 & -1 \\ 1 & 0 \end{pmatrix},$$

$$\sigma_3 = \imath\sigma^1 = \begin{pmatrix} 0 & \imath \\ \imath & 0 \end{pmatrix}$$

and the inner product

$$\langle \sigma_i, \sigma_j \rangle := \langle 1/2tr(\sigma_i^* \sigma_j) \rangle = \begin{pmatrix} 1 & 0 & 0 \\ 0 & 1 & 0 \\ 0 & 0 & 1 \end{pmatrix}.$$

We shall, of course, be particularly interested in the algebra, $\mathrm{End}_{\tilde{H}}(\Theta_{\tilde{H}})$, with the action of $\mathfrak{g}$, given for $\lambda \in \mathfrak{g}$, by

$$\lambda(\psi) := [\lambda, \psi], \ \ \psi \in \mathrm{End}_{\tilde{H}}(\Theta_{\tilde{H}}).$$

For this purpose, we need the details of the adjoint action of $\mathfrak{g}$.

9.4 Adjoint Actions of $\mathfrak{g}$

Let's have a look at the *Boson Fields*, and let us start with the adjoint action of the family $\mathfrak{g}$. Consider the basis, used above, h, e, f, u, r_1, r_2, given by the basis for the $\mathfrak{sl}(2)$,

$$h = \begin{pmatrix} 0 & 0 & 0 \\ 0 & 1 & 0 \\ 0 & 0 & -1 \end{pmatrix}, \quad e = \begin{pmatrix} 0 & 0 & 0 \\ 0 & 0 & 1 \\ 0 & 0 & 0 \end{pmatrix}, \quad f = \begin{pmatrix} 0 & 0 & 0 \\ 0 & 0 & 0 \\ 0 & 1 & 0 \end{pmatrix}$$

and the basis for the radical, $\mathfrak{r}$,

$$u = \begin{pmatrix} 0 & 0 & 0 \\ 0 & 1 & 0 \\ 0 & 0 & 1 \end{pmatrix}, \quad r_1 = \begin{pmatrix} 0 & 1 & 0 \\ 0 & 0 & 0 \\ 0 & 0 & 0 \end{pmatrix}, \quad r_2 = \begin{pmatrix} 0 & 0 & 1 \\ 0 & 0 & 0 \\ 0 & 0 & 0 \end{pmatrix},$$

where $u \notin [\mathfrak{g}, \mathfrak{g}]$, $[u, r_i] = -r_i$, $[r_1, r_2] = 0$, and the quotient,

$$\mathfrak{g}(\underline{t})/\mathfrak{r} = \mathfrak{sl}(2).$$

The adjoint action of $\mathfrak{g}$, in the above basis, is given as

$$\mathrm{ad}(h) = \begin{pmatrix} 0 & 0 & 0 & 0 & 0 & 0 \\ 0 & 2 & 0 & 0 & 0 & 0 \\ 0 & 0 & -2 & 0 & 0 & 0 \\ 0 & 0 & 0 & 0 & 0 & 0 \\ 0 & 0 & 0 & 0 & -1 & 0 \\ 0 & 0 & 0 & 0 & 0 & 1 \end{pmatrix},$$

$$\mathrm{ad}(e) = \begin{pmatrix} 0 & 0 & -1 & 0 & 0 & 0 \\ -2 & 0 & 0 & 0 & 0 & 0 \\ 0 & 0 & 0 & 0 & 0 & 0 \\ 0 & 0 & 0 & 0 & 0 & 0 \\ 0 & 0 & 0 & 0 & 0 & 0 \\ 0 & 0 & 0 & 0 & -1 & 0 \end{pmatrix},$$

$$\mathrm{ad}(f) = \begin{pmatrix} 0 & 1 & 0 & 0 & 0 & 0 \\ 0 & 0 & 0 & 0 & 0 & 0 \\ 2 & 0 & 0 & 0 & 0 & 0 \\ 0 & 0 & 0 & 0 & 0 & 0 \\ 0 & 0 & 0 & 0 & 0 & -1 \\ 0 & 0 & 0 & 0 & 0 & 0 \end{pmatrix},$$

$$\mathrm{ad}(u) = \begin{pmatrix} 0 & 0 & 0 & 0 & 0 & 0 \\ 0 & 0 & 0 & 0 & 0 & 0 \\ 0 & 0 & 0 & 0 & 0 & 0 \\ 0 & 0 & 0 & 0 & 0 & 0 \\ 0 & 0 & 0 & 0 & -1 & 0 \\ 0 & 0 & 0 & 0 & 0 & -1 \end{pmatrix},$$

$$\mathrm{ad}(r_1) = \begin{pmatrix} 0 & 0 & 0 & 0 & 0 & 0 \\ 0 & 0 & 0 & 0 & 0 & 0 \\ 0 & 0 & 0 & 0 & 0 & 0 \\ 0 & 0 & 0 & 0 & 0 & 0 \\ 1 & 0 & 0 & 1 & 0 & 0 \\ 0 & 1 & 0 & 0 & 0 & 0 \end{pmatrix},$$

$$\mathrm{ad}(r_2) = \begin{pmatrix} 0 & 0 & 0 & 0 & 0 & 0 \\ 0 & 0 & 0 & 0 & 0 & 0 \\ 0 & 0 & 0 & 0 & 0 & 0 \\ 0 & 0 & 0 & 0 & 0 & 0 \\ 0 & 0 & 1 & 0 & 0 & 0 \\ -1 & 0 & 0 & 1 & 0 & 0 \end{pmatrix}.$$

From this we deduce the Killing form of $\mathfrak{g}$, given, in the above basis, as

$$\langle h,h\rangle = 10,\ \langle h,e\rangle = 0,\ \langle h,f\rangle = 0\ \langle h,u\rangle = 0,\ \langle h,r_1\rangle = 0,\ \langle h,r_2\rangle = 0$$
$$\langle e,e\rangle = 0,\ \langle e,f\rangle = -3,\ \langle e,u\rangle = 0,\ \langle e,r_1\rangle = 0,\ \langle e,r_2\rangle = 0$$
$$\langle f,f\rangle = 0,\ \langle f,u\rangle = 0,\ \langle f,r_1\rangle = 0,\ \langle f,r_2\rangle = 0$$
$$\langle u,u\rangle = 2,\ \langle u,r_1\rangle = 0,\ \langle u,r_2\rangle = 0$$
$$\langle r_i,r_i\rangle = 0,\ \langle r_i,r_j\rangle = 0,$$

showing that $\mathfrak{g}^{\perp} = \{r_1, r_2\} \subset \mathfrak{g}$, and that $rad(\mathfrak{g})^{\perp} \simeq \mathfrak{sl}_2$.

Consider now the σ^i for $i = 1, 2, 3$, and put

$$\sigma^4 := u, \quad \sigma^5 := r_1, \quad \sigma^6 := r_2.$$

Then on this basis, the Killing form looks like

$$\langle \sigma^i, \sigma^j\rangle = \begin{pmatrix} 10 & 0 & 0 & 0 & 0 & 0 \\ 0 & 6 & 0 & 0 & 0 & 0 \\ 0 & 0 & 6 & 0 & 0 & 0 \\ 0 & 0 & 0 & 2 & 0 & 0 \\ 0 & 0 & 0 & 0 & 0 & 0 \\ 0 & 0 & 0 & 0 & 0 & 0 \end{pmatrix}.$$

Now, use Section 4.2, and see that we can, outside the sub-scheme $\underline{M}(B)$, identify $\mathfrak{sl}(2)$ with $\tilde{\Delta}$, such that h, e, f correspond to d_3, d_1, d_2, respectively.

Chapter 10

Summing Up the Model

Recall the philosophy of this model-building. We are assumed to have a moduli space, $\mathfrak{M}$, the points of which represent the different aspects of a physical phenomenon that we want to study. $\mathfrak{M}$ should be outfitted with a metric g, the time or clock of our model. Its dynamical extension, $\mathrm{Ph}^\infty(\mathfrak{M})$ with the canonical Dirac derivation δ, and together with its moduli space of representations, is the *non-commutative algebraic geometric space*, representing all possible *measurable* changes of the physical phenomena modeled by the objects of $\mathfrak{M}$.

Let us focus on an affine covering of $\mathfrak{M}$, the general object of which is the k-algebra C. To be able to work with this model, we need a choice of a dynamical structure, i.e. a δ-stable ideal (σ) of $\mathrm{Ph}^\infty(C)$, such that the dynamical system $A := C(\sigma) := \mathrm{Ph}^\infty(C)/(\sigma)$ becomes finitely generated.

We may also have global and local gauge groups, $\mathfrak{g}_0$ and $\mathfrak{g}_1$. The global one, $\mathfrak{g}_0$, containing the isometry Lie algebra of the metric g, and the local one, the Lie algebra $\mathfrak{g}_1$, of inessential automorphisms of the relevant representations of this dynamical system.

We should find conditions for the derivation, see Section 4.2,

$$\mathfrak{D} : \mathfrak{g}_0 \to \mathrm{Der}_k(\mathfrak{g}_1)$$

to be flat, i.e. have vanishing curvature, so that the direct sum

$$\mathfrak{g}* := \mathfrak{g}_0 \oplus \mathfrak{g}_1$$

becomes a Lie algebroid.

The goal is to classify the representations of $C(\sigma)/\mathfrak{g}*$, i.e. compute the *non-commutative algebraic quotient space*, the set of representations "insensitive" to the action of the gauge groups, thereby classifying the possible outcomes of measurements of the observables, i.e. of the elements of $C(\sigma)$, as time, "clocked" by the Dirac derivation δ, or its extension, $[\delta]$.

Recall also the definition, and the significance, of the notion of quotient in non-commutative algebraic geometry, see Lemma 4.1. Given a representation $\rho : C(\sigma) \to \mathrm{End}_k(V)$, there is the exact sequence

$$0 \to \mathrm{End}_{C(\sigma)}(V) \to \mathrm{End}_k(V) \to \mathrm{Der}_k(C(\sigma),$$
$$\mathrm{End}_k(V)) \overset{\kappa}{\to} \mathrm{Ext}^1_{C(\sigma)}(V, V) \to 0.$$

Any derivation $\xi \in \mathrm{Der}_k(C(\sigma))$ induces a derivation

$$\rho(\xi) := \xi\rho \in \mathrm{Der}_k(C(\sigma), \mathrm{End}_k(V))$$

and an extension of V, i.e. an element of the tangent space, at the point ρ, of $\mathrm{Rep}(C(\sigma))$,

$$\kappa(\rho(\xi)) \in \mathrm{Ext}^1_{C(\sigma)}(V).$$

If $\kappa(\rho(\xi))$ is 0, then this means that the infinitesimal change of ρ produced by the action of ξ is trivial.

Therefore, if the combined global and local gauge group, the Lie algebra of inessential derivations $\mathfrak{g}* = \mathfrak{g}_0 \oplus \mathfrak{g}_1$, is given, with its Lie–Cartan structure, i.e. with the existence of a (not necessarily flat) connection,

$$\mathfrak{D} : \mathfrak{g}_0 \to \mathrm{Der}_k(\mathfrak{g}_1),$$

the non-commutative quotient "space"

$$\mathrm{Rep}(C(\sigma))/\mathfrak{g}*$$

is the category of representations,

$$\mathrm{Rep}(C(\sigma))/\mathfrak{g}* := \{(\rho, \mu) : \rho : C(\sigma) \to \mathrm{End}_k(V),\ \mu : \mathfrak{g}_1 \to \mathrm{End}_C(V)\},$$

where, $\forall \gamma \in \mathfrak{g}_0$, the Kodaira–Spencer morphism,

$$ks(\rho) : \Theta_C = \mathrm{Der}_k(C, C) \to \mathrm{Ext}^1_C(V, V)$$

kills γ, i.e. $ks(\gamma) = 0$, and μ is a ρ-connection or, rather, a ρ-Lie-bundle morphism.

We shall use the notation

$$\rho \in \mathrm{Simp}(C(\sigma))/\mathfrak{g}*$$

to mean that the representation ρ is a simple object, in this category.

We shall also, in the sequel, often restrict the notion of Ext to extensions in the category $\mathrm{Rep}(C(\sigma))/\mathfrak{g}*$, demanding that the $\mathfrak{g}*$-structure lifts along with the representation ρ; see Section 12.2.

10.1 Metrics, Particles, and the Furniture

In our Toy Model situation, $\mathfrak{M} = \underline{\tilde{H}}/Z_2$, we have introduced a metric g in $\underline{\tilde{H}}$, taking care of time, via the dynamical structure, $\sigma_g = \langle [dt_i, t_j] - g^{i,j} \rangle$, the Dirac derivation $\delta := \mathrm{ad}(g - T)$ of $\tilde{H}(\sigma_g)$, and the Dirac derivation $[\delta]$ of the Furniture.

The global gauge group $\mathfrak{g}_0$ contains the isometry Lie algebra $\mathfrak{o}(g)$. The local gauge group $\mathfrak{g}_1$ is split into two parts, the $\underline{\tilde{H}}$-bundle of Lie algebras, above called $\mathfrak{g}$, and the Lie algebra bundle $\mathfrak{sgl}(3)$, (or, $\mathfrak{g}$ and $\mathfrak{su}(3)$), both acting on $\Theta_{\tilde{H}}$. One problem that we have to take care of in this situation is that $\underline{\tilde{H}}$ is not an affine space, so we cannot, uncritically use the results from Chapter 4. However, H is a polynomial algebra, and the scheme $\underline{H} - \underline{\Delta}$, is an open dense subset of $\underline{\tilde{H}}$. This helps us applying the above results, but we have to be careful.

Moreover, we do not have a unique dynamical system, furnishing a unique model for all forces and fields. But we have a very special vector space of states, $\Theta_{\tilde{H}}$, containing the states of our elementary particles, and we may, for every choice of a non-singular metric g, consider the Levi-Civita connection ∇ given as

$$\nabla_{\delta_i}(\delta_j) = \sum_k \Gamma^k_{i,j} \delta_k,$$

where we have put $\delta_i := \delta_{t_i}$, and the Christoffel symbols are given as

$$\Gamma^k_{i,j} = 1/2 \sum_l g^{k,l}(\delta_{t_j} g_{l,i} + \delta_{t_i} g_{l,j} - \delta_{t_k} g_{i,j}).$$

Its associated representation,

$$\rho_{lc} : \tilde{H}(\sigma_g) \to \mathrm{End}_k(\Theta_{\tilde{H}}),$$

is defined by

$$\rho_{lc}(dt_i) = \nabla_{\xi_i} = \xi_i + D_{\xi_i},$$

where, of course,

$$D_{\xi_i} \in \operatorname{End}_{\tilde{H}}(\Theta_{\tilde{H}}).$$

Recall that this last statement has sense, since when g is non-singular, the vector fields $\xi_i, i = 1, \ldots, 6$ form a $\tilde{H}$-basis for $\Theta_{\tilde{H}}$. Note also that in physics the Levi-Civita connection is usually just denoted $\nabla_{e_i} e_j$ for a given basis $\{e_1, \ldots, e_d\}$ of the bundle $\Theta_{\tilde{H}}$.

Definition 10.1. We propose that the notion of *particle* should be reserved for those representations ρ_1 of $\operatorname{Ph}(\tilde{H})$ induced by representations of $\tilde{H}(\sigma_g)/\mathfrak{g}$, on iterated extensions, $\mathfrak{B}$ of sub-representations of $\Theta_{\tilde{H}} = \tilde{c} \oplus \tilde{\Delta}$, with an extension given by elements in

$$\operatorname{Ext}^1_{\tilde{H}(\sigma_g),\rho_0}(\Theta_{\tilde{H}}, \Theta_{\tilde{H}})$$

such that the action of the metric, as a quadratic form on the $\tilde{H}$ modules $\mathfrak{B}$, lifts. Moreover, we shall promote ρ_{lc} to our base-point in the set of representations. Recall that the Levi-Civita connection is torsion-free, and compatible with the metric, i.e. for $\zeta, \eta, \mu \in \Theta_{\tilde{H}}$, we have

$$[\eta, \mu] = \nabla_\eta(\mu) - \nabla_\mu(\eta), \;\; \zeta(g(\eta, \mu)) = g(\nabla_\zeta(\eta), \mu) + g(\eta, \nabla_\zeta(\mu)).$$

This immediately implies that we have to restrict our metrics g to those that are invariant under $\mathfrak{g}$.

Note that we do not demand that the family of particles

$$\rho_i^1 : \operatorname{Ph}(\tilde{H}) \to \operatorname{End}_k(\mathfrak{B}_i)$$

form a swarm of any sort, see Section 4.1. But recall from Section 4.7 that the formula

$$\operatorname{Ext}^1_{\tilde{H}(\sigma_g),\rho_0}(\rho, \rho) = \mathcal{P}/\operatorname{Triv}$$

can easily be generalized to

$$\operatorname{Ext}^1_{\tilde{H}(\sigma_g),\rho_0}(\rho_i, \rho_j) = \mathcal{P}/\operatorname{Triv},$$

where

$$\mathcal{P}(\rho_i, \rho_j) = \operatorname{Hom}_{\tilde{H}}(\mathfrak{B}_i, \mathfrak{B}_j)^6/\operatorname{Triv}.$$

This will, in our case where $\mathfrak{B}_i$ are iterated extensions of simple subquotient representations of $\Theta_{\tilde{H}}$, help us to find a way of computing the possible particles of this theory.

Of course, the above definition is bold, since it implies non-locality for all particles. Any elementary particle, or a state of an elementary particle, exists everywhere, and may be observed at any point in space, i.e. a state is a section of a bundle on $\underline{\tilde{H}}$ marked as an eigenstate of an element of the Cartan sub-algebra of the local gauge group $\mathfrak{g}_1 = \mathfrak{g} \oplus \mathfrak{sgl}(3)$. Moreover, the particle is, by definition, provided with a "momentum," ρ_1, of ρ_0, the induced $\tilde{H}$-bundle.

Note that this is close to Penrose's idea of "wave functions" as physical waves, experiencing wave function collapse, and where release of gravitational energy also comes in. The "wild" ideas about "consciousness," as an independent entity (particle) that can survive the human body, seem to stem from the work of Penrose and Hameroff (2011). According to our Toy Model, they cannot be mathematically excluded.

In the next chapters, we shall come back to the question of how to construct (or, rather create) new particles, from known ones. Here is where the full power of non-commutative algebraic geometry has to be used!

We now find that the system we have got is very close to the usual set-up for QM, of QFT. We have an algebra of "Observers," several representations,

$$\rho : \tilde{H}(\sigma_g) \to \mathrm{End}_k(\mathfrak{B}),$$

representing our "Furniture." Each one of them is an iterated extension, trivial as extensions of $\tilde{H}$-modules, of the finite number of elementary particles, defined by the action of the gauge groups on $\Theta_{\tilde{H}}$. Consequently, the metric g induces quadratic forms on any such $\tilde{H}$-module $\mathfrak{B}$. The metric also defines a dynamical structure, a Dirac operator, taking care of the time development of all states of all particles. The time development of the particles, themselves, is more complicated and will be treated later. We, therefore, have a situation close to the classical quantum mechanics. This situation leads to the linked models of Chapter 4, which we shall now describe in detail for our purpose, in the situation of the Toy Model. Note first the following easy result.

Lemma 10.1. *Consider a representation*

$$\rho : \tilde{H}(\sigma_g) \to \mathrm{End}_k(\mathfrak{B})$$

with $\mathfrak{g}_0 = \mathrm{Der}_k(\tilde{H}_{\mathfrak{m}})$ *and* $\mathfrak{g}_1 = \mathfrak{g} \oplus \mathfrak{sgl}(3) \subset \mathrm{End}_{\tilde{H}}(\Theta_{\tilde{H}})$. *Let the* $\mathfrak{g}_0$*-connection*

$$\mathfrak{D} : \mathfrak{g}_0 \to \mathrm{Der}_k(\mathfrak{g}_1),$$

be defined as

$$\mathfrak{D}(\xi_i) = \mathrm{ad}(\nabla_{\xi_i} + \psi_i), \ \psi_i = g^{i,l}\gamma_l, \ \gamma_i \in \mathfrak{g}_1,$$

then we find that the curvature of $\mathfrak{D}$ *is given as*

$$F_{i,j} = \mathrm{ad}([\psi_i, \psi_j]).$$

Proof. See Sections 4.7 and 4.8. □

For the Weyl spinors, $\mathfrak{B} := B_o$ or $\mathfrak{B} := B_p$, and the Dirac spinors, $\mathfrak{B} := B_o \otimes B_p$, the Pauli operators, $\sigma^i, i = 1, 2, 3$ as well as the Dirac operators, γ^i, $i = 1, \ldots, 6$, generate the action of the local Gauge group $\mathfrak{g}_1$, but the curvature does not vanish. In fact, as we shall see in Section 10.3, we have the formulas

$$\sum_{j=1}^{3}[[\sigma^i, \sigma^j], \sigma^j] = 8\sigma^i, \quad \sum_{j=1}^{6}[[\gamma^i, \gamma^j], \gamma^j] = 10\gamma^i.$$

In this case, $\mathfrak{g}_0 \oplus \mathfrak{g}_1$ is therefore not a Lie algebroid, but the connection, $\mathfrak{D}$, extends to a connection,

$$\mathfrak{D} : \mathfrak{g} \to \mathrm{End}_k(\mathfrak{B}).$$

The results of Chapter 4, in particular Section 4.4, will in this Toy Model case take the form closely related to the general Theorem 9.1.

Recall the notations

$$T := \sum_j T_j dt_j = -1/2 \sum_{i,j,l} \delta_{t_l}(g_{i,j}) g^{l,i} dt_j = 1/2 \sum_{i,j,l} \delta_{t_j}(g^{i,j}) g_{i,l} dt_l$$

and the computations

$$T_l = -1/2\left(\sum_k \Gamma^k_{k,l} + \sum_{k,p,q} g^{k,q}\Gamma^p_{k,q}g_{p,l}\right) = -1/2\left(\sum_j \Gamma^j_{j,l} + \bar{\Gamma}^j_{j,l}\right),$$

where $\bar{\Gamma}^j_{j,l} := \sum_{p,q} g^{j,q}\Gamma^p_{j,q}g_{p,l}$, and the classical curvature came out as

$$\rho(F_{i,j}) = [\rho(dt_i), \rho(dt_j)] - \sum_p (\Gamma^{j,i}_p - \Gamma^{i,j}_p)\rho(dt_p) = [\nabla_{\xi_i}, \nabla_{\xi_j}] - \nabla_{[\xi_i,\xi_j]},$$

where

$$R_{i,j} := [dt_i, dt_j],\ F_{i,j} := R_{i,j} - \sum_p (\Gamma^{j,i}_p - \Gamma^{i,j}_p)dt_p.$$

We shall, when it is clear where we are, write $F_{i,j}$ for $\rho(F_{i,j})$, the obstruction for ρ inducing a representation of the Lie algebra, $\Theta_{\tilde{H}}$.

Theorem 10.1 (The generic time-action). *Fix a metric g, for $\underline{\tilde{H}}$. Consider a representation,*

$$\rho_0 : \tilde{H} \to \mathrm{End}_k(\mathfrak{B}), \rho_0 \in \mathrm{Rep}(\tilde{H})/\mathfrak{g}*,$$

and an extension

$$\rho : \tilde{H}(\sigma_g) \to \mathrm{End}_k(\mathfrak{B}),$$

considered as a momentum *of ρ_0, given by $\rho(dt_i) = \nabla_{\xi_i} = \sum_j g^{i,j}\nabla_{\delta_j}$, then, see Section 4.4, any other momentum of ρ_0 will look like*

$$\rho_\psi(dt_i) = \nabla_{\xi_i} + \psi_i,\ \psi_i \in \mathrm{End}_{\tilde{H}}(\mathfrak{B})$$

for some tangent direction *(i.e. some potential), $\underline{\psi} \in \mathbf{T}_{\rho_0} := \mathrm{End}_{\tilde{H}}(\mathfrak{B})^6$. By construction, the Dirac derivation in the generic dynamical system related to the metric g is the square of the Time operator. Here it is given by* the tangent direction *$\psi := \underline{\psi}$, as the* *action in* $\mathrm{End}_k(\mathfrak{B})$,

$$[\delta] = \mathrm{ad}(\rho_\psi(g - T)) = \mathrm{ad}(Q_h + [\psi] + Q_v),$$

meaning that

$$\rho_\psi(d^{n+1}t_i) = [(Q_h + [\psi] + Q_v),\ \rho_\psi(d^n t_i)],$$

where

$$Q_h := \rho(g - T) = Q := 1/2 \sum_{i,j} g^{i,j} \nabla_{\delta_i} \nabla_{\delta_j},$$

$$[\psi] := \sum_i \psi_i \nabla_{\delta_i},$$

$$Q_v := 1/2 \sum_{i,j} g_{i,j} \psi_i \psi_j + 1/2 \left(\sum_{j,l} \Gamma^j_{j,l} + \bar{\Gamma}^j_{j,l} \right) \psi_l + 1/2 \nabla.\psi,$$

where $\bar{\Gamma}^j_{j,l} := \sum_{p,q} g^{j,q} \Gamma^p_{j,q} g_{p,l}$. *This takes care of the complete time development of an operator* $\rho(a), a \in \tilde{H}$, *in* $\mathfrak{B}$, *as well as the time development of a state* $\phi \in \mathfrak{B}$ *for the dynamical structure* (σ_g). *But note that here we keep the representation* ρ_1 *"fixed" in its moduli space over* $\mathrm{Ph}^1(\tilde{H})$.

The second-order time-development, with a given metric g, *is also expressed in terms of the Force Laws in* $\mathrm{Ph}(\tilde{H})$, (*see Section* 4.4),

$$\begin{aligned} d^2 t_i = &- \sum_{p,q} \Gamma^i_{p,q} dt_p dt_q - \sum_{p,q} g_{p,q} F_{i,p} dt_q + 1/2 \sum_{l,p,q} g_{p,q} [F_{i,q}, dt_p] \\ &+ 1/2 \sum_{l,p,q} g_{p,q} [dt_p, (\Gamma^{i,q}_l - \Gamma^{q,i}_l)] dt_l + [dt_i, T]. \end{aligned}$$

Proof. See Section 4.5. □

Remark 10.1. Given the gauge group $\mathfrak{g}* = \mathfrak{g}_0 \oplus \mathfrak{g}_1$, there are some particular tangent directions in the space of potentials $\mathcal{P}$, given by $\psi := \mathfrak{v} = \{\mathfrak{v}_i\}$, with $\mathfrak{v}_i = \sum_j \gamma_j g^{j,i}$, and $\gamma_j \in \mathfrak{g}_1$, "normal" in $\mathfrak{B}$, naturally related to a derivation

$$\mathfrak{D} : \mathfrak{g}_0 \to \mathrm{Der}_k(\mathfrak{g}_1).$$

Put

$$[\mathfrak{v}] := \sum_i \mathfrak{v}_i \frac{\partial}{\partial t_i} = \sum_j \gamma_j \nabla_{\xi_j},$$

then with this choice of momentum, the time operator becomes

$$[\delta] = \mathrm{ad}(Q_h + [\mathfrak{v}] + Q_v) \in \mathrm{Der}(\mathrm{End}_k(\tilde{\mathfrak{B}}))$$

with the corresponding first-order time-action in the state-space being defined by

$$Q + [\mathfrak{v}] \in \mathrm{End}_k(\mathfrak{B}),$$

and the curvature equal to

$$F_{i,j} = \mathrm{ad}([\mathfrak{v}_i, \mathfrak{v}_j]).$$

Of course, since $\tilde{H}$ is not a polynomial algebra, and the metric is not non-degenerate, we have to be careful. We may, as we have seen, reduce to the subspace $\underline{\tilde{H}} - \underline{\Delta} = \mathbf{A}^3 \times \mathbf{A}^3 - \underline{\Delta}$.

In our Kepler-example, see Chapter 5, the case where we reduce the space by omitting the coordinate θ, reducing $\underline{\tilde{H}} = \mathbf{R}^3 \times S^2 \times \mathbf{R}_+$ to $R^3 \times S^1 \times \mathbf{R}_+$ as a quotient of $R^3 \times \mathbf{R} \times \mathbf{R}_+$, we may also safely use it, by "rolling up" ϕ.

Recall also that our four-dimensional subspace $\underline{M}(B)$ of $\underline{\tilde{H}}$, introduced in Chapter 7, to relate to cosmology is, topologically, the product $\underline{c}(\underline{0}) \times \mathbf{R}_+$, where $\underline{c}(\underline{0})$ is the light-space of the Big Bang. Here, the metric is singular, but given the form of the metric, called the Conformally Trivial one, in Chapter 8,

$$g = (1 - h(\rho\lambda))^2(d\rho^2 + d\lambda^2 + \rho^2 d\underline{\omega}_c^2 + \lambda^2 d\underline{\omega}_\lambda^2),$$

restricted to $\underline{c}(\underline{0})$, it is conformally equivalent to $\mathbf{A}^3$, with the Euclidean metric, where the $\mathbf{R}_+$-coordinate in $\underline{M}(B)$ is the uniquely defined zero-velocity coordinate, which we have called λ, the cosmological time, corresponding to the direction d_3. If we make a picture of this, with λ pointing upwards, we find, as we shall see better at the end of Section 10.2, the usual cosmological four-dimensional space–time picture, with the not so unimportant difference that the classical time coordinate is now replaced by *proper time*.

10.2 Time, Gravitation, and Einstein's Equation

We may consider the results above as introducing a fusion of General Relativity and Quantum Field Theory. We may also apply Theorem 4.8, relating to the creation of new Furniture, i.e. representations of $\mathrm{Ph}(\tilde{H})/\mathfrak{g}*$, and the change of metrics.

Theorem 10.2 (Time and gravitation). *Let* $\mathbf{M}$ *be the space of (isomorphism classes of) metrices on* $\tilde{H}$*. For every point* $g \in \mathbf{M}$*, consider the diagram,*

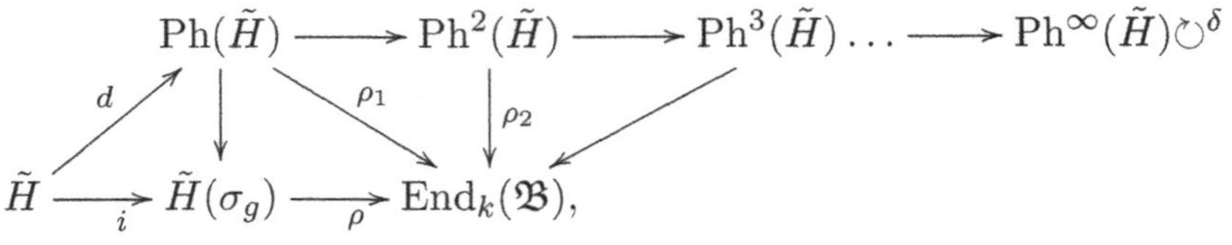

$$\begin{array}{ccccccc} & & \mathrm{Ph}(\tilde{H}) & \longrightarrow & \mathrm{Ph}^2(\tilde{H}) & \longrightarrow & \mathrm{Ph}^3(\tilde{H})\ldots \longrightarrow \mathrm{Ph}^\infty(\tilde{H})\circlearrowright^{\delta} \\ & {}^{d}\nearrow & \downarrow & \searrow^{\rho_1} & \downarrow{\scriptstyle \rho_2} & \swarrow & \\ \tilde{H} & \xrightarrow[i]{} & \tilde{H}(\sigma_g) & \xrightarrow[\rho]{} & \mathrm{End}_k(\mathfrak{B}), & & \end{array}$$

where ρ_1 *is the representation of* $\mathrm{Ph}(\tilde{H})$ *induced by the representation* ρ *of* $\tilde{H}(\sigma_g)$*, the momentum of* $\rho_0 = i \circ \rho$*. Any extension* ρ_2 *of* ρ_1 *corresponds to a tangent of* ρ_1*, therefore, to an acceleration of* $\rho_0 := i \circ \rho$*. Consider the family of* k*-algebras*

$$\{\tilde{H}(\sigma_g)|g \in \mathbf{M}\},$$

and denote by $\mu : \tilde{H}(\sigma) \to \mathbf{M}$ *the correspondence associating* g *to* $\tilde{H}(\sigma_g)$*, see what follows. Let,* $\mathbf{T}_{\mathbf{M},g}$ *be the tangent space to* $\mathbf{M}$*, at* g*, i.e.*

$$\mathbf{T}_{\mathbf{M},g} = \{(h_{i,j}\},\ h_{i,j} = h_{j,i} \in \tilde{H}\}.$$

Define, $h^{i,j}$ *by*

$$h_{i,j} = -\sum_{p,q} g_{i,p} h^{p,q} g_{q,j},\ \ h = \{h_{i,j}\} \in \mathbf{T}_{\mathbf{M}}$$

and put

$$T = -1/2 \sum_{i,j,l} \frac{\partial g_{i,j}}{\partial t_l} g^{l,i} dt_j$$

and

$$T' = -1/2 \sum_{i,j,l} \left(\frac{\partial (g_{i,j} + h_{i,j}\epsilon)}{\partial t_l} \right) (g^{l,i} + h^{l;i}\epsilon) dt_j.$$

Consider now the first-order deformation of the metric $g \in \mathrm{Ph}(\tilde{H})$*,* $(g + \epsilon h) \in \mathrm{Ph}_{k[\epsilon]}(\tilde{H} \otimes k[\epsilon]) = \mathrm{Ph}(\tilde{H}) \otimes k[\epsilon]$*, and the corresponding Dirac derivation,* $\mathrm{ad}(g + \epsilon h - T')$ *in* $\mathrm{Ph}_{k[\epsilon]}(\tilde{H} \otimes k[\epsilon])$*.*

Put $d't_i := \mathrm{ad}(g + \epsilon h - T')(t_i)$. *Then we find, in* $\tilde{H}(\sigma_g) \otimes k[\epsilon]$,

$$[d't_i, t_j] = g^{i,j} - h^{i,j}\epsilon.$$

Moreover, the ρ_1*-derivation* $\eta : \mathrm{Ph}(\tilde{H}) \to \mathrm{End}_k(\mathfrak{B})$ *defined by*

$$\eta(t_i) = 0,\ \eta(dt_i) = \sum_{l,q} h^{i,l}\rho_1(g_{l,q}dt_q) = \sum_{l} h^{i,l}\nabla_{\delta_l} \in \mathrm{Diff}^1(\mathfrak{B}, \mathfrak{B}),$$

corresponds to a first-order derivative of ρ_1, *i.e. to the morphism*

$$\eta(\rho_1) := \rho_2 : \mathrm{Ph}^2(\tilde{H}) \to \mathrm{End}_k(\mathfrak{B}),$$

for which, $\rho_2(d^2t_i) = \eta(dt_i)$. *This induces an element*

$$\eta(h) \in \mathrm{Ext}^1_{\rho_1,\rho_0}(\mathfrak{B}, \mathfrak{B}),$$

where $\mathfrak{B}$ *is the representation* ρ_1, *see Section* 4.5.1. *We obtain an injective map*

$$\eta : \mathbf{T}_{\mathrm{M},g} \to \mathrm{Ext}^1_{\mathrm{Ph}(\tilde{H})}(\mathfrak{B}, \mathfrak{B}),$$

onto the linear subspace

$$\mathrm{Ext}^1_{\mathrm{Ph}(\tilde{H}),\rho_0}(\mathfrak{B}, \mathfrak{B})^{(1)} \subset \mathrm{Ext}^1_{\mathrm{Ph}(\tilde{H})}(\mathfrak{B}, \mathfrak{B}),$$

of the first-order *non-trivial tangent space of the* $\mathrm{Ph}(\tilde{H})$*-representation* $(\rho_1, \mathfrak{B})$, *defined by the derivations,* $\eta : \mathrm{Ph}(\tilde{H}) \to \mathrm{End}_k(\mathfrak{B})$, *where for all* i, $\eta(dt_i) \in \mathrm{Diff}^1(\mathfrak{B}, \mathfrak{B})$.

In particular, this implies that the map, μ *defined above, is* C^1*-differentiable!*

Proof. See Section 4.8. □

Remark 10.2. Note that even though $\mathrm{End}_{\tilde{H}}(\mathfrak{B})^6$ is unchanged with respect to the deformations of the metrics, as long as the metric stays non-singular, we do not know that $\mathrm{Ext}^1_{\mathrm{Ph}(\tilde{H}),\rho_0}(\mathfrak{B}, \mathfrak{B}) \simeq \mathrm{End}_{\tilde{H}}(\mathfrak{B})^6/W$ is unchanged, since W depends on the metric.

10.2.1 *Einsteins field equations*

We have seen that the scalar curvature defines a function, or functional,

$$R : \mathbf{M} \to \tilde{H},$$

and we may look at the Hilbert–Einstein action, defining a *Dirac vector field*, of the form

$$[\delta] \in \Theta_{\mathbf{M}}, \ [\delta(g)] = \mathfrak{G} := Ric - 1/2R\overline{g}.$$

Recall that we have used the notation $g = 1/2\sum_{i,j} g_{i,j}dt_i dt_j \in \mathrm{Ph}(\tilde{H})$, $\overline{g} = \mathrm{Sym}(g)$.

This vector field $[\delta]$ on $\mathbf{M}$ is uniquely defined at each point $g \in \mathbf{M}$, and for each representation

$$\rho : \tilde{H}(\sigma_g) \to \mathrm{End}_k(\mathfrak{B}),$$

by the extension, ρ_1, of ρ,

$$[\delta](g : \rho_1) \in \mathrm{Ext}^1_{\mathrm{Ph}(\tilde{H}),\rho_0}(\mathfrak{B}, \mathfrak{B})$$

defined by the ρ_1-derivation

$$[\mathfrak{G}] : \mathrm{Ph}^1(\tilde{H}) \to \mathrm{End}_k(\mathfrak{B})$$

given by

$$[\mathfrak{G}](dt_i) = \sum_j \mathfrak{G}_{i,j}\nabla_{\delta_j}, \nabla_{\delta_j} := \rho_1(dt_j) = \xi_j \in \mathrm{End}_k(\mathfrak{B}).$$

We are missing the "stress–energy–mass tensor" in this picture. However, this "tensor" should be related to the Furniture of our Universe. Suppose we find that part of this furniture, for one reason or another, can be isolated, say as a representation $\rho_0 : \tilde{H} \to \mathrm{End}_k(\mathfrak{B})$, of course related to the gauge groups involved. Assume moreover that we have given, at the "point" $g \in \mathbf{M}$, i.e. with respect to the time–space at hand, defined by the metric g, a momentum ρ of $(\rho_0, \mathfrak{B})$.

Consider now the corresponding representation $\rho_1 : \mathrm{Ph}(\tilde{H}) \to \mathrm{End}_k(\mathfrak{B})$. Any first-order deformation $[h(\rho_1)]$ of ρ_1 will, according to the above theorem, correspond to a tangent vector $[h](g)$ of $\mathbf{M}$ at g.

This might be said a little differently as: Any non-trivial "graviton," i.e. mass–time quanta, changes any one of the furniture objects, in its versal space.

Now, we may compute all deformations of ρ_1, and in particular all one-dimensional families, $\rho_1(\tau)$ of "first-order deformations," i.e. those for which $\frac{d}{d\tau}(\rho_1(\tau))$ corresponds to a first-order extension, as explained in Theorem 10.1. According to the theorem, this should correspond to a one-dimensional family of metrics $g(\tau)$, with $g(0) = g$.

We would then be tempted to consider the equation

$$[\delta](g(\tau)) = \frac{d}{d\tau}(\rho_1(\tau)),$$

as our Field Equation, replacing the Einstein Field Equation. A solution of these coupled differential equations would give us a "curve" of metrics $g(\tau)$ in $\mathbf{M}$, together with a "curve" of corresponding representations $\rho_1(\tau)$ in $\mathrm{Rep}(\tilde{H}(\sigma_{g(\tau)}))$, determining the dynamics of the *past and the future* of our present Universe,

$$(\tilde{H}, g, \rho : \tilde{H}(\sigma_g) \to \mathrm{End}_k(\mathfrak{B})).$$

The set-up here is formally of the same type as the one in Section 3.2. Given a primary moduli space, here represented by the "blown up" k-algebra $\tilde{H}$, we find the moduli space $\mathbf{M}$, of dynamical systems σ_g, coupled with the moduli space of relevant representations $\mathfrak{B}$, for the corresponding $\mathrm{Ph}(\tilde{H})$, and the time-evolution is given as a vector field in $\mathbf{M}$.

The relation to the classical stress–energy–mass tensor when we model our Universe as a "gas" or "liquid," should follow from our treatment of the Heat- and Navier–Stokes equations, see Section 5.2.

However, the Cosmos exists, as a result of the Big Bang. To predict the future, we just need to know the present state of the time–space and its furniture, together with the momentum of the Universe, just now!

Of course, this depends upon the reasonableness of our Toy Model.

Example 10.1. Let us go back to Section 5.1, and our simple example,

$$g = \left(\frac{\rho - h}{\rho}\right)^2 d\rho^2 + (\rho - h)^2\, d\phi^2 + d\lambda^2.$$

We found the formulas

$$\Gamma^1_{1,1} = \frac{h}{\rho(\rho - h)},$$

$$\Gamma^1_{2,2} = -\frac{\rho^2}{(\rho - h)},$$

$$\Gamma^2_{1,2} = \frac{1}{(\rho - h)},$$

$$\Gamma^2_{2,1} = \frac{1}{(\rho - h)},$$

from which it follows that the non-trivial curvature is given by

$$R^1_{2,1,2} = \frac{h\rho}{(\rho - h)^2},$$

$$R^2_{1,2,1} = \frac{h}{\rho(\rho - h)^2},$$

$$Ric_{1,1} = \frac{h}{\rho(\rho - h)^2},$$

$$Ric_{2,2} = \frac{h\rho}{(\rho - h)^2},$$

$$R = 2\frac{h\rho}{(\rho - h)^4},$$

and we have the equation

$$\mathfrak{G} := Ric - 1/2Rg = -\frac{h\rho}{(\rho - h)^4} d\lambda^2,$$

so $\mathfrak{G} = 0$ on the light-space, but it is not a singularity of $[\delta]$ in $\mathbf{M}$. Note that in the book Eriksen *et al.* (2017), there is a simple mistake of the summation, making the formula on page 173 non-sensical!

Note that the right-hand part of the formula above looks like the time component of the classical ESM-tensor, since in a global situation, far outside the Horizon, $\frac{h\rho}{(\rho - h)^4}$ is very close to $h/(\rho - h)^3$ and therefore to the three-dimensional density of mass, classically denoted by $T^{0,0}$.

The space is therefore not a vacuum. The furniture is provided by the iterated representations of the simple $\mathfrak{g}$ sub-representations of the massy components of $\Theta_{\tilde{H}}$.

Moreover, we find that, for the metric of this simple example, the ring of observables is equal to

$$O := k[\lambda, \phi, \rho]\langle d\lambda, d\phi, d\rho\rangle / R \subset \tilde{H}(\sigma_g),$$

where the only relations are

$$R = \{[d\rho, \rho] = \left(\frac{\rho}{\rho - h}\right)^2, [d\phi, \phi] = (\rho - h)^{-2}, [d\lambda, \lambda] = 1\}.$$

But, what happens if, "suddenly," the furniture, i.e. the representation,

$$\rho : O \to \mathrm{End}_k(\mathfrak{B}),$$

changes, and according to Theorem 10.2 above, provokes a first-order change of the metric, $(h_{i,j})$? Since the Christoffel symbols are given as

$$\Gamma^k_{i,j} = 1/2 \sum_l g^{k,l}(\delta_{t_j} g_{l,i} + \delta_{t_i} g_{l,j} - \delta_{t_k} g_{i,j}),$$

and the Levi-Civita connection is given as

$$D_{\delta_i}(\delta_j) = \sum_k \Gamma^k_{i,j} \delta_k,$$

where we have put $\delta_i := \delta_{t_i}$, which gives us formulas resembling the formula of Einstein for gravitational Lens Effect, and the formulas of Section 5.2, i.e.

$$D_{\delta_\rho}(\delta_\phi) = \Gamma^2_{1,2} \delta_\phi = \frac{1}{(\rho - h)} \delta_\phi.$$

The corresponding first-order changes of the connection are given as

$$\Delta\Gamma^k_{i,j} = 1/2 \sum_l h^{k,l}(\delta_{t_j} g_{l,i} + \delta_{t_i} g_{l,j} - \delta_{t_k} g_{i,j})$$
$$+ 1/2 \sum_l g^{k,l}(\delta_{t_j} h_{l,i} + \delta_{t_i} h_{l,j} - \delta_{t_k} h_{i,j}),$$

which in this simple case, will be

$$\Delta\Gamma^1_{1,1} = -h_{\rho,\rho}\frac{h\rho}{(\rho-h)^3} + 1/2\frac{\rho^2}{(\rho-h)^2}\delta_\rho(h_{\rho,\rho}),$$

$$\Delta\Gamma^1_{2,2} = -\left(\sum_l h^{1,l}\right)(\rho-h) - 1/2\frac{h\rho}{(\rho-h)^3}\delta_\rho(h_{2,2})),$$

and

$$\Delta\Gamma^1_{1,2} = -h_{2,2}\frac{1}{(\rho-h)^3} + 1/2\frac{1}{(\rho-h)^2}\delta_\rho(h_{2,2}).$$

10.3 Energy, Dirac, and Maxwell

We have now got examples illustrating what we have termed a *Quantum Field Theory*, see Section 10.1 and see also Chapter 11.

Given a metric g, and a representation, i.e. a momentum,

$$\rho : \tilde{H}(\sigma_g) \to \mathrm{End}_k(\mathfrak{B}),$$

the objects in our QFT are the quantum fields, $\psi \in \mathrm{End}_{\tilde{H}}(\mathfrak{B})^6$, or the associated deformations ρ_ψ of ρ, given by

$$\rho_\psi(dt_i) = \rho(dt_i) + \psi_i,$$

together with the Dirac, or Heisenberg, operator acting on $\mathrm{End}_k(\mathfrak{B})$, see Section 10.3.1, given by

$$[\delta] = \mathrm{ad}(Q_h + [\psi] + Q_v)$$

acting as a derivation on $\mathrm{End}_k(\mathfrak{B})$, where $Q = (g-T) \in \tilde{H}(\sigma_g)$, and

$$*Q_h := \rho(g-T) = 1/2\sum_{i,j} g^{i,j}\nabla_{\delta_i}\nabla_{\delta_j},$$

$$[\psi] := \sum_i \psi_i\nabla_{\delta_i},$$

$$Q_v := \psi(g-T) = 1/2\sum_{i,j} g_{i,j}\psi_i\psi_j$$

$$+ 1/2\left(\sum_{j,l}\Gamma^j_{j,l} + \bar{\Gamma}^j_{j,l}\right)\psi_l + 1/2\nabla.\psi.$$

But, beware, everything should now also be related to "families of metrics," gravitational waves, in popular language, defined in terms of the time-evolution of the furniture $\mathfrak{B}$, considered as a representation of $\mathrm{Ph}(\tilde{H})$.

10.3.1 *Energy*

Recall that the notion of **Energy** is usually related to a **Property**, or some kind of "content," of an object in physics. The Einstein Equation

$$E_0 = mc^2$$

relates the **Energy** to the **Mass** of the object.

This is a formula most people know about, with the historical consequences we all know. However, the philosophical problems related to the "Explanation" of the formula has generated a large literature, see Fernflores (2019), for a recent and very informative paper on the subject. We shall come back to the difficulties, and "some possible" mathematical solutions, in Section 10.3.2.

Here, we propose a rather well-known definition of the notion energy/mass, and then we shall see how it works out, in relation to other variants.

By definition, time, is the metric chosen for our space $\underline{\tilde{H}}$. Since the metric is given by $g = \sum_i g_{i,j} dt_i dt_j \in \mathrm{Ph}(\underline{\tilde{H}})$, it looks like the square of the "length," defined by the metric. Time should therefore look like an operator q, in $\tilde{H}(\sigma_g)$, the square root of $Q := g - T$. This is what Dirac exploited, cooking up the Dirac equation for the electrons.

Classical General Relativity, see Sachs and Wu (1977), defines a point particle of rest mass m as a future pointing curve

$$\gamma : \mathbf{A}^1 \to \underline{\tilde{H}}$$

which, in algebraic geometry, is equivalent to a homomorphism, $\pi : \tilde{H} \to k[t]$, where we put $x_i := \pi(t_i) \in k[t]$. See Section 4.4.1, Example 4.1. The mass, $m(\gamma)$, is defined as

$$m(\gamma)^2 = -g(\gamma_*, \gamma_*) = -\sum_{k,l} g_{k,l} \frac{\partial x_k}{\partial t} \frac{\partial x_l}{\partial t} = -6h_{1,1},$$

where here (contrary to the notations in Section 4.4.1), g is the metric of $\underline{\tilde{H}}$, and h is the corresponding metric of $\mathbf{A}^1$. Moreover, the condition that the curve is future pointing, in the relativistic context of Sachs and Wu, loc.cit, means that the metric, restricted to the curve, given by $h_{1,1}$, is negative. However, in the above theory, time is

$$\delta = \mathrm{ad}(Q) \in \mathrm{Der}(\tilde{H}(\sigma_g)), or [\delta] \in \mathrm{End}_k(\mathfrak{B}),$$

and for every $\psi \in \mathrm{End}_{\tilde{H}}(\mathfrak{B})$, we have

$$[\delta](\psi) = [Q, \psi] = \sum_l \delta_l |\psi| dt_l,$$

where $|\psi|$ is ψ given as an $\tilde{H}$-matrix in the bases $\{\underline{c}, \underline{d}\}$, where we have assumed that all elements are unit vectors. This follows from the formula

$$[[\delta](\psi), t_k] = [[Q, \psi], t_k] = [[Q, t_k], \psi] = [dt_k, \psi] = \sum_l g^{k,l} \delta_l |\psi|.$$

In analogy with the classical situation, one might reasonably define the Energy/Mass of the Potential ψ as

$$m^2(\psi) = g([\delta](\psi), [\delta](\psi)) = \sum_{l,j} g^{i,j} \delta_l |\psi| \delta_j |\psi| \in \mathrm{End}_{\tilde{H}}(\mathfrak{B}).$$

Of course, here we have not obtained real values for the mass/energy, but we could hope for real values if we put, in perfect analogy with the classical formula,

$$m^2(\psi) = \sum_{l,j} g^{i,j} \delta_l |\psi| \delta_j |\psi|$$
$$m^2 = \det(m^2(\psi)).$$

For representations build from sub-modules of $\tilde{c}$ or of $\tilde{\Delta}$, see Section 10.3.4. Recall that our aim is to make a reasonable definition of Energy for particles, taking care of the fundamental Law of conservation of energy.

Before we go further in this direction, let us consider the possible Dirac variants, i.e. trying out the problem of finding a square root of Q.

Now, see the Scholie Theorem 9.1, and consider, the Pauli matrices,

$$\tilde{\sigma}_1 = e + f = \begin{pmatrix} 0 & 1 \\ 1 & 0 \end{pmatrix} =: \sigma_1, \tag{10.1}$$

$$\tilde{\sigma}^2 = ie - if = \begin{pmatrix} 0 & i \\ -i & 0 \end{pmatrix} =: i\sigma_2, \tag{10.2}$$

$$\tilde{\sigma}^3 = h = \begin{pmatrix} 1 & 0 \\ 0 & -1 \end{pmatrix} =: \sigma_3, \tag{10.3}$$

and the formula

$$\forall i = 1, 2, 3, \ \sum_{p=1}^{3} ad[\sigma^i, \sigma^p](\sigma^p) = 8\sigma^i$$

(and see also the discussion in Section 10.3.2). Recall from Section 9.3 that the parity operation P, and the charge conjugation C, defined in the product bundle $\Theta_{\tilde{H}} = \tilde{c} \times \tilde{\Delta}$ come out as

$$P = \begin{pmatrix} -1 & 0 \\ 0 & 1 \end{pmatrix}, C = \begin{pmatrix} 1 & 0 \\ 0 & -1 \end{pmatrix}$$

so that, with the time inversion operation, which here is

$$\mathbf{T} = \begin{pmatrix} -1 & 0 \\ 0 & -1 \end{pmatrix},$$

we find

$$C\mathbf{T}P = I.$$

Moreover, the parity operator P induces the Dirac matrices

$$\gamma^0 = \begin{pmatrix} 1 & 0 \\ 0 & -1 \end{pmatrix}, \quad \gamma^k = \begin{pmatrix} 0 & \sigma^k \\ \sigma^k & 0 \end{pmatrix}, \quad k = 1, 2, 3,$$

as well as the new operators

$$\gamma^{k+3} = \begin{pmatrix} \sigma^k & 0 \\ 0 & -\sigma^k \end{pmatrix}, \quad k = 1, 2, 3,$$

acting on $B_o \oplus B_p$, the Dirac Spinors (and our candidate for the electron), such that,

$$\forall p \neq q,\ \gamma^p\gamma^q = -\gamma^q\gamma^p, \gamma^p\gamma^p = 1,\ p,q = 1,2,3,4,5,6,$$

and the formula

$$\forall i = 1,\ldots,6,\ \sum_{p=1}^{6} ad[\gamma^i,\gamma^p](\gamma^p) = 10\gamma^i.$$

The corresponding $[\mathfrak{v}] = \sum_i \gamma_i\xi_i$, the "spin part" of the equation of motion (or energy), comes out like

$$[\mathfrak{v}](\phi) = \sum_{i,j} \gamma_i g^{i,j} \frac{\partial}{\partial t_j}(\phi) = E\phi, \phi \in \mathfrak{B}.$$

10.3.2 *Dirac*

This energy–mass equation looks very much like an equation of Dirac type, and in fact, assuming the metric Euclidean, (or Minkowski) we find that Theorem 10.1 actually produces the Weyl equation, for Weyl spinors,

$$[\mathfrak{w}](\phi) := \sum_i \tilde{\sigma}^i \frac{\partial}{\partial t_i}(\phi) = E\phi,\ \phi \in B_o,$$

and the Dirac equation, for Dirac spinors,

$$[\mathfrak{v}](\phi) = E\phi,\ \phi \in B_o \oplus B_p.$$

But beware, the parity operator P does not preserve the Weyl spinors.

As we have seen in Section 7.3, and shall explain again in Section 10.4, there are many good reasons to promote $B_o \oplus B_p$ to the rank of "Electron." The existence of the Dirac equation above is just one of them.

Now, for a given metric g, fixing a particle, i.e. a $\tilde{H}(\sigma_g)$-representation

$$\rho : \tilde{H}(\sigma_g) \to \mathrm{End}_k(\mathfrak{B}),$$

we have seen in section PCGISection that the set of isomorphism classes in the space of $\mathrm{Rep}(\tilde{H}(\sigma_g))/\mathfrak{g}*$ is given as a quotient $\mathbf{P}$, of the infinite dimensional affine space $\mathfrak{P} := \mathrm{End}_{\tilde{H}}(\mathfrak{B})^6$, the set of Potentials, by the action of the Lie algebra $\mathfrak{h} := \mathrm{End}_{\tilde{H}}(\mathfrak{B})$.

We observe that our Toy Model has furnished three nicely interrelated models:

- a Field Theory (including a YM model, see Section 4.6 for connections on $\tilde{H}$-bundles $\mathfrak{B}$, i.e. for representations,

$$\rho : \tilde{H}(\sigma_g) \to \mathrm{End}_k(\mathfrak{B})$$

with vanishing Dirac derivation for representations, and Hamiltonian $Q = \rho(g - T)$.

- a Quantum Field Theory, see Section 11.3, for "gauge" fields, i.e. defined for representations

$$\rho : \tilde{H}(\sigma_g) \to \mathrm{End}_{\mathbf{P}}(\tilde{\mathfrak{B}}),$$

where the restriction of $\mathrm{End}_{\mathbf{P}}(\tilde{\mathfrak{B}})$ to any $\psi \in \mathcal{P}$ is the representation

$$\rho_\psi : \tilde{H}(\sigma_g) \to \mathrm{End}_k(\mathfrak{B}),$$

defined by,

$$\rho_\psi(dt_i) = \nabla_{\xi_i} + \psi_i, \ \nabla_{\xi_i} := \rho_1(dt_i) = \sum_j g^{i,j} \frac{\partial}{\partial t_j}.$$

Here, we have a Dirac derivation $[\psi] = \sum_i \psi_i \frac{\partial}{\partial t_i}$ and a Hamiltonian $Q := \rho(g - T)$, and time development is given, as in Section 4.6, by

$$[\delta] = \mathrm{ad}((Q_h + [\psi] + Q_v)).$$

In case we consider $[\psi] := [\mathfrak{v}] = \sum_{p,q} g^{p,q} \gamma_p \frac{\partial}{\partial t_q}$, we find that the curvature comes out as

$$F_{i,j} = \mathrm{ad}([\psi_i, \psi_j]).$$

- Einstein's General Relativistic model, see Sections 4.7 and 10.2 for the scheme $\mathrm{Spec}(\mathrm{Ph}(\tilde{H})_{\mathrm{com}})$, i.e. for representations,

$$\rho : \tilde{H}(\sigma_0) \to \mathrm{End}_{\tilde{H}(\sigma_0)}(\tilde{H}(\sigma_0)) \simeq \tilde{H}(\sigma_0),$$

where we have $\sigma_0 = \langle\{[dt_i, t_j]\}\rangle$, so that $\tilde{H}(\sigma_0) = \mathrm{Ph}(\tilde{H})_{\mathrm{com}} = k[t_1, \ldots t_6, \xi_1 \ldots \xi_6]$, where ξ_i here is the class of dt_i in $\tilde{H}(\sigma_0)$, and where the Dirac derivation takes the form

$$[\delta] = \sum_{i,p,q} \left(-\Gamma^i_{p,q} \xi_p \xi_p \frac{\partial}{\partial \xi_i} + \xi_i \frac{\partial}{\partial t_i} \right)$$

and Hamiltonian $Q = 0$, i.e. see Section 4.3.

- The General Force Law. The exterior force fields in physics are in our Field Theory Model induced by representations like $\rho : \tilde{H}(\sigma_g) \to \mathrm{End}_k(\mathfrak{B})$. In fact, the dynamics of ρ is given by the time derivative $\dot{\rho}$ of ρ, being defined by

$$\begin{aligned}
\dot{\rho}(dt_i) = \rho(d^2 t_i) =& \rho(-1/2 \sum_{p,q} (\bar{\Gamma}^i_{p,q} + \bar{\Gamma}^i_{q,p}) dt_p dt_q \\
&+ 1/2 \sum_{p,q} g_{p,q}(R_{p,i} dt_q + dt_p R_{q,i}) + [dt_i, T]), \\
=& - \sum_{p,q} \Gamma^i_{p,q} \nabla_p \nabla_q - 1/2 \sum_{p,q} g_{p,q}(F_{i,p} \nabla_q + \nabla_p F_{i,q}) \\
&+ 1/2 \sum_{l,p,q} g_{p,q} [\nabla_p, (\Gamma^{i,q}_l - \Gamma^{q,i}_l)] \nabla_l + [\nabla_i, T], \\
=& - \sum_{p,q} \Gamma^i_{p,q} dt_p dt_q - \sum_{p,q} g_{p,q} F_{i,p} dt_q \\
&+ 1/2 \sum_{p,q} g_{p,q} [F_{i,q}, dt_p] \\
&+ 1/2 \sum_{l,p,q} g_{p,q} [dt_p, (\Gamma^{i,q}_l - \Gamma^{q,i}_l)] dt_l + [dt_i, T],
\end{aligned}$$

where $\nabla_p := \rho(dt_p)$. Recall the definition

$$\Gamma^k_{i,j} = 1/2 \sum_l g^{k,l} (\delta_{t_j} g_{l,i} + \delta_{t_i} g_{l,j} - \delta_{t_k} g_{i,j}),$$

and from Section 10.1 the definition

$$T := \sum_j T_j dt_j = -1/2 \sum_{i,j,l} \delta_{t_l}(g_{i,j}) g^{l,i} dt_j = 1/2 \sum_{i,j,l} \delta_{t_j}(gi, j) g_{i,l} dt_l,$$

and the computation

$$T_l = -1/2\left(\sum_k \Gamma^k_{k,l} + \sum_{k,p,q} g^{k,q}\Gamma^p_{k,q}g_{p,l}\right) = -1/2\left(\sum_j \Gamma^j_{j,l} + \bar{\Gamma}^j_{j,l}\right),$$

where $\bar{\Gamma}^j_{j,l} := \sum_{p,q} g^{j,q}\Gamma^p_{j,q}g_{p,l}$.

10.3.3 *Classical Maxwell equations*

In QFT, the Dirac derivation [$\mathfrak{v}$] gives us

$$\nabla_p = \gamma_p\xi_p = \sum_s \gamma_p g^{p,s}\frac{\partial}{\partial t_s}.$$

For the Euclidean or the Minkowski metric, in the classical case, we find

$$\nabla_p = \xi_p = g^{p,p}\frac{\partial}{\partial t_p},$$

and the Hamiltonian $Q = \rho(g - T)$ is reduced to the Laplace differential operator,

$$\Delta = \sum_p g^{p,p}\frac{\partial}{\partial t_p}\frac{\partial}{\partial t_p}.$$

Since in both cases $\Gamma^i_{p,q} = 0$, and $T = 0$, the Force Law, or the equation of motion, reduces to

$$\dot{\rho}(dt_i) = \rho(d^2t_i) = -\sum_p \left(F_{i,p}\nabla_p + 1/2\frac{\partial}{\partial t_p}(F_{i,p})\right).$$

It is not difficult to see that this last equation, restricted to the subspace $M(B)$, contains Maxwell's equation. Interpreting,

$$\mathfrak{B} = \theta_{M(B)},\ v_p := \rho(dt_p) = \nabla_p = \left(\frac{\partial}{\partial t_p} + \psi_p\right), \psi_p \in \mathrm{End}_{M(B)}(\Theta),$$

$$\dot{v}_p := \rho(d^2t_i),$$

we find that the observed electromagnetic fact, i.e. the Lorentz Force Law,

$$\dot{v}_i = -\sum_p F_{i,p}v_p$$

implies that the tangent to $\dot{\rho}$, given by the potential $\{(1/2\sum_p \frac{\partial}{\partial t_p}(F_{i,p})), i = 1,\ldots,4\}$, must be zero, (or as physicists might say, *gauge*). This means, see Chapter 4, that there exist $\Phi \in \mathrm{End}_{M(B)}(\Theta_{M(B)})$, such that

$$1/2\sum_p \frac{\partial}{\partial t_p}(F_{i,p}) = \frac{\partial}{\partial t_i}(\Phi) + \left[1/2\sum_p \frac{\partial}{\partial t_p}(F_{i,p}), \Phi\right], \quad i = 1,\ldots,4.$$

Assuming that the electromagnetic field potentials $\psi_i = A_i$ are *scalars*, the curvatures $F_{i,p} = \frac{\partial A_i}{\partial t_p} - \frac{\partial A_p}{\partial t_i}$ are also scalars, so all commutators $[1/2\sum_p \frac{\partial}{\partial t_p}(F_{i,p})), \Phi]$ vanish, implying,

$$1/2\sum_p \frac{\partial F_{i,p}}{\partial t_p} = \frac{\partial \Phi}{\partial t_i}, \quad i = 1,\ldots,4.$$

Now, let us choose coordinates such that $t_4 =: t_0 = \lambda$ (the proper time for physicists) is the only zero-velocity coordinate of $M(B)$, and use the classical notations

$$E_i = F_{i,0} = \frac{\partial A_i}{\partial t_0} - \frac{\partial A_0}{\partial t_i}, \quad B_{p\times q} = F_{p,q} = \frac{\partial A_p}{\partial t_q} - \frac{\partial A_q}{\partial t_p}.$$

This is, classically, written as

$$F^{\mu,\nu} = \begin{pmatrix} 0 & E_x & E_y & E_z \\ -E_x & 0 & B_z & -B_y \\ -E_y & -B_z & 0 & B_x \\ -E_z & B_y & -B_x & 0 \end{pmatrix}.$$

A simple computation then shows

$$1/2\sum_p \frac{\partial F_{0,p}}{\partial t_p} = -1/2\nabla_s E,$$

$$-1/2\sum_p \frac{\partial F_{1,p}}{\partial t_p} = -1/2\left(\frac{\partial F_{1,0}}{\partial t_0} + \frac{\partial F_{1,2}}{\partial t_2} + \frac{\partial F_{1,3}}{\partial t_3}\right)$$

$$= -1/2\frac{\partial E_1}{\partial t_0} - 1/2(\nabla_s \times B)_1,$$

$$-1/2\sum_p \frac{\partial F_{2,p}}{\partial t_p} = -1/2\left(\frac{\partial F_{2,0}}{\partial t_0} + \frac{\partial F_{2,1}}{\partial t_1} + \frac{\partial F_{2,3}}{\partial t_3}\right)$$
$$= -1/2\frac{\partial E_2}{\partial t_0} - 1/2(\nabla_s \times B)_2,$$
$$-1/2\sum_p \frac{\partial F_{3,p}}{\partial t_p} = -1/2\left(\frac{\partial F_{3,0}}{\partial t_0} + \frac{\partial F_{3,1}}{\partial t_1} + \frac{\partial F_{3,2}}{\partial t_2}\right)$$
$$= -1/2\frac{\partial E_3}{\partial t_0} - 1/2(\nabla_s \times B)_3,$$

where ∇_c is the (light-) space part of ∇. From this, and from the vanishing of the tangent to $\dot{\rho}$, i.e. the equation above, the Maxwell equations follow, with $\Phi = \nabla \cdot A$, *electric current* $\mathfrak{J} = \nabla_c(\Phi)$, and *electric charge* $\nu = \frac{\partial \Phi}{\partial t_0}$. Together, $(\mathfrak{J}, \nu)$ is a zero-tangent to the representation ρ, the first component in $\tilde{c}$, i.e. in space (or light) direction, and the second in $\tilde{\Delta}$, or in zero-velocity direction. The result is, therefore,

$$\nabla_c \wedge E = -\frac{\partial A}{\partial t_0}, B = \nabla_c \wedge A, \nabla_c \cdot E = \nu, \nabla_s \wedge B = \mathfrak{J} + \frac{\partial E}{\partial t_0},$$

where $t_0 = \lambda$. Note that if we use the classical Minkowski, relativistic metric, in the same space we find exactly the same result, although the time derivative, $\dot{\rho}$, now must be thought of as the derivative with respect to *proper time.*

Note also that we might have looked at the perfectly symmetric situation, where we concentrate on the subspace of $\underline{\tilde{H}}$ corresponding to $\{c_3, d_1, d_2, d_3\}$. We would then have got a *massy* force with properties analogous to the electromagnetic force, and with equations just like Maxwell's.

In case we put the electric current and the charge density both $=0$, what we get is simply the field of photons and the field of down quarks.

However, neither the above example, nor the dual case just mentioned have philosophical sense. The only time–space that we have at hand is the Toy Model, and this is the quotient space $\tilde{H}/(Z_2)$ where the generator of Z_2 is the parity operator P, operating on $B_o \oplus B_p$, given by the matrix

$$\gamma^5 = \begin{pmatrix} 0 & id \\ id & 0 \end{pmatrix},$$

in the base $\{e_1 = (c_1 + d_1), e_2 = (c_2 + d_2), e_1* = (-c_1 + d_1), e_2* = (-c_2 + d_2)\}$, which turns left handedness to right handedness, with respect to the direction (o, p), respectively, (p, o).

Now, in this general situation, let coordinates t_i be chosen such that

$$dt_0, dt_1, dt_2, dt_3, dt_4, dt_\infty$$

are dual to

$$c_3, e_1, e_2, e_1*, e_2*, d_3,$$

and pick potentials A_i symmetric with respect to P. Then, just as above, put

$$E_i = F_{0,i}, H_i = F_{\infty,i}, B_1 = F_{3,2}, B_2 = F_{4,1}, B_3 = F_{1,2}, B_4 = F_{2,3}$$

and assume a reasonable Lorentz Force Law holds, i.e.

$$\dot{v}_i = E \cdot \sum_{j=0}^{\infty} F_{i,j} v_j$$

and that the tangent to this vanishes, like above. Then there exists an element $\Phi \in \mathrm{End}_{\tilde{H}}(\Theta)$, such that for $i = 0, 1, \ldots, 4, \infty$,

$$1/2 \sum_p \frac{\partial F_{i,p}}{\partial t_p} = \frac{\partial \Phi}{\partial t_i},$$

and we find

$$\frac{\partial F_{3,2}}{\partial t_2} = \frac{\partial F_{3,4}}{\partial t_4},$$

$$\frac{\partial F_{4,1}}{\partial t_1} = \frac{\partial F_{4,3}}{\partial t_3},$$

$$\frac{\partial F_{1,2}}{\partial t_2} = \frac{\partial F_{1,4}}{\partial t_4},$$

$$\frac{\partial F_{2,1}}{\partial t_1} = \frac{\partial F_{2,3}}{\partial t_3},$$

and

$$\frac{\partial\Phi}{\partial t_0} = 1/2\sum\frac{\partial E_p}{\partial t_p} + \frac{\partial H_0}{\partial t_\infty},$$
$$\frac{\partial\Phi}{\partial t_\infty} = 1/2\sum\frac{\partial H_p}{\partial t_p} + \frac{\partial E_\infty}{\partial t_0},$$
$$\frac{\partial\Phi}{\partial t_1} = \frac{\partial B_3}{\partial t_2} - 1/2\left(\frac{\partial E_1}{\partial t_0} + \frac{\partial H_1}{\partial t_\infty}\right),$$
$$\frac{\partial\Phi}{\partial t_2} = \frac{\partial B_4}{\partial t_3} - 1/2\left(\frac{\partial E_2}{\partial t_0} + \frac{\partial H_2}{\partial t_\infty}\right),$$
$$\frac{\partial\Phi}{\partial t_3} = \frac{\partial B_1}{\partial t_4} - 1/2\left(\frac{\partial E_3}{\partial t_0} + \frac{\partial H_3}{\partial t_\infty}\right),$$
$$\frac{\partial\Phi}{\partial t_4} = \frac{\partial B_2}{\partial t_1} - 1/2\left(\frac{\partial E_4}{\partial t_0} + \frac{\partial H_4}{\partial t_\infty}\right).$$

We have found that any particle, given by a representation,

$$\rho_\psi : \tilde{H}(\sigma_g) \to \mathrm{End}_k(\mathfrak{B})$$

induces a force field. In fact, the dynamics of ρ_ψ is given by the time derivative $\dot{\rho}_\psi$ of ρ_ψ, being defined just like above,

$$\begin{aligned}\dot{\rho}_\psi(dt_i) = &-\sum_{p,q}\Gamma^i_{p,q}\nabla_p\nabla_q - 1/2\sum_{p,q} g_{p,q}(F_{i,p}\nabla_q + \nabla_p F_{i,q})\\ &+ 1/2\sum_{l,p,q} g_{p,q}[\nabla_p, (\Gamma^{i,q}_l - \Gamma^{q,i}_l)]\nabla_l + [\nabla_i, T],\end{aligned}$$

where now $\nabla_p := \rho_\psi(dt_p)$.

Going back to Chapter 4, Theorem 9.1, and the summarizing above, we find a general Force Law, containing the classical Lorentz Force Law, and also equations for the dynamics of generalized *Spin Structures*.

Theorem 10.3. *Put*

$$\mathfrak{v}_p := \sum_q g^{p,q}\gamma^q, \psi_p := \mathfrak{v}_p + A_p$$

with $A_p \in \tilde{H}$ *and let* $\gamma^p \in \mathfrak{g}$ *be the Dirac matrices. Consider the representations*

$$\rho_A,\ \rho_\psi : \tilde{H}(\sigma_g) \to \mathrm{End}_k(B_o \oplus B_p),$$

defined by

$$\rho_A(dt_i) = \nabla_{\xi_i} + A_i,\ \rho_\psi(dt_i) = \nabla_{\xi_i} + \psi_i.$$

Then the curvature of ρ_ψ *is equal to*

$$F_{i,j} := \mathrm{ad}([\mathfrak{v}_i, \mathfrak{v}_j]) + ([\nabla_{\xi_j}, A_i] - [\nabla_{\xi_i}, A_j]),$$

where the second summand is part of the curvature of the representation ρ_A. *We find the following* Force Law, *for* ρ_ψ,

$$\dot{\mu}_p = \sum_q F_{p,q}\mu_q$$

for $\mu_q = \rho_\psi(dt_q)$, *and* $\dot{\mu}_p = \rho_\psi(d^2t_i)$.

Proof. Already done. Note also that this Force Law is modeled on the notion that our object, a point-object, is being exposed to a Force. It gives the relation between the (present, starting) velocity of the object, and the result, due to the acceleration transmitted by the Force. □

10.3.4 *Photons, tenebrons, and electrons*

Consider the conformally trivial metric of Section 8.3,

$$g = \left(\frac{\rho - h(\lambda)}{\rho}\right)^2 d\rho^2 + \left(\frac{\lambda - h(\rho)}{\lambda}\right)^2 d\lambda^2 + (\rho - h(\lambda))^2 d\underline{\omega}_c^2$$
$$+ (\lambda - h(\rho))^2 d\underline{\omega}_\lambda{}^2 = \Omega(d\rho^2 + d\lambda^2 + \rho^2 d\underline{\omega}_c{}^2 + \lambda^2 d\underline{\omega}_\lambda{}^2)$$

where $\Omega = (1 - (h_0/\rho\lambda))$, and $h(\lambda) = (h_0/\lambda)$, $h(\rho) = (h_0/\rho)$.

Since the direction of light (towards us), at the point $(o,p) = \underline{t} = (\underline{\lambda}, \omega_c, \rho)$ is given by $-c_3 = -\frac{\partial}{\partial\rho}$, i.e. by the choice of $\underline{\lambda}, \omega_c, \rho$, we find that the spin operator $\psi_\nu^0 := \nu(e - f) \in \mathfrak{g}$, is "positive" with respect to the light direction, i.e. $-c_3$, and a tangent in the potential space of the **Photon**, i.e. of the $\mathfrak{g}$-representation

$$\rho : \tilde{H}(\sigma_g) \to \mathrm{End}_k(\mathfrak{B}),$$

where $\mathfrak{B} = \langle c_1, c_2 \rangle$ is "normal" to c_3. (In fact, any point (o, p) of $\underline{\tilde{H}}$ might have been better denoted by (opsis), the greek name for a *visual object*.)

The photon should therefore turn out to be the **orbits** related to the transverse light-waves of the potential

$$\sigma_\nu = \exp(\nu\rho \cdot (e - f)) = \begin{pmatrix} \cos(\nu\rho) & \sin(\nu\rho) & 0 & 0 & 0 & 0 \\ -\sin(\nu\rho) & \cos(\nu\rho) & 0 & 0 & 0 & 0 \\ 0 & 0 & 0 & 0 & 0 & 0 \\ 0 & 0 & 0 & 0 & 0 & 0 \\ 0 & 0 & 0 & 0 & 0 & 0 \\ 0 & 0 & 0 & 0 & 0 & 0 \end{pmatrix}.$$

By plugging in $\phi := \nu\rho$, we find the "wave length," $\mathfrak{l}_\rho = 2\pi/\nu$. And, obviously, the 2π measured in our metric clock would be $\hbar$, the Planck constant (see the consequences of this for the cosmological red-shift, in the next chapters). Since one easily checks that $\delta_\rho|\sigma_\nu| = (e - f)\sigma_\nu$, we find, as we might have expected,

$$[\delta](\sigma_\nu) = \nu(e - f)\sigma_\nu d\rho.$$

Now, using the definition of mass/energy

$$m^2(\psi) = g([\delta](\sigma_\nu), [\delta](\sigma_\nu)) = \sum_{l,j} g^{i,j}\delta_l|\psi|\delta_j|\psi| \in \mathrm{End}_{\tilde{H}}(\mathfrak{B}),$$

we obtain

$$m^2(\sigma_\nu) = -\Omega^{-2}\nu^2{\sigma_\nu}^2,\ m^2 = \nu^2\Omega^{-2}.$$

Moreover, the orbits of σ_ν,

$$\sigma_\nu^+ = \cos(\nu\rho)c_1 - \sin(\nu\rho)c_2,\ \sigma_\nu^- = \sin(\nu\rho)c_1 + \cos(\nu\rho)c_2,$$

are eigenstates of $(e - f)\delta_\rho$, and $\sigma_\mu^- = -(e - f)\sigma_\mu^+$. (This is usually just expressed as a *state function*, $\psi(\tau) = \exp(\imath \cdot \nu\tau)$).

Since the parity operator, P, identifies σ_ν, and $-\sigma_\nu$, the eigenvalue comes out as $-\nu^2$, or as we wish ν^2, so that one usually puts

- Energy of the Photon: $\nu = 2\pi/\mathfrak{l}$
- $(e-f)(\sigma^+) + \sigma^- = 0$

And, of course, the velocity must be one.

But now there is an obvious symmetric object, that we shall, in spite of the mix of greek and latin, call the **Tenebron**, i.e. the $\mathfrak{g}$-representation,

$$\rho : \tilde{H}(\sigma_g) \to \mathrm{End}_k(\mathfrak{D}),$$

where $\mathfrak{D} = \langle d_1, d_2 \rangle$ is "normal" to d_3, and given by the potential

$$\tau_\mu = \exp(\mu\lambda \cdot (e-f)) = \begin{pmatrix} 0 & 0 & 0 & 0 & 0 & 0 \\ 0 & 0 & 0 & 0 & 0 & 0 \\ 0 & 0 & 0 & 0 & 0 & 0 \\ 0 & 0 & 0 & \cos(\mu\lambda) & \sin(\mu\lambda) & 0 \\ 0 & 0 & 0 & -\sin(\mu\lambda) & \cos(\mu\lambda) & 0 \\ 0 & 0 & 0 & 0 & 0 & 0 \end{pmatrix}.$$

By plugging in $\phi = \mu\lambda$, the energy (the rest-energy = rest-mass) comes out as μ, the wave-length, $\mathfrak{l}_\lambda = 2\pi/\mu$, and the velocity must be 0, since we are in $\tilde{\Delta}$. As above, τ_μ and the orbits,

$$\tau_\mu^+ = \cos(\mu\lambda)d_1 - \sin(\mu\lambda)d_2, \;\; \tau_\mu^- = \sin(\mu\lambda)d_1 + \cos(\mu\lambda)d_2$$

are eigenvectors of $(e-f)\delta_\lambda$, and $\tau_\mu^- = -(e-f)\tau_\mu^+$, so in particular:

- Energy of the Tenebron: μ
- $(e-f)(\tau^+) + \tau^- = 0$

Moreover, the charge of this object is "obviously" $-2/3$. But as the name of the object is supposed to refer to (ténèbres; absence de lumière) this object cannot be "seen"! It is part of the "dark furniture" of our Universe. (Recall that $\tilde{\Delta}$ are the directions in which the observer o and the observed p are moving with no relative geometric change.)

The comparable situation for the **Electron**, the canonical representation on $B_o \oplus B_p$, or on $\langle c_1, c_2, d_1, d_2 \rangle$, see Section 9.3.1, is given by

$$e(\kappa, \mu) := \begin{pmatrix} \cos(\kappa\rho) & \sin(\kappa\rho) & 0 & 0 & 0 & 0 \\ -\sin(\kappa\rho) & \cos(\kappa\rho) & 0 & 0 & 0 & 0 \\ 0 & 0 & 0 & 0 & 0 & 0 \\ 0 & 0 & 0 & \cos(\mu\lambda) & \sin(\mu\lambda) & 0 \\ 0 & 0 & 0 & -\sin(\mu\lambda) & \cos(\mu\lambda) & 0 \\ 0 & 0 & 0 & 0 & 0 & 0 \end{pmatrix}.$$

The orbits are the electrons e^- and the positron e^+

$$e^-(\kappa\mu) := \sigma_\kappa + \tau_\mu, e^+(\kappa\mu) := \sigma_\kappa - \tau_\mu.$$

Now, again using the definition of mass/energy,

$$m^2(\psi) = \sum_{l,j} g^{i,j} \delta_l |\psi| \delta_j |\psi| \in \mathrm{End}_{\tilde{H}}(\mathfrak{B}).$$

Plugging in our metric, we find that the formula gives us

$$m^2(e(\kappa, \mu)) = -\Omega^{-2}(\kappa^2 {\sigma_\kappa}^2 + \mu^2 \tau_\mu)^2.$$

Moreover, the charge of this object is "obviously" $-2/3$ for the electron, and $+2/3$, for the positron, according to the choice of the action $\mathfrak{h}_2^\pm$, see Section 9.3. Recall that the Charge Conjugation Operator C restricted to $\tilde{\Delta}$ is $T := -\,\mathrm{id}$.

The velocity comes out as $v^2 = \kappa^2/(\kappa^2 + \mu^2)$. Put the "massy energy"

$$E = \sqrt{(\kappa^2 + \mu^2)} = \mu\gamma, \ \gamma := 1/\sqrt{(1 - v^2)},$$

where γ is the Lorentz factor, and denote by $m_e := \mu$, the mass, and by K_e, the kinetic energy of the electron. With this, we find well-known formulas,

$$E = \gamma m_e, \ K_e = E - \mu = (\gamma - 1) m_e,$$

and for a given velocity v there is obviously a corresponding wavelength, comparable to the de Broglie wavelength $\mathfrak{l}_{db} = 2\pi/E = \mathfrak{l}_\rho v$, known to physicists.

Moreover, we find that

$$e^-(\kappa\mu) + e^+(\kappa\mu) = 2\sigma_\kappa^+ = 2(\cos(\kappa\rho)c_1 + \sin(\kappa\rho)c_2.$$

It is interesting to compare this with the so-called "dark matter annihilation,"

$$e^+ + e^- = \gamma + \gamma.$$

See Nima Arkani-Hamed and Weiner (2009), and Fernflores (2019), for recent papers on this subject. Clearly, the mass has disappeared, since the result is light, but the energy, the sum of the square energies of the reactant two electrons, will be equal to $2\kappa^2$, i.e. to the sum of the square energies of the two photons.

We may construct objects, like "axions," and other WIMPs, and their wave-like phonons, the "black" representations onto $\langle d_1, d_2, d_3\rangle$, and its dual, the "extended photon," $\langle c_1, c_2, c_3\rangle$. See the paper Gooth (2017) for a very well-documented exposition of the present theory/experiments.

More complex is the mathematical reason for the proposal (and discussion) concerning the "rest-mass" of systems of so-called "non-parallel photons." This is related to the fact that two photons corresponding to different light-spaces, $\underline{c}(\underline{\lambda}_i, i = 1, 2)$, therefore to different elements $\underline{\lambda}_i \in \underline{\Delta}, i = 1, 2$, are connected by, say a curve, corresponding to a zero-velocity movement, i.e. to some "rest mass."

10.4 Black Mass and Energy

Recall that the Toy Model is actually about the quotient space $\mathrm{Hilb}^{(2)}(\mathbb{A}^3) = \underline{\tilde{H}}/(Z_2)$, where $\underline{\tilde{H}}$ is the space of pairs of ordered points in $\mathbb{A}^3$, and the generator of Z_2 is the parity operator P.

Recall also that at each point of $\underline{\tilde{H}}$, there is a unique decomposition of the tangent bundle $\Theta_{\underline{\tilde{H}}}$, into $\underline{\tilde{c}}$, and $\underline{\tilde{\Lambda}}$ with, at every point (o, p), (projective) bases, usually referred to as the generators of the light-directions, therefore called $\langle c_1, c_2, c_3\rangle$, and the generators of the null-, or the dark-directions, therefore called $\langle d_1, d_2, d_3\rangle$. With the introduction of a metric, g, we obtain real bases, unit vectors in these directions with obvious orientations, denoted as above, $\langle c_1, c_2, c_3, d_1, d_2, d_3\rangle$.

We shall assume that the metric chosen makes $\underline{\tilde{c}}$ and $\underline{\tilde{\Lambda}}$ normal, implying that any tangent-direction ξ at the point $(o,p) \in \underline{\tilde{H}}$ corresponds to a momentum, with velocity, or speed, $\sin(\phi)$ where ϕ is the geometric angle between ξ and $\underline{\tilde{\Lambda}}$.

Consider a metric that is non-singular on an open subset of $\underline{\tilde{H}}$, like in the examples we have treated above, where the metrics are non-trivial outside the "Horizon." Then, obviously, in this open space, the vector fields are generated by the $\langle \xi_p \rangle, p = 1, \ldots, 6$, and the Force Laws, see Theorem 10.1, will give us the time-development of the states (base elements) of $\mathfrak{B}$), following from the formula.

$$[\delta](\rho(dt_i)) = \rho(d^2 t_i),$$

and from

$$d^2 t_i = -\sum_{p,q} \Gamma^i_{p,q} dt_p dt_q - \sum_{p,q} g_{p,q} F_{i,p} dt_q + 1/2 \sum_{l,p,q} g_{p,q} [F_{i,q}, dt_p]$$
$$+ 1/2 \sum_{l,p,q} g_{p,q} [dt_p, (\Gamma^{i,q}_l - \Gamma^{q,i}_l)] dt_l + [dt_i, T].$$

As we have seen, P operates on $B_o \oplus B_p$, with the base $\{e_1 = (c_1 + d_1), e_2 = (c_2 + d_2), e_1* = (-c_1 + d_1), e_2* = (-c_2 + d_2)\}$, as the matrix,

$$\gamma^5 = \begin{pmatrix} 0 & id \\ id & 0 \end{pmatrix},$$

turning left handedness to right handedness, with respect to the direction (o,p), respectively, (p,o).

The parity operator, P, therefore induces the Dirac matrices,

$$\gamma^0 = \begin{pmatrix} 1 & 0 \\ 0 & -1 \end{pmatrix}, \quad \gamma^k = \begin{pmatrix} 0 & \sigma^k \\ \sigma^k & 0 \end{pmatrix}, \quad k = 1, 2, 3,$$

as well as the new operators,

$$\gamma^{k+3} = \begin{pmatrix} \sigma^k & 0 \\ 0 & -\sigma^k \end{pmatrix}, \quad k = 1, 2, 3,$$

acting on $\Theta_{\tilde{H}}$, producing the Dirac equation, for Dirac spinors $B_o \oplus B_p$, see Section 10.3.2.

$$[\mathfrak{v}](\phi) := \sum_i \gamma^i \xi_i(\phi) = E\phi.$$

As we have seen above, and shall explain here, there are many good reasons to promote the particle $B_o \oplus B_p$ to the rank of "Electron." The existence of the Dirac equation above is just one of them. For any object $\mathfrak{B}$, the states $\phi \in \mathfrak{B}$ are uniquely provided with a "velocity," given in terms of its content of zero-velocity and light-velocity terms, respectively d_i and c_i's.

Clearly, all classical particles are visible, and therefore extensions of combinations of zero-velocity and light velocity states, so in principle they can move with all velocities strictly between zero and one. Photons are the only states of objects that have maximal velocity.

However, to explain the furniture of our Universe, we may have to include particles composed only of extensions of sub-representations of $\tilde{\Delta}$. Such objects have zero-velocities, and would be completely *Dark*, like the *Tenebrons*, treated above. They will have energies or mass, of some sort, that would be called Dark Mass, and this scenario is what we would like to unravel. The literature on the subject of Black Mass and Energy is becoming difficult to overview, to put it conservatively, so let me just refer to one paper that seemed to be understandable, for a mathematician, Nima Arkani-Hamed and Weiner (2009).

The search for WIMPs, the weakly interacting massive particles, assumed to be responsible for the problems with the "amount" of positrons and electrons, observed by cosmic ray detectors, is very intense, and one would believe that any mathematical model, vaguely resembling the proposed physical theories, might be of interest.

Obviously, our way of "unraveling" this subject should be considered as part of the present "mathematical Toy Model." The subject is not ready for a realistic "clean-up," considering the enormous experimental material obtained in the last 40 years. This said, let us take an example.

Consider the conformally trivial metric of $\underline{\tilde{H}}$, of Chapter 8,

$$g = (1 - h(\rho\lambda))^2(d\rho^2 + d\lambda^2 + \rho^2 d\underline{\omega}_c^2 + \lambda^2 d\underline{\omega}_\lambda^2),$$

with $h(x) = h_0 x^{-1}$.

Obviously, the metric can, as an operator in $\tilde{H}(\sigma_g)$, be cut up into the sum of

$$g_c = (1 - h(\rho\lambda))^2(d\rho^2 + \rho^2 d\underline{\omega}_c^2)$$

and

$$g_\lambda = (1 - h(\rho\lambda))^2(d\lambda^2 + \lambda^2 d\underline{\omega}_\lambda^2).$$

The eigenvectors of the λ-Hamiltonian $Q_\lambda := \rho(g_\lambda - T)$ in the state space for certain elementary particles (i.e. for representations ρ of $\tilde{H}(\sigma_g)$) may be linearly independent, giving birth to the three families of elementary particles, distinguished by their different mass, or λ-energies. This may therefore explain the origin of the three families, or generations, of otherwise identical elementary particles. Recall that the cylindrical coordinates, introduced in Section 5.1, and used here, correspond to the coordinates of the space

$$\underline{\tilde{H}} = \underline{\Delta} \times S^2 \times \mathbb{A}^1,$$

where $\underline{\Delta} \subset \mathbb{A}^3 \times \mathbb{A}^3$ is the diagonal, and S^2 is the blow-up of a point in $\underline{\Delta}$, in $\underline{H}$.

In complete generality, we now have a way of distinguishing types of energy in the Dirac equation. Go back to Section 8.3, and consider the formulas of Remark 8.1, relating our coordinates $(\lambda, \underline{\omega}_\lambda, \rho, \underline{\omega}_c)$ to the coordinates $\underline{t} = (t_1, \ldots, t_6)$, and see that the derivation

$$[\nu] := \sum_{p,q=1,2,3} g^{p,q}\gamma^p \frac{\partial}{\partial t_q} + \sum_{p,q=4,5,6} g^{p,q}\gamma^p \frac{\partial}{\partial t_q}$$

may be cut up into two parts,

$$[\kappa] := \sum_{p,(q=1,2,3)} g^{p,q}\gamma^p \frac{\partial}{\partial t_q}, \; [\mu] := \sum_{p,(q=4,5,6)} g^{p,q}\gamma^p \frac{\partial}{\partial t_q}.$$

The above is, in our model, a kind of combined Einstein, Klein–Gordon, and Dirac equation. And we have got a way of defining mass and charge in the same way, for all our representations, i.e. everywhere! That seems to be the only justification for introducing the ad hoc Higgs mechanism, with its Higgs particle, and eventual relation to the graviton. Observe that we have the (light) Photon $\langle c_1, c_2 \rangle$, the (black) Tenebron $\langle d_1, d_2 \rangle$, the action of the weak bosons, Z, W^+, W^-, together with the action of the graviton, given by non-zero elements in $\mathrm{Ext}^1_{\mathrm{Rad}(\mathfrak{g})}(\mathfrak{B})$, respectively by the element of $\mathrm{Ext}^1_{\mathrm{Ph}(\tilde{H})}(\mathfrak{B},\mathfrak{B})$ induced by $\mathfrak{G}$, producing the most elementary atomic structures, see Section 12.2. Note that, for the trivial metric, i.e. for constant $g^{i,j}$

$$[\mathfrak{v}]^2 = \sum_p {\delta_p}^2 = \Delta,$$

is the Laplace operator.

In light of this, some of the problems mentioned by, Nima Arkani-Hamed and Weiner (2009), can, maybe, be understood. Above we mentioned the role of the **Lorentz factor**.

It was defined as $\gamma := 1/\sqrt{(1-v^2)}$, and we found

$$E = \gamma m_e,\ K_e = E - \mu = (\gamma - 1)m_e.$$

Moreover, we found that

$$e^-(\kappa\alpha\beta) + e^+(\kappa\alpha\beta) = 2\sigma_\nu^+ = \cos(\nu\rho)c_1 + \sin(\nu\rho)c_2,$$

which is a kind of "dark matter annihilation,"

$$e^+e^- = 2\sigma_\nu^+,$$

where the photon σ_ν^+ is mass-less.

Note also that for $\rho \gg 0$ the metric in $\tilde{\Lambda}$-directions tend to be trivial, so this could explain that Dark Matter is found to be gravitationally neutral, far away from the Big Bang.

10.5 Ensembles, Bi-Algebras, and Quantum Groups

The elementary particles of our QFT-model are, see Section 10.4, representations of the form

$$\rho : \tilde{H}(\sigma_g) \to \mathrm{End}_k(\mathfrak{B}),$$

where $\mathfrak{B}$ is an $\tilde{H}(\sigma_g)$-sub-quotient of $\Theta_{\tilde{H}}$ with the fixed global gauge groups $\mathfrak{g}_0 \subset \Theta_{\tilde{H}}$, and a fixed local gauge group $\mathfrak{g}_1 \subset \mathrm{End}_{\tilde{H}}(\mathfrak{B})$. All other particles will be assumed to be iterated extensions of these. And, as we shall see, we should also extend the notion of particle, using non-commutative deformation of Modules, to obtain more general Ensembles of particles, like structures containing many identical particles, but being bound together with Forces defined via the general O-construction introduced in Section 4.1, which we shall now recall for use in our Toy Model.

But, first in physics, *sums* or *ensembles* of particles are often represented by tensor products of the representations defining the set of particles. For this to fit into the philosophy we have followed here, we must give reasons for why these tensor products pop up, seen from our moduli point of view.

It seems to me that the most natural point of view might be the following: Suppose, in all generalities, A is the moduli algebra parametrizing some objects $\{X\}$, and B is the moduli algebra for some other objects $\{Y\}$, then considering the *product*, or rather, the pair, (X, Y), one would like to find the moduli space of these *pairs*. A good guess would be $A \otimes_k B$ since it algebraically defines the product of the two moduli spaces. However, this is, as we know, too simplistic. There are no reasons why the pair of two objects should deform independently, unless we assume that they do not fit into any *ambient space*, i.e. unless the two objects are considered to sit in totally separate universes, and then we have done nothing but doubling our model in a trivial way. In fact, we should, for the purpose of explaining the role of the *product*, assume that our entire Universe is parametrized by the moduli algebra A, and accepting, for two objects X and Y in this Universe, that the pair (X, Y) correspond to a new object parametrized by A.

If A is a Hopf algebra, for example if $\mathrm{Spec}(A)$ is an algebraic group, then we find that there are canonical homomorphisms,

$$A \to A \otimes_k A,$$

so that the category $\mathrm{Rep}(A)$ has natural tensor products, used heavily in physics, in forming new models of physically interesting objects, like ensembles, from a given family of such.

For A commutative, there exists a natural homomorphism of algebras,

$$\iota : A \to \mathrm{Ph}(A) \otimes_A \mathrm{Ph}(A)$$

and an extension

$$\nabla : \mathrm{Ph}(A) \to \mathrm{Ph}(A) \otimes_A \mathrm{Ph}(A), \nabla(da) = 1/2(da \otimes 1 + 1 \otimes da)$$

given by the ι-derivation, γ, defined by $\gamma(a) = 1/2(da \otimes 1 + 1 \otimes da)$, with the property that

$$\begin{aligned}\gamma(ab) &= 1/2(da \cdot b + a \cdot db) \otimes 1 + 1 \otimes 1/2(da \cdot b + a \cdot db) \\ &= a1/2(db \otimes 1 + 1 \otimes db) + 1/2(da \otimes 1 + 1 \otimes da)b,\end{aligned}$$

since $a(1 \otimes db) = 1 \otimes a \cdot db$, and $(da \otimes 1)b = (da \cdot b \otimes 1)$.

But then, the algebra of observables, $\tilde{H}(\sigma_g)$, in our Toy Model is also a natural bi-algebra over $\tilde{H}$ (for non-trivial g). In fact, one easily computes,

$$\begin{aligned}\nabla([dt_i, t_j]) &= 1/2[(dt_i \otimes 1 + 1 \otimes dt_i), t_j] \\ &= 1/2([dt_i, t_j] \otimes 1 + 1 \otimes [dt_i, t_j]) \\ &= 1/2(g^{i,j} \otimes 1 + 1 \otimes g^{i,j}) = g^{i,j}.\end{aligned}$$

Therefore,

$$\nabla : \tilde{H}(\sigma_g) \to \tilde{H}(\sigma_g) \otimes_{\tilde{H}} \tilde{H}(\sigma_g)$$

is well defined.

However, $\tilde{H}(\sigma_g)$, is not a k-bi-algebra, so we need a different way of representing "sums" or "ensembles" of particles. Given a family of particles,

$$\rho_i : \tilde{H}(\sigma_g) \to \mathrm{End}_k(\mathfrak{V}_i), \;\; i = 1, 2, \ldots, n, \mathfrak{V} = \{\mathfrak{V}_1, \ldots, \mathfrak{V}_n\}$$

we have seen in Section 4.1 that the "sum" of the family of representations is given by the *O-construction*,

$$\tilde{H}(\sigma_g) \to O^{\tilde{H}(\sigma_g))}(\mathfrak{V}) \to \mathrm{End}_H(\mathfrak{V}) := (H_{i,j} \otimes_k \mathrm{Hom}(\mathfrak{V}_i, \mathfrak{V}_j)),$$

where $H := H(\mathfrak{V})$ is the prorepresenting hull of the deformation functor $\mathsf{Def}_{\mathfrak{V}}$. There are natural projections,

$$\pi_i : \mathrm{End}_H(\mathfrak{V}) \to \mathrm{End}_k(\mathfrak{V}_i),$$

inducing ρ_i. Note that for a chosen metric g, and for representations as above, we have the composite representations,

$$\rho_i^1 : \mathrm{Ph}(\tilde{H}) \to \tilde{H}(\sigma_g) \to \mathrm{End}_k(\mathfrak{V}_i),$$

and the corresponding O-construction will give us morphisms as above,

$$\mathrm{Ph}(\tilde{H}) \to O^{\mathrm{Ph}(\tilde{H})}(\mathfrak{V}) \to \oplus_{i=1}^n \mathrm{End}(\mathfrak{V}_i).$$

The idea is then, from families of elementary particles, to use the non-commutative theory of Chapter 3 to construct all other objects constituting our Furniture of the Universe. Using Theorem 10.1 we propose another way of looking at superpositions and relations between superposed mass-/space time separation and coherence time (see Penrose and Hameroff, 2011).

We shall, in Chapter 11, come back to this, but let us first see that the physicists Ensembles, i.e. the tensor products, of $A(\sigma_g)$-representations,

$$\mathrm{End}_k(V_1) \otimes_k \mathrm{End}_k(V_2) \otimes_k \cdots \otimes_k \mathrm{End}_k(V_n),$$

have a reasonable time-operator, given by $Q \in A(\sigma_g)$. In fact, the Dirac derivation $\delta := \mathrm{ad}(g-T) := \mathrm{ad}(Q)$ of $A(\sigma_g)$ gives us derivations $\eta_i := \mathrm{ad}(\rho_i(Q))$ of each factor in the tensor product, and therefore a corresponding derivation,

$$(\eta_1 \otimes_k 1 \otimes_k \cdots \otimes_k 1) + (1 \otimes_k \eta_2 \otimes_k 1 \cdots \otimes_k 1) + \cdots + (1 \otimes_k 1 \otimes_k \cdots \otimes_k \eta_n)$$

of the tensor product. Moreover, if like in Chapters 4 and 7, we have a local gauge group $\mathfrak{g}$ acting on all V_i, we obtain in the same way an action of $\mathfrak{g}$ on the tensor product of the $\mathrm{End}_k(V_i)$.

In particular, we shall use this for marking ensembles of particles in our Toy Model. As an example, suppose $v_i^p \in V_i$ are eigenvectors for the action of $\mu \in \mathfrak{h} \subset \mathfrak{g}$ with eigenvalues $h_i^p \in k$, then $v_i^p \otimes v_j^q \in V_1 \otimes_k V_2$ is an eigenstate of μ with eigenvalue $h_i^p + h_j^q$. But a sum like $v_1^p \otimes v_2^q + v_1^{p'} \otimes v_2^{q'}$ would not necessarily be an eigenstate of μ. It would, however, be reasonable to call it a state of μ. If $h_1^p + h_2^q = h_1^{p'} + h_2^{q'} = 0$, this is an example of a possibly *entangled* state, in the language of physicists. Even though μ is a representation that conserves the two (maybe identical) particles, it is clear that the read outs of measurements for the different particles may be different, with some interesting consequences. Note, however, that our model is by construction "non-local," so the EPR-paradox is in our sense not a mathematical paradox.

We shall not elaborate on this here, since the technique used would be identical to that used by physicists, in the vast literature on entanglement, quantum computation, dots, etc. See also Laudal (2011), Example 4.14, for an effort to use this in explaining some interactions in physics.

10.6 Black Mass and Gravitational "Waves"

Given the metric g above, let the representation

$$\rho : \tilde{H}(\sigma_g) \rightarrow \mathrm{End}_k(\mathfrak{B})$$

correspond to a black particle. It is therefore an iterated extension of the simple sub-representations of $\tilde{\Delta}$. The gauge group being the simple Lie algebra $\mathfrak{sgl}(3)$, it is reasonable that all potential extensions are trivial. Any non-trivial deformation (change) of ρ must therefore be a gravitational extension, corresponding to an infinitesimal action

$$\mathfrak{h} \in \mathrm{Ext}^1_{\mathrm{Ph}(\tilde{H})}(\mathfrak{B}, \mathfrak{B}).$$

As we know, h would act on all "nearby located" objects, as a gravitational wave, with de Broglie wavelength $(M)^{-1}$, where M would be the change of mass, i.e. the change of dark energy corresponding to h, and therefore related to the change in area of the horizon $E(\underline{\lambda})$ at the point $\underline{\lambda} \in \underline{\Lambda}$.

Note also that for $\rho \gg 0$, the metric in $\tilde{\Lambda}$-directions tend to be trivial, so this should explain that Dark Matter is gravitationally neutral.

Note also the possible explanation for: deflection of an electron by, say a proton; the acceleration of the electron resulting in the emission of *Bremsstrahlung*; and Compton scattering: An inelastic collision between a photon and a solitary (free) electron; resulting in a transfer of momentum and energy between the particles, which modifies the wavelength of the photon by an amount called the *Compton shift*.

Chapter 11

Particles, Fields, and Probabilities

The elementary particles of the Standard Model are, in our QFT model, identified with the simple gauge representations of the form

$$\rho : \tilde{H}(\sigma_g) \to \mathrm{End}_k(\mathfrak{B}),$$

where $\mathfrak{B}$ is any $\tilde{H}$-flat sub-quotient of the $\mathfrak{g}$-representation $\Theta_{\tilde{H}}$. The states in $\Theta_{\tilde{H}}$ or in $\mathfrak{B}$ are classified according to being particular eigenvectors, for the different *markers*, the elements of the Cartan sub-algebra $\mathfrak{h}$ of the local gauge group, $\mathfrak{g}_1 := \mathfrak{g} \oplus \mathfrak{su}(3)$, or $\mathfrak{g}_1 := \mathfrak{g} \oplus \mathfrak{sgl}(3)$, see Section 9.3.

If the particle $\rho : \tilde{H}(\sigma_g) \to \mathrm{End}_{\tilde{H}}(\mathfrak{B})$ is defined in terms of $\rho(dt_p) = \xi_p + \psi_p$, $\psi_p \in \mathrm{End}_{\tilde{H}}(\mathfrak{B})$, then the $\tilde{H}$ sub-module of $\mathfrak{B}$ generated by the eigenvectors of the element $h' \in \mathfrak{h}$ will be a sub-representation of ρ if (and only if ?) $[h', \psi_p] = \lambda_p \psi_p$, for some $\lambda_p \in \tilde{H}$ and all p.

Consider now the list of such markers used by physicists:

The ordinary spin: h
The isospin: $1/2h_1$
The electric charge: h_2
The weak isospin: $I_3 = 3/4h_2 \pm 1/2h_1$
Weak hypercharge: $Y_W = 2(h_2 - I_3) = 1/2h_2 \pm h_1$.

Since the direction d_3 is universally determined, the gauge group $\mathfrak{g}_1$ is basically reduced to $\langle h, h_1, h_2 \rangle$. Moreover, the change of sign in the last formula is, as we shall see later, related to what the physicists are calling *Chirality*, choosing a *Left-hand* or a *Right-hand* orientation of the coordinate system for $\tilde{\Delta}$. In the choice made above, the coordinate system $\{d_1, d_2, d_3\}$ is a right-handed one, and exchanging d_1 and d_2, make the system left-handed. It turns out that the concept of left–right in physics is the opposite, as the following display will show.

11.1 Elementary Particles

The elementary particles are characterized by the following display:

Markers:	$1/2 \cdot h$	$\mathrm{ad}(u)$	$h_1^{\pm}$	h_2	$I_3^{\pm}$	$Y_W^{\pm}$
c_1	$1/2$	1	0	0	0	0
c_2	$-1/2$	1	0	0	0	0
c_3	0	0	0	0	0	0
d_1	$1/2$	1	$\pm 1/2$	$-1/3$	$0(-1/2)$	$-2/3(1/3)$
d_2	$-1/2$	1	$\mp 1/2$	$-1/3$	$-1/2(0)$	$1/3(-2/3)$
d_3	0	0	0	$2/3$	$1/2$	$1/3$

We have the following (intrinsic) spin operator:

$$e - f = \begin{pmatrix} 0 & 1 & 0 \\ -1 & 0 & 0 \\ 0 & 0 & 0 \end{pmatrix}$$

which is the rotation in $\tilde{c}$ about the axis c_3. It is also defining a rotation, or spin, in $\tilde{\Delta}$ about the axis d_3. Moreover, we have two *folds*, the first,

$$f - r_2 = \begin{pmatrix} 0 & 0 & 0 \\ 1 & 0 & 0 \\ 0 & -1 & 0 \end{pmatrix}$$

in $\tilde{c}$, (as well as in $\tilde{\Delta}$) mapping c_1 to c_2 and c_2 to c_3, killing c_3, and finally,

$$e - r_1 = \begin{pmatrix} 0 & 1 & 0 \\ 0 & 0 & 0 \\ -1 & 0 & 0 \end{pmatrix},$$

in $\tilde{c}$, (as well as in $\tilde{\Delta}$) mapping c_2 to c_1 and c_1 to c_3 and killing c_3.

Remark 11.1. Given these tables, we should explain why it seems that physicists use the *marker* $1/2 \cdot h$ for the (classical) angular momentum J, and why J and the isospin measured by $\mathfrak{h}_1$ coincide for d_1, d_2. The ordinary spin J, induced by action of the global gauge group $\mathfrak{g}_0$, see Chapter 5, for the metrics of interest here, i.e. our metric g of the Toy Model (as well as the Euclidean, the Minkowski, or the Schwarzschild metrics), was seen to be given as a rotation in ϕ, the component of the parameter $\underline{\omega}$. Now, if we consider a vector field ψ of $\mathbf{H} := \mathrm{Hilb}^2(\mathbb{A}^3)$, it is clear that, pulled up to a vector field ψ' of $\underline{\tilde{H}}$, a rotation of ψ' by $\phi = 2\pi$ corresponds to a double rotation of ψ, since the parity operator, the generator of Z_2, is the rotation by $\phi = \pi$, in the "parameter" $\underline{\omega}$, of $\underline{c}(\underline{\lambda})$.

This seems to be the reason why physicists introduce the notion of *fermions*, for globally defined states of $\Theta_{\mathbf{H}}$. These are, after all, the only physically well-defined states of our Toy Model. In this sense, all representations of the ring of observables $\mathrm{Ph}^{\infty}(\tilde{H})$, invariant under $Z_2 = \langle P \rangle$ and *vectors*, are Fermions, and thus of Spin $1/2$. As in Laudal (2011), (4.6), we show that the parity operator P cuts up the bundle $\Theta_{\underline{\tilde{H}}}$ into the sum of two pieces,

$$\Theta^L_{\underline{\tilde{H}}} := \mathrm{im}(1/2(1 - P)) \simeq \tilde{c}$$

and

$$\Theta^R_{\underline{\tilde{H}}} := \mathrm{im}(1/2(1 + P)) \simeq \tilde{\Delta}.$$

In physics, they are referred to as the *left-hand* states and *right-hand* states, respectively. This is the basis for the notion of *chirality*, to which we shall return.

Since, evidently, the metric must be invariant under the Parity operator, we find that any representation

$$\rho : \tilde{H}(\sigma_g) \to \mathrm{End}_k(\Theta_{\tilde{H}})$$

has a Hamiltonian $Q = \rho(g - T) \in \mathrm{End}_k(\Theta_{\tilde{H}})$ that respects the decomposition

$$\Theta_{\underline{\tilde{H}}} = \Theta^R_{\underline{\tilde{H}}} \oplus \Theta^L_{\underline{\tilde{H}}}.$$

The eigenvectors of $\mathrm{ad}(Q)$ in $\mathrm{End}_k(\Theta_{\tilde{H}})$ are therefore divided into two groups,

$$\mathrm{Bosons} \subset \mathrm{End}_k(\tilde{\Delta}) \oplus \mathrm{End}_k(\tilde{c})$$

and

$$\mathrm{Fermions} \subset \mathrm{Hom}_k(\tilde{\Delta}, \tilde{c}) \oplus \mathrm{Hom}_k(\tilde{c}, \tilde{\Delta}),$$

or, considering the obvious graded structure on $\Theta_{\underline{\tilde{H}}}$, Bosons being the degree 0 endomorphisms, and Fermions the degree ± 1 endomorphisms.

The corresponding *Elementary Bosons*, corresponding to the gauge group $\mathfrak{g} \subset \mathrm{End}_{\tilde{H}}(\Theta_{\underline{\tilde{H}}})$ are given by the following list:

Markers	$1/2 \cdot h$	$\mathrm{ad}(u)$
h	0	0
e	1	0
f	-1	0
u	0	0
r_1	$-1/2$	-1
r_2	$1/2$	-1

Here, there is also a spin structure. The action of

$$\mathrm{ad}(e-f) = \begin{pmatrix} 0 & 0 & 0 \\ 0 & 0 & 1 \\ 0 & -1 & 0 \end{pmatrix}$$

induces a spin structure in $\mathrm{rad}(\mathfrak{g})$ about the axis u. Moreover, as we shall see in Chapter 12, the Weak Force could be seen as a consequence of the extensions given by the Potentials $\mathfrak{P}$, generated by the subset, $\mathrm{rad}(\mathfrak{g})$ of $\mathrm{End}_{\tilde{H}}(\Theta_{\tilde{H}})$, i.e. generated by $\langle u, r_1, r_2 \rangle$.

This takes care of a kind of *Super Symmetry*, and makes it possible to introduce the usual structures related to Planck's constant, vacuum state, creation, and annihilation operators. Thus, we find what we have termed a Quantum Field Theory, see Laudal (2011), Section 4.6, see also Section 11.3.

We should, at this point, try to model the most elementary particles we meet in physics. But first, what is a *particle* in say, the Standard Model? Is it point-like, or is it a bumpy field defined in our space? The literature concerned with this question is formidable, and the points of views are quite different.

We have already, in Section 10.3, proposed well-defined models for the simplest possible objects, like the quarks, the photons, their dark cousins, the tenebrons, the electrons, and their anti-particles. Via the Weak Force, we shall also find models for neutrons and the hydrogen atom, and simple explanations for the different decays, like the β-decay, and so also for neutrons, and their relations to the Weak Force.

Going further we shall have to use the extension-technique introduced in Section 10.5, and this will be postponed to Chapter 12.

Let us first have a look at what comes out of the philosophy above, with respect to notions just defined, Bosons and Fermions, and the place of probability in this context.

In all three models, we have a vector space of states, $\Theta_{\tilde{H}}$, producing the elements of the extension-spaces $\mathfrak{B}$, for all the iterated representations,

$$\rho : \mathrm{Ph}(\tilde{H}) \to \mathrm{End}_k(\mathfrak{B})$$

of the simple ones, those we would like to call particles.

Moreover, we have a Hamiltonian, $Q = \rho(g - T)$, acting on this space, and therefore acting on the ring of fields, $\mathrm{End}_k(\mathfrak{B})$, by the adjoint action. Since $\mathrm{ad}(Q)$ is a derivation in $\mathrm{End}_k(\mathfrak{B})$, the k-spectrum must be an additive monoid $\Lambda \subset k^+$. Suppose further that there exists a (positive) generator $\hbar$ of Λ. In (Laudal (2011), (4.6)) this was called our Planck's constant, and we let $f_\hbar$ be the corresponding eigenvector, or maybe eigenfield, of $\mathrm{ad}(Q)$. The corresponding spectrum of Q in $\mathfrak{B}$ would be contained in the set of differences of Λ, see (Laudal (2011), (4.6)).

So a Bosonic particle, obviously corresponding to a point in the moduli space of representations, turns out to be related with just one field, $f_\hbar$, carrying a bag of discrete energies $n\hbar$. If $f_\hbar^2 = 0$, we have a Fermionic representation, and no states carry more than one energy quantum (at a time!), i.e. the Pauli principle, see again (Laudal (2011), (4.6)). But first, let us fill in the part of Probabilities, so important in modern physics.

11.2 Time as a Source for Probabilities

We have created a theory that relates any particle to a representation

$$\rho : \tilde{H}(\sigma_g) \to \mathrm{End}_k(\mathfrak{B})$$

and any state of the particle, to an element of the representation space, $\phi \in \mathfrak{B}$, an iterated extension of the elementary particles representation spaces, i.e. of the different sub-bundles of our universal bundles of vector fields, $\tilde{\Delta}$ and $\tilde{c}(\underline{\lambda})$.

But the observables $\alpha \in \tilde{H}(\sigma_g)$, acting on this representation space might have many (possible) values, meaning that the observable α might have a series of possible eigenvalues, κ_i for eigenvectors, $\phi_i \in \mathfrak{B}$, for $i = 1, \dots, n$, such that,

$$\rho(\alpha)(\phi_i) = \kappa_i \phi_i.$$

Now, contrary to the situation in classical quantum mechanics, our measures of space-coordinates, of an observed, p, are not aleatory. In fact, if the object is observed by an observer, o, in the only three-dimensional space she has at hand, namely in the space $\underline{c}(\underline{\lambda}) = E(\underline{\lambda}) \times \mathbb{R}$, here $\underline{\lambda} \in \underline{\Delta}$, is the element named $c(o,p)$, see Section 5.1. We have only one set of space-coordinates, namely $(\underline{\omega}, \rho)$, the direction of (our) sight towards the observed, and the distance, i.e. the time since the light-wave passed the observed on its way to (me), the observer.

(Recall here that $\mathfrak{B}$ is an iterated extension, trivial with respect to the $\tilde{H}$-module structure, so the corresponding Space is a union of subspaces of Spec($\tilde{H}$).)

However, other observables might not be equally well determined. So, which ones should we relate to, in practice? Of course, this has to be expressed differently, in a dynamical context; having "measured" that the value of the observable $\alpha \in \tilde{H}(\sigma_g)$ is κ, i.e. that there is an eigenstate $\phi \in \mathfrak{B}'$ for the representation ρ such that $\rho(\alpha)(\phi) = \kappa\phi_i$, what comes "next"? Well, now the dynamics enters the game, and our equations of motion indicate that ρ will change to ρ', $\mathfrak{B}$ to $\mathfrak{B}'$, and ϕ to $\phi' \in \mathfrak{B}'$.

In $\mathfrak{B}'$, the observable α will have a series of values κ_i and eigenvectors ϕ_i, for $i = 1, \dots, n$. What will the "probability" be for finding that κ has changed to κ_i? Or, does this question have any meaning?

Consider the situation of Section 10.1. Our time, the metric g, gives us a quadratic form, denoted $\langle\ \rangle$, on all the representation spaces. Assume first that the motion has not changed the representation, i.e. $\rho_0 = \rho_0'$, but time has changed ϕ into ϕ'. Then this gives us a way of introducing probabilities associated to the different eigenvalues we obtain, and this for every point (o, p) in the Universe, $\underline{\tilde{H}}$. In fact, an observable $\alpha \in \tilde{H}(\sigma_g)$ acts on the finite rank $\tilde{H}(\sigma_g)$-bundle $\mathfrak{B}$. We should suppose that its eigenvectors, ϕ_i, all assumed to be of norm 1, with corresponding eigenvalues κ_i, now form a complete, separable structure, such that any state $\phi' \in \mathfrak{B}$, of norm 1, is given on the form $\phi' = \sum \langle \phi', \phi_i \rangle \phi_i$. Then we might define the probability of finding that the measurement κ of α on ϕ has changed to κ_i with eigenvector ϕ_i to be

$$\text{prob}(\alpha, \phi; \phi', \phi_i) = \langle \phi', \phi_i \rangle^2.$$

The sum being $(\phi', \phi') = 1$, by definition, so formally it looks OK, the sum of all probabilities comes out as 1. However, to make this reasonable, we should use the fact that the norm introduced is, Time, defined by the metric g. And here, Time is pushing ϕ into a grid, that we might suppose is orthonormal, and the measurement picks out the different projections. We shall come back to this philosophically important point, at the end of Section 11.3.

The measurement of a state of the particle, defined by a representation $\rho : \tilde{H}(\sigma_g) \to \text{End}_k(\mathfrak{B})$, that was known to look like $\phi \in \mathfrak{B}$ (for the observer, therefore, at any point of the (light) space of the observer, (some "time" before) having undergone time-development according to the equations of Section 10.3, of length τ, is now assumed to correspond to the calculation of $\phi := \exp(\tau[\delta](\phi)$ and its eigenvalues, therefore, given as a set of probabilities for the different outcomes. The state ϕ rests as a state, there is no "collapse" of the "wave-packet," but to obtain information about what happens "next," the only thing we know after the measurement is that we have a series of probabilities for ϕ, to start from. This may be fine if we believe that the first computation was OK, and if there is nothing else troubling the dynamics. We obviously have to start with something we believe is a real information about our system, therefore, certainly about the state we saw it in. But the representation may also have changed radically. And with it, the metric, and so the whole system of dynamics, of $\mathrm{Rep}(\tilde{H}(\sigma_g))$.

In particular, the energy of the system changes form; we may see processes nurtured by borrowing from the gravitation, and lending it back. Here is where I suppose the ORCH OR "calculus" of Penrose–Hameroff comes in, see Penrose and Hameroff (2011). Of course, we know, from the prominent physicist Laurence Krauss, that this is total non-sense; but my simple conclusion is that these proposals are not easy to contradict from a purely mathematical point of view. See also the paper of Volk (2018).

11.3 Quantum Field Theory, Wightman's Axioms

Now we have a situation in which we might consider the Wightman axioms for Quantum Field Theory, say along the lines of Deligne (1999). We defined

$$\Theta^L_{\underline{\tilde{H}}} := \mathrm{im}(1/2(1-P)) \simeq \tilde{c},\ \Theta^R_{\underline{\tilde{H}}} := \mathrm{im}(1/2(1+P)) = \tilde{\Delta}.$$

Recall that $\underline{H}$ is an affine space of dimension six, so putting $V = \underline{H}$, $G = \mathrm{Der}(U(o,p))$ with its action on

$$R := T_{\underline{\tilde{H}},(o,p)} \simeq k^6, P = \mathfrak{g},$$

and

$$\mathfrak{R} = \Theta_{\tilde{H}} = \tilde{c} \oplus \tilde{\Delta},\ \mathfrak{S} := \mathfrak{S}^R = \mathfrak{S}^R_{even} \oplus \mathfrak{S}^R_{odd} \simeq \Theta_{\underline{\tilde{H}}},$$

the axioms look like

(1) There is a finite-dimensional real representation $\rho : G \to \mathrm{Aut}(R)$.
(2) There is a positive unitary representation $U : P \to \mathrm{Aut}(\mathfrak{H})$, where $\mathfrak{H}$ is a super Hilbert space.
(3) There exists a dense P-invariant subspace $\mathfrak{D} \subset \mathfrak{H}$, and a P-invariant vector $\Omega \in \mathfrak{H}$.
(4) There is a linear map $\phi : \mathfrak{S}^R \to \mathrm{End}(\mathfrak{D})$, called the field map, such that:

(α) ϕ is P-equivariant, where P acts on $\mathrm{End}(\mathfrak{D})$ by conjugation.
(β) $\phi(f)$ is super-Hermitian symmetric.
(γ) For any $\psi_1, \psi_2 \in \mathfrak{D}$, the map $f \longmapsto \langle \psi_1, \phi(f)\psi_2 \rangle$ is continuous, $\mathfrak{S}$ equipped with the Schwartz topology.
(δ) The space $\mathfrak{D}$ is generated by vectors $\phi(f_1)\phi(f_2)\ldots\phi(f_n)\Omega$, for $f_i \in \mathfrak{S}$.
(ϵ) (space-locality) If $f_1, f_2 \in \mathfrak{S}$ are such that for any $v_1 \in \mathrm{Supp}(f_1), v_2 \in \mathrm{Supp}(f_2)$ we have $(v_1 - v_2) \in V_{space}$, then $[\phi(f_1), \phi(f_2)] = 0$, where [-] stands for super-commutator.

Reducing to the corresponding Lie algebras for all Lie groups involved, we observe that for any metric g on our Toy Model space, $\underline{\tilde{H}}$, the set-up above, with particles defined as representations

$$\rho : \tilde{H}(\sigma_g) \to \mathrm{End}_k(\mathfrak{B})$$

and ρ any iterated extension of the elementary particles defined above, is a Wightman QFT. The homomorphism ρ composed with $d : \tilde{H} \to \mathrm{Ph}(\tilde{H}) \to \tilde{H}(\sigma_g)$ is a derivation $\tilde{H} \to \mathrm{End}_k(\mathfrak{D})$, inducing the trivial extension, therefore, P acts via the conjugation on $\mathrm{End}_k(\mathfrak{D})$.

Of course, we are not in the Minkowski situation, so we do not need point (ϵ) of Axiom (4). The rest is pretty obvious.

Chapter 12

Interactions

Return to the treatment of finite dimensional representations of finitely generated associative k-algebras A of Chapter 3. Given a derivation ξ of A, and the induced dynamical system $A(\sigma)$, we considered the moduli space of simple n-dimensional representations,

$$\rho : A(\sigma) \to \mathrm{End}_k(V),$$

and found a versal space, $U(n) \subset \mathrm{Spec}(C(n))$, where the commutative k-algebra $C(n)$ could be "effectively" computed. The dynamical structure induced a vector-field $[\xi]$ in $C(n)$, see Theorems 12.2 and 12.3, such that any representation ρ corresponding to a point $\underline{t} \in U(n)$ would change along the integral curve $\mathbf{c}$ of $[\xi]$, through $\underline{t}$. Recall that the point $\underline{t} \in U(n)$ corresponds to an A-representation ρ_0, together with a formal curve of representations through ρ_0, We are, of course, assuming that the field k is contained in the real numbers. Now, this point $\underline{t}$ may well be a point on the border of $\mathrm{Simp}_n(A(\sigma))$, i.e. in $\Gamma_n = \mathrm{Simp}(C(n)) - U(n)$, where it must *decay* into a decomposable representation, i.e. into an iterated extension of two or more new particles $\{V_i \in \mathrm{Simp}_{n_i}(A(\sigma), n = \sum n_i\}$. See Laudal (2011), Section 4.1, and Eriksen *et al.* (2017), Section 3.7. What happens next is dependent on the dynamics of these iterated extensions. And, according to our philosophy, this is again dependent on the geometry of the moduli space of all such iterated extensions.

Now, let us go back to our Toy Model, treated above. Assume that the world is populated by perhaps an infinite family of elementary particles $\mathfrak{B} = \{\mathfrak{B}_i\}_{i=1}^{\infty}$, such that the Furniture of the Cosmos is composed of the finitely iterated extensions of these elementary particles, forming atoms, molecules, and the myriads of different stuff we observe around us, then the changing world would be described as a dynamical structure defined on the moduli space of finitely iterated extensions of the particles in $\mathfrak{B}$.

Since by definition, see Section 10.1, all the state spaces, $\mathfrak{B}_i$, of our elementary objects in our Toy Model, $\mathrm{Rep}](\tilde{H})/\mathfrak{g}*$, have "canonical" norms, quadratic forms given by the metric, and since any extension of these conserve this norm, we may introduce probability structures and classical measurements, as shown in Chapter 11. Since measurements "clearly" are dependent on some sort of "memory," we shall have to be interested in what memory is, in our model.

Obviously, a big particle is an iterated extension of many particles, in a certain succession, each one either given by extensions of the form

$$\psi \in \mathrm{Ext}^1_{\tilde{H},\rho_0}(\mathfrak{B}_i, \mathfrak{B}_j), \psi \in \mathfrak{P}$$

or reflected in a possible change of the metric. "Pealing off" the elementary (or less complicated) sub-particles could be a "recollection," and gives sense to the notion of memory.

Before we continue this line of philosophy, we should return to the general theory of non-commutative deformation.

12.1 Interaction and Non-Commutative Deformations

The purpose of this section is to sketch a mathematical model for the above scenario. Recall from Section 3.3, and see Laudal (2011), (3.3), that for any associative k-algebra A, a family of A-representations $\mathbf{V} = \{V_i\}_{i=1}^{r}$ is called a swarm, if

$$\dim_k \mathrm{Ext}^1_{\mathbf{A}}(V_i, V_j) < \infty \quad \text{for all} \quad i, j = 1, \ldots, r.$$

Consider now, for any swarm, $\mathbf{V} := \{V_i\}, i = 1, \ldots, r$, of $\mathbf{A}$-representations, the deformation functor

$$\mathsf{Def}_{\mathbf{V}} : \underline{a}_r \longrightarrow \mathbf{Sets},$$

and its formal moduli

$$H(\mathbf{V}) := \begin{pmatrix} H_{1,1} & \dots & H_{1,r} \\ & \dots & . \\ H_{r,1} & \dots & H_{r,r} \end{pmatrix},$$

together with the versal family, and the O-construction, i.e. the essentially unique homomorphism of k-algebras,

$$\tilde{\rho} : A \longrightarrow O(\mathbf{V}) := \begin{pmatrix} H_{1,1} \otimes \mathrm{End}_k(V_1) & \dots H_{1,r} \otimes \mathrm{Hom}_k(V_1, V_r) \\ . & \dots & . \\ H_{r,1} \otimes \mathrm{Hom}_k(V_r, V_1) \dots & H_{r,r} \otimes \mathrm{End}_k(V_r) \end{pmatrix}.$$

Recall the form of this homomorphism. Let $\Psi_{i,j}$ be the k-dual of $\mathrm{rad}(H)_{i,j}$, and let $\{\psi^s_{i,j}\}$ and $\{h^s_{i,j}\}$ be dual k-bases of $\Psi_{i,j}$ and $\mathrm{rad}(H)_{i,j}$, then for $v_i \in V_i$,

$$\tilde{\rho}(a)(v_i) = v_i a + \sum_{j,s} h^s_{i,j} \otimes \psi^s_{i,j}(a)(v_i), \psi^s_{i,j}(a) \in \mathrm{Hom}_k(V_i, V_j).$$

This is, in an obvious sense, the *universal interaction*. The elements $\hat{h}_{i,j}$ of the k-dual, $\mathrm{rad}(H)^*$, of $\mathrm{rad}(H)$, correspond to linear maps $A \to \mathrm{Hom}_k(V_i, V_j)$, (which we should consider *forces mediated* by elements of A), acting by mapping the states of V_i into states of V_j.

12.2 The Weak and Strong Interactions

Given the metric g on $\underline{\tilde{H}}$, and let $\mathfrak{B} = \{\mathfrak{B}_i\}$, $i \geq 1$, be the family of elementary sub-quotients of the $\tilde{H}(\sigma_g)$-module $\Theta_{\tilde{H}}$. Then

$$\rho_i : \tilde{H}(\sigma_g) \to \mathrm{End}_k(\mathfrak{B}_i)$$

is not a swarm. The k-dimension of $\mathrm{Ext}^1_{\tilde{H}(\sigma_g),\rho_0}(\mathfrak{B}_i, \mathfrak{B}_j)$ is not necessarily finite. We must therefore be careful if we want to use the technique of the O-construction above.

However, the k-Lie algebra generated by the extended Gauge Group $\mathfrak{g}*$, the union of $\mathfrak{g}$ acting on $\Theta_{\tilde{H}}$, and $\mathfrak{sgl}(3)$ acting on $\tilde{\Delta}$, is of course of finite dimension over k.

Consider now the force fields $\mathfrak{F}$ defined by

$$\mathfrak{F}(\mathfrak{B}_i, \mathfrak{B}_j) \subset \mathcal{P}(\mathfrak{B}_i, \mathfrak{B}_j)/\operatorname{Triv} \simeq \operatorname{Ext}^1_{\tilde{H}(\sigma_g),\rho_0}(\mathfrak{B}_i, \mathfrak{B}_j)$$

generated by $\mathfrak{g}*^6 \subset \mathcal{P}(\mathfrak{B}_i, \mathfrak{B}_j) = \operatorname{Hom}_{\tilde{H}}(\mathfrak{B}_i, \mathfrak{B}_j)^6$. Here, the inclusion

$$\mathfrak{F}(\mathfrak{B}_i, \mathfrak{B}_j) \subset \operatorname{Ext}^1_{\tilde{H}(\sigma_g),\rho_0}(\mathfrak{B}_i, \mathfrak{B}_j)$$

is computed via $\psi = \{\psi_l\} \in \mathfrak{F}(\mathfrak{B}_i, \mathfrak{B}_j)$, corresponding to the derivation

$$\psi \in \operatorname{Der}_k(\tilde{H}(\sigma_g), \operatorname{Hom}_{\tilde{H}}(\mathfrak{B}_i, \mathfrak{B}_j)),$$

defined by $\psi(t_l) = 0$, and $\psi(dt_l) = \psi_l$, for all l, see Section 4.5.1.

The gauge fields, generated as above by a representation of the principal Lie bundle $\mathfrak{g}$, into $\operatorname{End}_{\tilde{H}}(\mathfrak{B})$, the corresponding extensions, and therefore to tangent directions in $\mathbf{T}_\rho$, are quite easy to understand.

Recall from Section 4.5 that if ρ_i is given by

$$\rho_i(t_j) = 0, \; \rho(dt_j) = \xi_j + \phi^i_j, \; \phi^i_j \in \operatorname{End}_{\tilde{H}}(\mathfrak{B}_i),$$

then an element $\psi^{1,2} \in \mathcal{P}(\mathfrak{B}_1, \mathfrak{B}_2)$ is 0 in $\operatorname{Ext}^1_{\tilde{H}(\sigma_g),\rho_0}(\mathfrak{B}_1, \mathfrak{B}_2)$ if and only if there exists $\Phi^{1,2} \in \operatorname{Hom}_{\tilde{H}}(\mathfrak{B}_1, \mathfrak{B}_2)$ such that

$$\psi^{1,2}_i = \xi_i(\Phi^{1,2}) + \phi^1_i \Phi^{1,2} - \Phi^{1,2}\phi^2_i.$$

Since $\mathfrak{F}$ is of finite dimension over k, we should be able to copy the O-construction of the family $\mathfrak{B}$ using only the Force Fields, see Eriksen *et al.* (2017), Chapter 3, in particular Section 3.7.

These Force Fields contain our version of the Weak and the strong Forces, and the outcome would be a way of computing all Physical Particles that are iterated extensions of the elementary ones.

Let us consider just the Weak Force. We have the representation of $\operatorname{rad}(\mathfrak{g})$ in $\operatorname{End}_{\tilde{H}}(\Theta_{\tilde{H}})$, computed in Section 9.2, in the universal basis $\langle c_1, c_2, c_3, d_1, d_2, d_3 \rangle$.

The formulas there tell us that u is the identity on $\langle c_1, c_2, d_1, d_2 \rangle$, and that r_1 maps c_1 to c_3, and d_1 to d_3, while r_2 maps c_2 to c_3, and d_2 to d_3. But this is exactly the definition, in the Standard Model, of the actions of Z, and $W_i, i = 1, 2$.

Let us now show that this is related to the obvious lifting problems for representation $\rho_0 : \tilde{H} \to \mathrm{End}_k(\mathfrak{B})$ where $\mathfrak{B}$ is a representation of the principal Lie bundle $\mathfrak{g} \oplus \mathfrak{su}(3)$, or $\mathfrak{g}*$. If we stick to just the real part of the story, we should consider $\mathrm{Ext}^1_{\tilde{H}(\sigma_g),\rho_0}(\mathfrak{B},\mathfrak{B})$, or self-interactions of representations

$$\rho : \mathrm{Ph}(\tilde{H}(\sigma_g)) \to \mathrm{End}_{\tilde{H}}(\mathfrak{B}),$$

consistent with extensions of the representation of the gauge group, and of course, keeping ρ_0 constant.

In this case, there are other forces to be considered. The $\tilde{H}$ module $\mathfrak{B}$ is free, and $\mathfrak{sl}(2)$ and $\mathfrak{su}(3)$ being semi-simple, do not have non-trivial extensions. But the $\mathrm{rad}(\mathfrak{g})$ part of $\mathfrak{g} = \mathfrak{sl}(2) \oplus \mathrm{rad}(\mathfrak{g})$ has non-trivial extension modules, and as we shall see, this give us a model for the *Weak Interaction.*

Lemma 12.1. *Consider the action of* $\mathfrak{g} = \mathfrak{sl}(2) \oplus \mathrm{rad}(\mathfrak{g})$ *on* $\tilde{c}$ *and* $\tilde{\Delta}$, *then there is an isomorphism,*

$$\mathrm{Ext}^1_{\mathfrak{g}}(\Theta,\Theta) \simeq \mathrm{Ext}^1_{\mathrm{rad}(\mathfrak{g})}(\Theta,\Theta)$$

and,

$$\begin{aligned}\mathrm{Ext}^1_{\mathrm{rad}(\mathfrak{g})}(\Theta,\Theta) = {} & \mathrm{Ext}^1_{\mathrm{rad}(\mathfrak{g})}(\tilde{c},\tilde{c}) \oplus \mathrm{Ext}^1_{\mathrm{rad}(\mathfrak{g})}(\tilde{\Delta},\tilde{\Delta}) \\ & \oplus \mathrm{Ext}^1_{\mathrm{rad}(\mathfrak{g})}(\tilde{c},\tilde{\Delta}) \oplus \mathrm{Ext}^1_{\mathrm{rad}(\mathfrak{g})}(\tilde{\Delta},\tilde{c}).\end{aligned}$$

Moreover,

$$\mathrm{Ext}^1_{\mathrm{rad}(\mathfrak{g})}(\tilde{c},\tilde{c}) \simeq \mathrm{Ext}^1_{\mathrm{rad}(\mathfrak{g})}(\tilde{\Delta},\tilde{\Delta})$$

and, any $\psi \in \mathrm{Ext}^1_{\mathrm{rad}(\mathfrak{g})}(\tilde{\Delta},\tilde{\Delta})$ *is given by the derivation*

$$\psi : \mathrm{rad}(\mathfrak{g}) \to \mathrm{End}_{\tilde{H}}(\tilde{\Delta})$$

in the bases, $\{d_1, d_2, d_3\}$ *of* $\tilde{\Delta}$, *the quarks, and* $\{u, r_1, r_2\}$ *of* $\mathrm{rad}(\mathfrak{g})$, *modulo the trivial derivations, given by an element* $a := (a_{i,j}) \in$

$\mathrm{End}_{\tilde{H}}(\tilde{\Delta}, \tilde{\Delta})$. *We find*

$$\psi(u) = \begin{pmatrix} \alpha_{1,1} & 0 & \alpha_{1,3} \\ 0 & \alpha_{2,2} & \alpha_{2,3} \\ \alpha_{3,1} & \alpha_{3,2} & \alpha_{3,3} \end{pmatrix} \mathrm{mod} \begin{pmatrix} 0 & 0 & -a_{1,3} \\ 0 & 0 & -a_{2,3} \\ a_{3,1} & a_{3,2} & 0 \end{pmatrix}$$

and

$$\psi(r_1) = \begin{pmatrix} -\alpha_{1,3} & 0 & 0 \\ -\alpha_{2,3} & 0 & 0 \\ \alpha^1_{3,1} & \alpha^1_{3,2} & \alpha_{1,3} \end{pmatrix} \mathrm{mod} \begin{pmatrix} a_{1,3} & 0 & 0 \\ a_{2,3} & 0 & 0 \\ a_{3,3} - a_{1,1} & -a_{1,2} & -a_{1,3} \end{pmatrix},$$

$$\psi(r_2) = \begin{pmatrix} 0 & -\alpha_{1,3} & 0 \\ 0 & -\alpha_{2,3} & 0 \\ \alpha^2_{3,1} & \alpha^2_{3,2} & \alpha_{2,3} \end{pmatrix} \mathrm{mod} \begin{pmatrix} 0 & a_{1,3} & 0 \\ 0 & a_{2,3} & 0 \\ -a_{2,1} & a_{3,3} - a_{2,2} & -a_{2,3} \end{pmatrix}.$$

In particular, we observe that $\alpha_{2,1} = \alpha_{1,2} = 0$. *Choose* $(a_{i,j}) = (\alpha_{i,j})$, *except for* $a_{3,1} = -\alpha_{3,1}, a_{3,2} = -\alpha_{3,2}$, *and add the terms, to find the derivation*

$$\psi(u) = \begin{pmatrix} \alpha & 0 & 0 \\ 0 & \alpha & 0 \\ 0 & 0 & \alpha \end{pmatrix}$$

and

$$\psi(r_1) = \begin{pmatrix} 0 & 0 & 0 \\ 0 & 0 & 0 \\ \alpha^1_{3,1} & \alpha^1_{3,2} & 0 \end{pmatrix},$$

$$\psi(r_2) = \begin{pmatrix} 0 & 0 & 0 \\ 0 & 0 & 0 \\ \alpha^2_{3,1} & \alpha^2_{3,2} & 0 \end{pmatrix},$$

and see that the particular values of ψ

$$\psi(u) = \begin{pmatrix} \alpha & 0 & 0 \\ 0 & \alpha & 0 \\ 0 & 0 & \alpha \end{pmatrix}$$

and

$$\psi(r_1) = \begin{pmatrix} 0 & 0 & 0 \\ 0 & 0 & 0 \\ \beta & 0 & 0 \end{pmatrix},$$

$$\psi(r_2) = \begin{pmatrix} 0 & 0 & 0 \\ 0 & 0 & 0 \\ 0 & \gamma & 0 \end{pmatrix},$$

correspond to a non-zero element of

$$\mathrm{Ext}^1_{\mathrm{rad}(\mathfrak{g})}(\tilde{c}, \tilde{c}) \simeq \mathrm{Ext}^1_{\mathrm{rad}(\mathfrak{g})}(\tilde{\Delta}, \tilde{\Delta}).$$

If we want to consider the full extension over $\mathfrak{g}$, *with*

$$\psi_* \in \mathrm{Der}(\mathfrak{g}, \mathrm{End}_{\tilde{H}}(\tilde{\Delta})$$

or the obvious restriction

$$\psi_* \in \mathrm{Der}(\mathfrak{g}, \mathrm{Hom}_{\tilde{H}}(\langle c_1, c_2, d_1, d_2 \rangle, \langle c_3, d_3 \rangle))$$

for which ψ_* *restricted to* $\mathfrak{sl}_2$ *is trivial, then we find the condition* $\beta = \gamma$.

Proof. Recall the structure of rad($\mathfrak{g}$). We have $[u, r_i] = \pm r_i$, $\hat{E}[r_1, r_2] = 0$, and the actions on Θ, in the basis $\{c_1, c_2, c_3, d_1, d_2, d_3\}$, given by

$$u = \begin{pmatrix} 1 & 0 & 0 & 0 & 0 & 0 \\ 0 & 1 & 0 & 0 & 0 & 0 \\ 0 & 0 & 0 & 0 & 0 & 0 \\ 0 & 0 & 0 & 1 & 0 & 0 \\ 0 & 0 & 0 & 0 & 1 & 0 \\ 0 & 0 & 0 & 0 & 0 & 0 \end{pmatrix},$$

$$r_1 = \begin{pmatrix} 0 & 0 & 0 & 0 & 0 & 0 \\ 0 & 0 & 0 & 0 & 0 & 0 \\ 1 & 0 & 0 & 0 & 0 & 0 \\ 0 & 0 & 0 & 0 & 0 & 0 \\ 0 & 0 & 0 & 0 & 0 & 0 \\ 0 & 0 & 0 & 1 & 0 & 0 \end{pmatrix},$$

$$r_2 = \begin{pmatrix} 0 & 0 & 0 & 0 & 0 & 0 \\ 0 & 0 & 0 & 0 & 0 & 0 \\ 0 & 1 & 0 & 0 & 0 & 0 \\ 0 & 0 & 0 & 0 & 0 & 0 \\ 0 & 0 & 0 & 0 & 0 & 0 \\ 0 & 0 & 0 & 0 & 1 & 0 \end{pmatrix}.$$

Now, $\mathrm{Ext}^1_{\mathrm{rad}(\mathfrak{g})}(\tilde{\Delta}, \tilde{\Delta}) = \mathrm{Der}(\mathrm{rad}(\mathfrak{g}), \mathrm{End}_k(\tilde{\Delta}))/\,\mathrm{Triv}$. The rest is verified by computing $\psi(r_i) = [r_i, \psi(u)] + [\psi(r_i), u]$ for $i = 1, 2$, and $0 = [r_1, \psi(r_2)] + [\psi(r_1), r_2]$. The particular case above follows by simply checking that it fits with the Lie relations in $\mathrm{rad}(\mathfrak{g})$, or more generally with those of $\mathfrak{g}$, where ψ vanish on the sub-Lie algebra $\mathfrak{sl}_2$. □

The particular case, above, may be interpreted as follows: $Z := \psi(u)$ mediates a force, mapping $\langle d_i \rangle$ to $\langle d_i \rangle$, $W_1 := \psi(r_1)$ maps $\langle d_1 \rangle$ to $\langle d_3 \rangle$, and $W_2 := \psi(r_2)$ maps $\langle d_2 \rangle$ to $\langle d_3 \rangle$.

Go back to Section 9.3, and see that W_1 may be identified with e^3_-, and W_2 with e^2_-, so we may consider the Weak Force as part of the Strong Force. We found the following explicit formulas for the action of $\mathfrak{sgl}(3)$ on $\tilde{\Delta}$, which we shall identify with the Strong Force, given first by the restriction of $h \in \mathfrak{g}$ to $\tilde{\Delta}$, and the charge operator h_2, which in the basis $\{d_1, d_2, d_3\}$ look like

$$h_1 = \begin{pmatrix} 1/2 & 0 & 0 \\ 0 & -1/2 & 0 \\ 0 & 0 & 0 \end{pmatrix}, \quad h_2 = \begin{pmatrix} -1/3 & 0 & 0 \\ 0 & -1/3 & 0 \\ 0 & 0 & 2/3 \end{pmatrix},$$

and then the operators $\mathfrak{e}^i_\pm, i = 1, 2, 3$, given by

$$\mathfrak{e}^1_+ = \begin{pmatrix} 0 & 1 & 0 \\ 0 & 0 & 0 \\ 0 & 0 & 0 \end{pmatrix}, \quad \mathfrak{e}^2_+ = \begin{pmatrix} 0 & 0 & 0 \\ 0 & 0 & 1 \\ 0 & 0 & 0 \end{pmatrix}, \quad \mathfrak{e}^3_+ = \begin{pmatrix} 0 & 0 & 1 \\ 0 & 0 & 0 \\ 0 & 0 & 0 \end{pmatrix},$$

and their "duals,"

$$\mathfrak{e}^1_- = \begin{pmatrix} 0 & 0 & 0 \\ 1 & 0 & 0 \\ 0 & 0 & 0 \end{pmatrix}, \quad \mathfrak{e}^2_- = \begin{pmatrix} 0 & 0 & 0 \\ 0 & 0 & 0 \\ 0 & 1 & 0 \end{pmatrix}, \quad \mathfrak{e}^3_- = \begin{pmatrix} 0 & 0 & 0 \\ 0 & 0 & 0 \\ 1 & 0 & 0 \end{pmatrix}.$$

We observe that the action of $\mathfrak{e}_+^1 = e$ and $\mathfrak{e}_-^1 = f$, $\mathfrak{e}_-^2 = r_2$ and $\mathfrak{e}_-^3 = r_1$, in $\mathfrak{g}$. Moreover,

$$[u, \mathfrak{e}_\pm^i] = \pm \mathfrak{e}_\pm^i$$

and $\mathfrak{e}_-^2, \mathfrak{e}_-^3$ are the "anti-operators" of $\mathfrak{e}_+^2$, respectively $\mathfrak{e}_+^3$.

We see that the weak interaction on quarks, mediated by the weak interaction bosons, Z, W_1, W_2, identified above with the Bosons $u, r_1, r_2 \in \mathfrak{g}$, respectively, have charges as they should. Clearly, Z is neutral, and both W_i have charge 1, since

$$C(Z) = (\mathfrak{h}_2^+ Z - Z\mathfrak{h}_2^+)(d_j) = 0, C(W_i)(\mathfrak{h}_2^+ W_i - W_i\mathfrak{h}_2^+)(d_i) = 1, \quad i = 1, 2.$$

Moreover, we see that the Charge Conjugation C, see Section 10.3, also changes the sign of the weak charge.

Knowing that the Weak Force, restricted to $\tilde{\Delta}$, is contained in the Strong Force, we find that the Force Fields, $\mathfrak{F}$, defined above

$$\mathfrak{F}(\Theta_{\tilde{H}}, \Theta_{\tilde{H}}) \subset \mathcal{P}(\Theta_{\tilde{H}}, \Theta_{\tilde{H}})/\operatorname{Triv} \simeq \operatorname{Ext}^1_{\tilde{H}(\sigma_g),\rho_0}(\Theta_{\tilde{H}}, \Theta_{\tilde{H}})$$

generated by $(\mathfrak{g}*)^6$, is the combined Weak and Strong Forces, acting upon the elementary particles, from which we should be able to reconstruct, by iterated extensions, all particles of our Toy Model.

Talking about "Mass for Bosons," one needs an introduction that will have to wait.

From the lemma above, we learn that there are two obvious extensions of the "electron =:e," in our sense, with the "proton=:p," also in our sense.

First, the natural "neutron=:n," i.e. the extension:

$$0 \to \langle c_3, d_3 \rangle \to \langle c_1, c_2, c_3, d_1, d_2, d_3 \rangle \to \langle c_1, c_2, d_1, d_2 \rangle \to 0,$$

or the same written in another way as

$$0 \to \begin{pmatrix} c_3 \\ d_3 \end{pmatrix} \to \begin{pmatrix} c_1 & c_2 & c_3 \\ d_1, & d_2 & d_3 \end{pmatrix} \to \begin{pmatrix} c_1 & c_2 \\ d_1 & d_2 \end{pmatrix} \to 0,$$

given by the weak force boson

$$W = W^{+,-} \in \operatorname{Hom}_{\tilde{H}}(\langle c_1, c_2, d_1, d_2 \rangle, \langle c_3, d_3 \rangle)$$

and its image in $\operatorname{Ext}^1_{\tilde{H}(\sigma_g),\rho_0}(e, p)$.

This should give birth to the $\beta - decay$

$$n \to e^- + p + \nu_e,$$

where the neutrino ν_e should be neutral and perhaps having light velocity, thus no mass.

Let us look at the extension above, at the level of sub-modules of $\Theta_{\tilde{H}}$, and let us try out the following definition of the neutrino:

$$\nu_e := \begin{pmatrix} c_1 & c_2 & c_3 \\ 0 & 0 & 0 \end{pmatrix}$$

as an extension of the photon

$$\gamma := \begin{pmatrix} c_1 & c_2 \\ 0 & 0 \end{pmatrix}$$

with

$$c_3 = \begin{pmatrix} c_3 \\ 0 \end{pmatrix}$$

corresponding to $W = W^{+,-} \in \mathrm{Ext}^1_{\tilde{H}(\sigma_g),\rho_0}(\gamma, c_3)$.

Note also that we have a dual, dark-matter particle, given in the same way as

$$\eta := \begin{pmatrix} 0 & 0 & 0 \\ d_1 & d_2 & d_3 \end{pmatrix}$$

as an extension of the tenebron

$$\gamma := \begin{pmatrix} 0 & 0 \\ d_1 & d_2 \end{pmatrix}$$

with

$$d_3 := \begin{pmatrix} 0 \\ d_3 \end{pmatrix}$$

corresponding to $W = W^{+,-} \in \mathrm{Ext}^1_{\tilde{H}(\sigma_g),\rho_0}(\gamma, c_3)$. For lack of a more fancy name, call it the *Hebron*.

It is now quite easy to translate the $\beta - decay$ formula above. The neutron and the kicked out proton must have different momenta, and even different velocities. These velocities are given as $\sin(\langle c_3, d_3\rangle)$. Observing the proton after the decay, keeping $\langle d_3\rangle$ constant, we obtain that the proton kicked out has a different form $(\langle c_3', d_3\rangle)$.

The (energy)-difference between the neutron and the evicted proton and electron must be the energy of the neutrino. Due to the fact that we assume the neutron has charge 0, the neutrino should be an object of the above form! Usually this "decay" is pictured as,

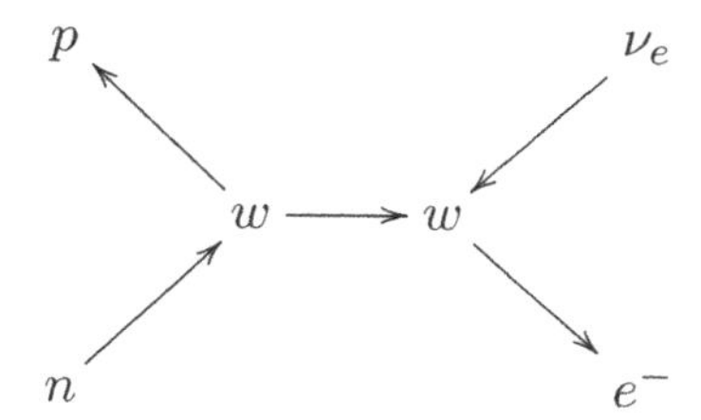

or as Pauli wrote it, $n_0 \to p_+ + e^- + \nu_{e^-}$.

We might try to use the technique of non-commutative deformations, to see the relations between the objects involved. The existence of the anti-operators $\mathfrak{e}^3_\pm$ in the strong force, coupled to the action of $\mathfrak{g}$, forces the family of particles $\mathfrak{B} := \{e, p\}$, defined as the representations of $\tilde{H}(\sigma_g)$, to have more structure, namely the non-trivial Ring of Observables $O(\mathfrak{B})$, defined on $\mathbf{a}_r$, for $r = 2$, with the defining morphism

$$\eta : \tilde{H}(\sigma_g) \to O(\mathfrak{B}).$$

From this, and from the corresponding construction for any set of particles, the usual diagrams of β-decay, "electron"-capture, and lots of other Decay diagrams might follow. As an example, we may compute the O-construction above, and find, for (as above), W restricted to $\tilde{c}$. We find,

$$O(\mathfrak{B}) =: \begin{pmatrix} H_{1,1} \otimes \mathrm{Hom}(p,p) & H_{1,2} \otimes \mathrm{Hom}(p,e) \\ H_{2,1} \otimes \mathrm{Hom}(e,p) & H_{2,2} \otimes \mathrm{Hom}(e,e) \end{pmatrix},$$

see Section 12.1, for the definitions of $H = \{H_i\}$, $i = 1, 2$, and of the morphism η.

From this, and from the corresponding construction, with $\mathfrak{F}$ replacing Ext^1, for any set of particles, the usual diagrams of β-decay, "electron"-capture, and lots of other Decay diagrams might follow.

As an example, we may compute the O-construction above. We find,

$$O(\mathfrak{B}) = \begin{pmatrix} H_{1,1} \otimes \mathrm{Hom}_k(p,p) & H_{1,2} \otimes \mathrm{Hom}_k(p,e) \\ H_{2,1} \otimes \mathrm{Hom}_k(e,p) & H_{2,2} \otimes \mathrm{Hom}_k(e,e) \end{pmatrix}$$

see Section 12.1, for the definitions of $H = \{H_{i,j}\}, i = 1, 2$, and of the morphism η.

Now, since, for the purpose of creating new particles, we should use our Forces $\mathfrak{F}(\mathfrak{B}_i, \mathfrak{B}_j) \subset \mathrm{Ext}^1_{\tilde{H}(\sigma_g),\rho_0}(\mathfrak{B}_i, \mathfrak{B}_j)$, instead of the full Ext^1, and since $\mathfrak{F}(\mathfrak{B}_i, \mathfrak{B}_j)$ is just a finite-dimensional k-Lie algebra, let us put

$$F^{i,j} := \mathfrak{F}(\mathfrak{B}_i, \mathfrak{B}_j),$$

and denote by $F_{i,j}$ the k-dual vector-space. The restricted non-commutative deformation of the family $\mathfrak{B}=\{e, p\}$ is then, at the first level, i.e. defined on the object, $F \in \mathbf{a}_2$, with $\mathrm{rad}(F)^2 = 0$,

$$O_{\mathfrak{F}}(\mathfrak{B}) = \begin{pmatrix} (k \oplus F_{1,1}) \otimes \mathrm{Hom}_k(p,p) & F_{1,2} \otimes \mathrm{Hom}_k(p,e) \\ F_{2,1} \otimes \mathrm{Hom}_k(e,p) & (k \oplus F_{2,2}) \otimes \mathrm{Hom}_k(e,e) \end{pmatrix},$$

which might be considered as the hydrogen atom. Built into it, one finds easily, both the neutron, as a consequence of the Weak Force, and the electron "fixed to the proton" by the Strong Force!

Instead of following this bumpy road further, we shall in Section 12.3, introduce a slightly different way of looking at the construction and decay of particles. See Chapter 3 of Eriksen *et al.* (2017) for a thorough introduction to this theme. The "classification" of the particles in our Cosmological Furniture will have to wait for its place in a future paper.

But let us here close the treatment of the Weak and the Strong Forces, by adding the following example.

Example 12.1. We have, above, omitted the full story, the action of Z, W_1, W_2 in $\mathrm{Hom}_{\tilde{H}(\mathfrak{g})}(\tilde{c}, \tilde{\Delta})$ and in $\mathrm{Hom}_{\tilde{H}(\mathfrak{g})}(\tilde{\Delta}, \tilde{c})$. This may be interesting, so without further comments, we add the corresponding matrices for Z_0 and r_1.

$$\psi(u) = \begin{pmatrix} \alpha_{1,1} & 0 & \alpha_{1,3} & \alpha_{1,4} & 0 & \alpha_{1,6} \\ 0 & \alpha_{2,2} & \alpha_{2,3} & 0 & \alpha_{1,4} & \alpha_{2,6} \\ \alpha_{3,1} & \alpha_{3,2} & \alpha_{3,3} & \alpha_{3,4} & \alpha_{3,5} & \alpha_{1,4} \\ \alpha_{4,1} & 0 & \alpha_{4,3} & \alpha_{4,4} & 0 & \alpha_{4,6} \\ 0 & \alpha_{4,1} & \alpha_{5,2} & 0 & \alpha_{5,5} & \alpha_{5,6} \\ \alpha_{6,1} & \alpha_{6,2} & \alpha_{4,1} & \alpha_{6,4} & \alpha_{6,5} & \alpha_{6,6} \end{pmatrix}$$

where $\alpha_{1,1} = \alpha_{2,2} = \alpha_{3,3} = \alpha_{4,4} = \alpha_{5,5} = \alpha_{6,6}$, and modulo matrices of the form

$$\begin{pmatrix} 0 & 0 & a_{1,3} & 0 & 0 & a_{1,6} \\ 0 & 0 & a_{2,3} & 0 & 0 & a_{2,6} \\ -a_{3,1} & -a_{3,2} & 0 & -a_{3,4} & -a_{3,5} & 0 \\ 0 & 0 & a_{4,3} & 0 & 0 & a_{4,6} \\ 0 & 0 & a_{5,2} & 0 & 0 & a_{5,6} \\ -\alpha_{6,1} & -\alpha_{6,2} & 0 & -a_{6,4} & -a_{6,5} & 0 \end{pmatrix}$$

and

$$\psi(r_1) = \begin{pmatrix} -\alpha_{1,3} & 0 & 0 & -\alpha_{1,6} & 0 & 0 \\ -\alpha_{2,3} & 0 & 0 & -\alpha_{2,6} & 0 & 0 \\ \alpha^1_{3,1} & \alpha^1_{3,2} & \alpha_{1,3} & \alpha^1_{3,4} & \alpha^1_{3,5} & \alpha_{1,6} \\ -\alpha_{4,3} & 0 & 0 & -\alpha^1_{4,6} & 0 & 0 \\ -\alpha_{5,3} & 0 & 0 & 0 - \alpha_{5,6} & 0 & 0 \\ \alpha^1_{6,1} & \alpha^1_{6,2} & \alpha_{4,3} & \alpha^1_{6,4} & \alpha^1_{6,5} & \alpha_{4,6} \end{pmatrix},$$

modulo matrices of the form

$$\begin{pmatrix} a_{1,3} & 0 & 0 & a_{1,6} & 0 & 0 \\ a_{2,3} & 0 & 0 & a_{2,6} & 0 & 0 \\ b_{3,1} & -a_{1,2} & -a_{1,3} & b_{3,4} & -a_{1,5} & -a_{1,6} \\ a_{4,3} & 0 & 0 & a_{4,6} & 0 & 0 \\ a_{5,3} & 0 & 0 & a_{5,6} & 0 & 0 \\ b_{6,1} & -a_{4,2} & -a_{4,3} & b_{6,4} & -a_{4,5} & -a_{4,6} \end{pmatrix},$$

where $b_{3,1} = a_{3,3} - a_{1,1}, b_{3,4} = a_{3,6} - a_{1,4}, b_{6,1} = a_{6,3} - a_{4,1}, b_{6,4} = a_{6,6} - a_{4,4}$, for some matrix $a = (a_{i,j})$.

This means that $\psi(u)$ is the identity, and $\psi(r_1), \psi(r_2)$ are given by the maps: $d_j \rightarrow c_3$, $d_j \rightarrow c_3$, $d_j \rightarrow d_3$, $c_j \rightarrow c_3$, $c_j \rightarrow c_3$, for, respectively j=2, and j=1.

12.3 Graphs and Sub-Categories Generated by a Family of Modules

Let A be any associative k-algebra, and assume given a swarm $\mathbf{V} = \{V_i\}$ of A-representations. Let Γ be an ordered graph with set of nodes $|\Gamma| = \{V_i\}$. Starting with a first node V_{i_1}, of Γ, we can construct, in many ways, an extension of the module V_{i_1} with the module V_{i_2}, corresponding to the end point of the first arrow of Γ, then continue, choosing an extension of the result with the module corresponding to the endpoint of the second arrow of Γ, etc. until we have reached the endpoint of the last arrow. Any finite length module can be made in this way, for some Γ corresponding to a decomposition of the module into simple constituencies, by peeling off one simple sub-module at a time, i.e. by picking one simple sub-module and forming the quotient, picking a second simple sub-module of the quotient and taking the quotient, and repeating the procedure until it stops.

The *ordered* k-algebra $k[\Gamma]$ of the ordered graph Γ is the quotient algebra of the usual algebra of the graph Γ by the ideal generated by all admissible words which are not "intervals" of the ordered graph. Say $\ldots\gamma_{i,j}(n-1)\gamma_{j,j}(n)\gamma_{j,k}(n+1)\ldots$ is an interval of the ordered graph, then $\gamma_{i,j}(n-1).\gamma_{j,k}(n+1) = 0$ in $k[\Gamma]$.

The main result in this context is now the following result, see Laudal (2003).

Proposition 12.1. *Let A be any associative k-algebra, $\mathbf{V} = \{V_i\}_{i=1}^r$ any swarm of right A-modules. Consider an iterated extension $\mathbf{E}$ of $\mathbf{V}$, with extension graph Γ. Then there exists a morphism of k-algebras*

$$\phi : H(\mathbf{V}) \rightarrow k[\Gamma]$$

such that, as right A-modules,

$$\mathbf{E} \simeq k[\Gamma] \otimes_\phi \tilde{V}.$$

Here, $\tilde{V}$ is the versal deformation, of the family $\mathbf{V}$*, as left* $H(\mathbf{V})$*, and right, A-module.*

Moreover, the set of equivalence classes of iterated extensions of $\mathbf{V}$ *with representation graph* Γ *is a quotient of the set of closed points of the affine algebraic variety*

$$\underline{A}[\Gamma] = \mathrm{Mor}(H(\mathbf{V}), k[\Gamma]).$$

There is a versal family $\tilde{V}[\Gamma]$ *of A-modules defined on* $\underline{A}[\Gamma]$*, containing as fibers all the isomorphism classes of iterated extensions of* $\mathbf{V}$ *with extension graph* Γ.

Let $\mathbf{Mod}_A^{\mathbf{V}}$ denote the full abelian sub-category of $\mathbf{Mod}_A$ generated by the iterated extensions of the objects in $\mathbf{V}$, and let $\mathbf{Mod}_{H(\mathbf{V})}$ be the category of *finite-dimensional* H-modules. Then we have the following structure theorem, generalizing a result, of Beilinson, see Beĭlinson (1978).

Theorem 12.1. *Let A be any k-algebra, and fix a swarm,* $\mathbf{V} = \{V_i\}_{i=1}^r$*, of A-modules, then there exists a functor*

$$\iota(\mathbf{V}) : \mathbf{Mod}_{H(\mathbf{V})} \to \mathbf{Mod}_A^{\mathbf{V}}$$

which is an isomorphism on equivalence-classes of objects, and injective on morphisms. If $\mathbf{V}$ *consists of simple modules, then* κ *is an equivalence.*

Proof. Any right $H(\mathbf{V})$-module M is a k^r-module, so it can be decomposed as $M = \oplus M_i$, where $M_i := Me_i$. The structure map is therefore given as

$$\rho_0 : H(\mathbf{V}) \to \mathrm{End}_k(M) = (\mathrm{Hom}_k(M_i, M_j)).$$

Here ρ_0 maps $H_{i,j}$ into $\mathrm{Hom}_k(M_i, M_j)$, and therefore the formal family may be decomposed to give us the following k-algebra homomorphisms,

$$\begin{aligned}\rho : A &\xrightarrow{\tilde{\rho}} (H_{i,j} \otimes \mathrm{Hom}_k(V_i, V_j)) \xrightarrow{\rho_0} (\mathrm{Hom}_k(M_i, M_j) \otimes_k \mathrm{Hom}_k(V_i, V_j)) \\ &= (\mathrm{Hom}_k(M_i \otimes_k V_i, M_j \otimes V_j)) = \mathrm{End}_k(W).\end{aligned}$$

Here, $W = \oplus_{i=1}^r (M_i \otimes V_i)$, and by definition,

$$\iota(\mathbf{V})(M) := W.$$

Since M by definition of $\mathbf{Mod}_{H(\mathbf{V})}$ is of finite dimension as k-vector space, and therefore is an iterated extension of the simple modules

k_i, it is clear that W is an iterated extension of the modules V_i in the family $\mathbf{V}$. Moreover, we have seen that for any object being an iterated extension, along a graph Γ, of the modules in $\mathbf{V}$, we can find a morphism

$$\phi : H(\mathbf{V}) \to k[\Gamma]$$

which obviously defines $k[\Gamma]$ as a finite-dimensional $H(\mathbf{V})$-module, such that

$$E \simeq k[\Gamma]\otimes_{\phi}\tilde{V} = \iota(\mathbf{V})(k[\Gamma]).$$

Thus, the first part of the theorem follows from Theorem 4.1. The rest is more or less evident. □

12.3.1 *Interactions and dynamics*

Go back to the start-scenario of this section, and to our assumption that all the stuff in our world is represented by iterated extensions of the modules representing the elementary particles $\mathbf{V} = \{V_i\}$. Assume that $\mathbf{V}$ is a swarm of $\mathbf{A}(\sigma)$ modules. Then $H(\mathbf{V}) = (H_{i,j})$ is finitely generated, but maybe not algebraic. To simplify the situation a little, we shall therefore assume there is an algebraization, $\mathbf{H}(\mathbf{V}) = (\mathbf{H}_{i,j})$, for which $H(\mathbf{V}) = (H_{i,j})$ is the formalization at the canonical module, given by the structure morphism $\pi : \mathbf{H}(\mathbf{V}) \to W_0 := k^r$. Moreover, we assume there exists an extension,

$$\tilde{\rho} : A \to O(\mathbf{V}) = (\mathbf{H}_{i,j} \otimes_k \mathrm{Hom}_k(V_i, V_j))$$

of the versal family.

Any elementary particle, represented by the simple representation

$$\rho : \mathbf{A}(\sigma) \to \mathrm{End}_k(V)$$

evolves, as time goes by, maybe ending up by decaying into a composite particle, represented by the module V_∞ which by our assumption must be an iterated extension of the simple modules in $\mathbf{V}$. Use the Beilinson theorem, above, and identify V_∞ with a $\mathbf{H}(\mathbf{V}) = (\mathbf{H}_{i,j})$ module, M_0. There may be many extension graphs corresponding to this module. Pick one, Γ, and consider the moduli space, $\underline{A}[\Gamma]$, of such iterated modules. The dynamics of $\underline{A}[\Gamma]$ is now taken care of by the same method that we applied at the outset of this study. We must look for a dynamical structure for $\mathbf{H}(\mathbf{V})$, i.e. look at δ-stable ideals,

$$(\sigma) \subset \mathrm{Ph}^{\infty}(\mathbf{H}(\mathbf{V})),$$

and we must be prepared to study the category of morphisms,

$$\mathbf{H}(\mathbf{V})(\sigma) := \mathrm{Ph}^{\infty}(\mathbf{H}(\mathbf{V}))/(\sigma) \to k[\Gamma].$$

Any such will, as we know, give us a formal curve of iterated modules with extension graph Γ. Then we may try to copy the technique of Chapter 4. In the general situation, this does not seem so easy, but in our Toy Model case, it should be possible, as we have had a glimpse of in Section 11.1, where something looking very much like a mathematical equivalent to the Weak Force popped up.

One question that comes up in relation to a more or less serious interpretation of our mathematical model is: can we have an interaction of several particles the outcome of which is an elementary particle? It is reasonable to conjecture the following result.

Theorem 12.2. *Suppose, in the situation above, that there is a deformation W of W_0, as $\mathbf{H}(\mathbf{V})$ module, such that*

$$\pi' : \mathbf{H}(\mathbf{V}) \to \mathrm{End}_k(W) \simeq M_r(k),$$

deforming π, is surjective. Then $\iota(\mathbf{V})(W)$ is a simple A module.

Note that if k is algebraically closed, any simple deformation of W_0 as $\mathbf{H}(\mathbf{V})$ module must necessarily be given by a surjective structure map,

$$\rho_W : \mathbf{H}(\mathbf{V}) \to \mathrm{End}_k(W) = M_r(k).$$

The existence of a surjective homomorphism of k-algebras, ρ_W, is now a problem we have to address. Consider for this purpose, the following definition, see Laudal (1986), and Eriksen *et al.* (2017),

Definition 12.1. Let Q be the quiver associated to the family $\mathbf{V}$ of A modules. A sub-graph, Γ of Q, is called *Massey-trivial* if all Massey products containing non-trivial sub-graphs of Γ vanish.

We may then prove the following result.

Theorem 12.3. *Suppose there is a* Massey-trivial *connected complete cycle, Γ, of Q. Then there is a simple deformation, W, of W_0, as $\mathbf{H}(\mathbf{V})$ module.*

Obviously, this result makes the interaction scenario sketched above quite involved, even though we may concentrate on a finite number of particles in the theory we have chosen.

Example 12.2 (Spontaneous interaction and evolution). Suppose given a sub-family of finite-dimensional simple modules, $\mathbf{S} \subset \mathbf{V}$, and suppose there is a surjective homomorphism of k-algebras,

$$\mu : \mathbf{H}(\mathbf{S}) \to M_s(k),$$

where s is the number of elements in $\mathbf{S}$. Consider now the composition

$$\begin{aligned} \tilde{\mu} : A \overset{\tilde{\rho}}{\to} O(\mathbf{S}) \overset{\mu \otimes id}{\to} M_n(k) \otimes \mathrm{End}_k \left(\bigoplus_{V \in \mathbf{S}} V) \right) &= \mathrm{End}_k \left(\bigoplus_{V \in \mathbf{S}} V \right) \\ &= \mathrm{End}_k(U), \end{aligned}$$

where $U = \iota(\mathbf{S})(k^s)$ and ι is defined as in Chapter 8. According to the results above, we see that this composition defines a new simple A module. The process of replacing $\mathbf{V}$ by $(\mathbf{V} - \mathbf{S}) \cup \{U\}$ is sometimes called a mutation.

12.4 Creating New Particles from Old Ones

Above we asked, given the way we have defined the elementary particles, how do we construct the more complicated constituents of our world, the atoms and molecules, in short the "Furniture of our Universe"? And what are the forces acting upon these (elementary) particles, performing the changes we see?

The most elementary of these kinds of questions (which all physicists seem to agree upon) would obviously be, what do we mean by (*a family of equal*), or an ensemble of particles? This, we have already treated above, in relation to the O-construction. The first one, how do (we) create new ones, we termed more demanding. In fact, it is not, once we learn about how to perform iterated extensions, and see the relations to the *O-construction*.

The technique of producing iterated extensions of a family of representations helps us to create new representations from the primeval ones. Before we shall try to produce our model of hydrogen and helium, let us recall the way we produce new particles out of two particles (or modules) by two different extension methods.

First, recall that we have the formula, see Remark 4.5.

Given a representation $\rho : \tilde{H}(\sigma_g) \to \mathrm{End}_k(\mathfrak{B})$, corresponding to $\psi \in \mathcal{P}$, i.e. such that $\rho(t_i) = t_i, \rho(dt_i) = \xi_i + \psi_i$, we find

$$\mathbf{T}_{\rho,\rho} = \mathrm{Ext}^1_{C(\sigma_g),\rho_0}(\rho, \rho) = \mathcal{P}/\,\mathrm{Triv},$$

where the trivial derivations $W \in \mathrm{Triv}$, mapping t_i to 0, are exactly those given by the n-tuples,

$$\left(\left(\sum_j^n g^{1,j}\left(\frac{\partial \Phi}{\partial t_j}\right) + [\psi_1, \Phi]\right), \ldots, \left(\sum_{j=1}^n g^{n,j}\left(\frac{\partial \Phi}{\partial t_j}\right) + [\psi_n, \Phi]\right)\right),$$

for some $\Phi \in \mathrm{End}_C(V)$, by

$$W(t_i) = 0,\ W(dt_i) = \left(\sum_j^n g^{i,j}\left(\frac{\partial \Phi}{\partial t_j}\right) + [\psi_i, \Phi]\right).$$

The expression,

$$\Phi(\psi) := (\xi_1(\Phi) + [\psi_1, \Phi], \ldots, \xi_n(\Phi) + [\psi_n, \Phi]),$$

therefore, corresponds to an infinitesimal gauge transformation,

$$\Phi \in \mathrm{Der}_k(\mathcal{P})$$

of the space $\mathcal{P}$ of representations of $C(\sigma_g)$, acting linearly like

$$\Phi(\rho + \psi) = (\rho + \psi) + \Phi(\psi)$$

The *physically* relevant tangent space is, therefore, the quotient

$$\mathbf{P} = \mathcal{P}/\mathfrak{h}$$

of $\mathcal{P}$ with respect to the action of the abelian Lie algebra $\mathfrak{h} := \mathrm{End}_C(V)$.

From this we shall construct new objects, from old ones.

With the Weak force: Hydrogens, Neutrons, etc.

And with the Strong force: Hadrons, Mesons, Baryons, Gluons, etc.

12.5 Entanglement, Consciousness

The present theoretical basis for entanglement in QT is build on the trivial notion of extensions of sub-systems (objects), defined as tensor products $V \otimes W$, of representations of the algebra of observables, and entanglement of states are explained by accepting that if $V \neq W$, and $\phi_1 \in V, \phi_2 \in W$, the action of an observable $a \in A$ on $\phi_1 \in V$ may introduce an action of a on $\phi_2 \in W$. I have to admit that I have never understood the arguments for only accepting tensor products as products in quantum theory. The recent works of among others Penrose (see Penrose and Hameroff, (2011)) tells me that the problems here come from the unlucky choice of just one representation space, the "Hilbert space."

In fact, if we have a family of representations, $\mathbf{B} = \{\rho_{0i} : \tilde{H} \to \mathrm{End}_k(\mathfrak{B}_i)\}$, and momenta,

$$\rho_i : \tilde{H}(\sigma_g) \to \mathrm{End}_k(\mathfrak{B}_i),$$

consider the O-construction, of the corresponding family over $\mathrm{Ph}(\tilde{H})$,

$$\eta : \mathrm{Ph}(H)(\sigma_g) \to O(\mathbf{B}),$$

then the family should be called entangled if $O(\mathbf{B})$ does not split up into a sum of the $\mathrm{End}_k(\mathfrak{B}_i)$.

Consider the case where we have just one representation ρ_0, and a momentum

$$\rho : \tilde{H}(\sigma_g) \to \mathrm{End}_k(\mathfrak{B}),$$

then we know from Section 4.5.1 that we have

$$\mathrm{Ext}^1_{\tilde{H}(\sigma_g)/\rho_0}(\mathfrak{B}, \mathfrak{B}) = \mathbf{P} := \mathfrak{P}/\mathfrak{h},$$

where $\mathfrak{P}$ is the space of potentials. Consider now a tangent to our metric,

$$h = (h_{i,j}) \in \mathbf{T}_{\mathbf{M},g}.$$

According to Theorem 10.2, there is also an injective morphism,

$$\eta : \mathbf{T}_{\mathbf{M},g} \to \mathrm{Ext}^1_{\mathrm{Ph}(\tilde{H})}(\mathfrak{B}, \mathfrak{B}),$$

onto the subspace of first-order extensions, $\mathrm{Ext}^1_{\mathrm{Ph}(\tilde{H}/\rho_0)}(\mathfrak{B},\mathfrak{B})$. For $\Phi \in \mathfrak{P}$, and $h = (h_{i,j}) \in \mathbf{T}_{\mathbf{M},g}$, the corresponding derivations,

$$\Phi : \tilde{H}(\sigma_g) \to \mathrm{End}_k(\mathfrak{B}), \textit{and } \eta(h) : \mathrm{Ph}(\tilde{H}) \to \mathrm{End}_k(\mathfrak{B}),$$

defined by $\Phi(dt_i) = \Phi_i$ and $\eta(h)(dt_i) = \sum_j h^{i,j}\delta_j$, changing the value of ρ from $\rho(dt_i) = \sum_j g^{i,j}\delta_j + \psi_i$ to $\rho(dt_i) = \sum_j g^{i,j}\delta_j + \psi_i + \epsilon\Phi_i$, and from $\rho(dt_i) = \sum_j g^{i,j}\delta_j$ to $\rho(dt_i) = \sum_j g^{i,j}\delta_j + \epsilon\sum_j h^{i,j}\delta_j\psi_i$. Recall that $\eta(h)$ maps $\tilde{H}$ to 0, so $(\phi + \eta(h)(a)(\phi))$ will look the same as ϕ considered as elements in the $\tilde{H}$ module. So, even though $\phi' := \phi + \eta(h)(a)(\phi)$ is different from ϕ, it will have the same values for all observables $a \in \tilde{H}$.

Going back to Theorem 10.2, we therefore find that the energy that brought about the superposed separation of ϕ and ϕ' is related to the inverse time, squared, taking place in $\tilde{\Delta}$, so that we find reasons to believe in the formulas of Penrose and Hameroff (2011),

$$E = h/T, E = Gm^2a^{-1},$$

where a is the so-called "displacement separation," which here should be related to the determinant of $(h^{i,j})$. Moreover, T, the time used for this separation, should be related to the square root of $(h_{i,j})$.

Note also that if there is an algebraic versal space, $\mathfrak{V}$, for some representation ρ_0, then we may go back to Chapter 6, and consider deformations $\rho_i \in \mathfrak{V}$, in the same moduli-room of the Moduli Suite of V_0. They are all entangled, since although they are isomorphic, the automorphism $\alpha : V \to V$ separates states, $\alpha(\phi) \neq \phi$, but all observables are perfectly correlated! The entropy of the ρ_is then becomes a kind of measure of entanglement, known to physicists.

12.5.1 *Self-reflection*

In line with Penrose and Hameroff's ideas about conscience, as a quantum phenomenon, loc.cit., one might wonder about the possibilities of an observer to observe itself, i.e. how to model, for an arbitrary metric g, the quantum representations of the algebra $\tilde{H}(\sigma_g)$ restricted to the point (o,o) or to the sphere $E(\underline{\lambda})$? For the simplified metric

$$g = (\rho - h)^2/\rho^2 d\rho^2 + (\rho - h)^2 d\phi^2 + \kappa d\lambda^2$$

we find

$$O := \tilde{H}(\sigma_g) = k[\lambda, \phi, \rho] < d\lambda, d\phi, d\rho > /R$$

where the only relations are

$$R = \{[d\rho, \rho] = \left(\frac{\rho}{\rho - h}\right)^2, [d\phi, \phi] = (\rho - h)^{-2}, [d\lambda, \lambda] = 1\},$$

as ρ goes to zero, i.e. the sub-algebra of O,

$$O_o := \tilde{H}(\sigma_g)_o = k[\lambda, \phi]\langle d\lambda, d\phi\rangle / R,$$

with the relations

$$R_o = \{[d\phi, \phi] = h(\lambda)^{-2}, [d\lambda, \lambda] = \kappa^{-1}\}.$$

Chapter 13

Comparing the Toy Model with the Standard Model

Before we end this book, let us take a bird's-eye view of the current state of the Standard Model (SM) and compare it with our Toy Model (TM) considered in Chapters 5–12.

But first, let us make a short review of some introductory books in mathematical physics, treating Quantum Theory (QT) and General Relativity theory (GRT). We may refer to the paper of Deligne and Freed, and also to the paper of Fadeev, in: *Quantum Fields and Strings: A Course for Mathematicians*, see Deligne (1999). This is a high-brow introduction to Field Theory, considered necessary to understand Quantum Fields and Strings. One may also look into the book by Sachs and Wu, (1977). This is my favorite book, treating General Relativity, and it has served as a model for the treatment of the theory above. However, mathematicians would maybe be more interested in the basic ideas and notions in modern physics, those that we have used to make our general constructions. And it may be interesting to most physicists to know how this basic physics is understood by mathematicians, like me, that try to understand their field. Therefore, let me refer to my favorite book on quantum field theory see, Mandl and Shaw (2010), and, of course, to the book of Weinberg (2005).

All of these books have contributed to the general feeling I have had that I finally understood how physicists thought. This, I am afraid, is not supported by the reactions of most of the listeners at seminars for mathematical physicists that I have endured. So, where are the problems, the obstructions, between the mathematical language, and the physicists way of understanding their field?

I do not know, and maybe I shall never know, so let us just take a superficial look at the way the Standard Model is presented in physics, and the relations I see between this and the mathematical theory presented above, i.e. my "Toy Model."

The ingredients are quite different. First, the SM is build upon the four-dimensional Minkowski space, where proper time must be compared to our $\lambda_3 =: \lambda$, and where our two other null-directions are missing. Then, our gauge groups, which turn out to contain the gauge group of the SM, are no longer only related to the metric. Our gauge groups are the Lie algebras responsible for the equivalence relation in our non-commutative algebraic space, of representations of the sheaf $\mathrm{Ph}(\tilde{H})$. These are all extensions of the $\mathfrak{g}$-invariant sub-modules of $\Theta_{\tilde{H}}$. Our basic elementary particles come out naturally; the weak bosons as the generators of $\mathfrak{g}$, and the gluons as the generators of $\mathfrak{sgl}(3)$, the massless states are $\{c_1, c_2, c_3\}$, and the massive and charged ones are $\{d_1, d_2, d_3\}$. The corresponding simple modules generate all particles, by tensor products, and by iterated extensions. The chirality, massiveness, charge, etc. all come out canonically, via reasonably canonical choices of markers.

There are also many similarities. Both the Weyl and the Dirac spinors have the same dimension and the same symmetries in SM as in our model.

We have the possibility to define three different mass terms, namely, via

$$\rho : \tilde{H}(\sigma_g) \to \mathrm{End}_k(\Theta_{\tilde{H}}),$$

the operators

$$\rho(d\lambda_i) \in \mathrm{End}_k(\Theta_{\tilde{H}}), \quad i = 1, 2, 3.$$

We already know that there are, at any point $(o, p) \in \underline{\tilde{H}}$, three tangent planes, called $B_o, B_p, A_{o,p}$, the two first "orthogonal" to the (op)-direction, and the third containing the canonical light and zero-velocity directions at that point. Note that the corresponding objects, i.e. the $\mathfrak{g}$-stable representations of $\tilde{H}$, are simple. And note that as objects they are visible in the sense of Section 10.4!

There are natural bases for these planes, $\{(c_1 + d_1), (c_2 + d_2)\}$, $\{(-c_1 + d_1), (-c_2 + d_2)\}$, and $\{c_3, d_3\}$, respectively, and the action of $\mathfrak{g}$ on these planes can be read from the values in Section 9.2.

We have also seen that the Pauli matrices turn up as

$$\sigma^1 = e + f = \begin{pmatrix} 0 & 1 \\ 1 & 0 \end{pmatrix},$$

$$\sigma^2 = ie - if = \begin{pmatrix} 0 & i \\ -i & 0 \end{pmatrix},$$

$$\sigma^3 = h = \begin{pmatrix} 1 & 0 \\ 0 & -1 \end{pmatrix},$$

and that the parity operator P induces the Dirac matrices

$$\gamma^0 = \begin{pmatrix} 1 & 0 \\ 0 & -1 \end{pmatrix}, \quad \gamma^k = \begin{pmatrix} 0 & \sigma^k \\ \sigma^k & 0 \end{pmatrix}, \quad k = 1, 2, 3,$$

as well as the new operators

$$\gamma^{k+3} = \begin{pmatrix} \sigma^k & 0 \\ 0 & -\sigma^k \end{pmatrix}, \quad k = 1, 2, 3,$$

acting on $B_o \oplus B_p$, such that

$$\forall p \neq q, \ \gamma^p \gamma^q = -\gamma^q \gamma^p, \gamma^p \gamma^p = 1, \quad p, q = 1, 2, 3, 4, 5, 6.$$

Chirality, in the physicist's language, was explained as follows. The morphism P, extended to $B_o \oplus B_p$, is in the basis chosen above, given by the matrix

$$\gamma^5 = \begin{pmatrix} 0 & \mathrm{id} \\ \mathrm{id} & 0 \end{pmatrix},$$

which turns left handedness to right handedness, with respect to the direction (o, p), respectively, (p, o). Recall also that there are faithful representations of $\mathfrak{g}$ in the bundle $B_o \oplus B_p$, and that $\mathfrak{g}$ kills $A_{o,p}$. We see that the representations B_o and B_p, the "points" of the non-commutative quotient of the moduli space $\tilde{H}(\sigma_g)$, by the gauge group $\mathfrak{g}$, are the two-component *Weyl Spinors* of physicists, and the space

$B_o \oplus B_p$ is the space of *Dirac Spinors*. The Hamiltonian, the operator representing time and energy Q, looks the same in both models.

Now, recall the basic ingredients of SM, the elementary particles, their organization into groups, and the way they are characterized. First, written up in the first three columns in the following diagram, the particles. The quarks, uq the up quark, usually just written u, and its more massive sisters c, charm, and t, topp. Then dq, the down quark, usually written d, and its brothers, s the strange, and b, the bottom. Below are the Leptons, the electron and its companions, μ and τ, and finally the corresponding neutrinos. The fourth column is reserved for the Forces represented by the "bosons," γ, the photon, g the very mysterious graviton, and then the bosons responsible for the weak force, Z, W^+, and W^-. Their actions on the quarks are specified separately.

$$\begin{array}{cccc} uq & c & t & \gamma \\ \\ dq & s & b & g \\ \\ e & \mu & \tau & W^{+/-} \\ \\ \nu_e & \nu_\mu & \nu_\tau & Z^0 \end{array}$$

We may now cook up a list of *particles* resembling the ingredients of the Standard Model from our arsenal of representations of the local gauge group, $\mathfrak{g}* = \mathfrak{g} \oplus \langle h_2 \rangle \subset \mathfrak{g} \oplus \mathfrak{su}(3)$. Let us start with the simple sub-representations of $\Theta_{\tilde{H}}$. They coincide with the elementary simple representations of $\mathfrak{sl}(2)$,

$$L(1) = \{\langle c_1, c_2 \rangle, \langle c_3 \rangle\}; D(1) = \{\langle d_1, d_2 \rangle, \langle d_3 \rangle\},$$

but, beware, these representations are, as we have seen, in particular, in Section 12.2, not invariant with respect to the full gauge group $\mathfrak{g}$ nor of the real part of the $\mathfrak{su}(3)$. As we have seen, we may consider "our" Hydrogen atom, as an extension of the electron with the proton, under the influence of the weak and strong forces.

The tensor products of the representations $L(1)$ and $D(1)$ will produce lots of simple, and non-simple, representations, writing $d_i d_j$ for $1/2(d_i \otimes d_j + d_j \otimes d_i)$, and in general, $d_i d_j \ldots d_k$ for the symmetric tensor product, the first tensor (or symmetric) products of $D(1)$, will give us the simple representations

$$D(2) = \langle d_1 d_1, d_1 d_2, d_2 d_2 \rangle, D(3) = \langle d_1 d_1 d_1, d_1 d_1 d_2, d_1 d_2 d_2, d_2 d_2 d_2 \rangle.$$

We find, of course, the same results for the tensor products of $L_c(1)$, and using elementary representation theory of the Lie algebras $\mathfrak{g}_0, \mathfrak{g}, \mathfrak{sgl}(3)$, or $\mathfrak{so}(3)$, we may, in principle, classify all the possible representations that would qualify for our Furniture of the Universe, as defined in the previous sections. Since the following is meant to show the relations between the Toy Model and the Standard Model, we shall concentrate our efforts on the "build up" of the so-called elementary particles, in the SM. They are defined in terms of quarks. The Bosons are already treated, and their definitions are dependent on the same type of markers as we have used.

Put Isospin=$\mathfrak{h}_1$, Charge=$\mathfrak{h}_2$, Weak Isospin= $I_3 = 3/4\mathfrak{h}_2 + 1/2\mathfrak{h}_1$, Weak Hyper-charge $= Y_W = 1/2\mathfrak{h}_2 - \mathfrak{h}_1$, and use these numbers, together with the "boson" $\mathbf{u}$, as markers. (Recall that $\mathbf{u}$, which as element in $\mathfrak{g}$, was called u, is our Z of the weak force.)

Then consider the diagram:

$Markers:$	$1/2 \cdot h$	u	$\mathfrak{h}_1$	$\mathfrak{h}_2$	I_3	Y_W
$L_d(0):$						
$uq = d_3$	0	0	0	$2/3$	$1/2$	$1/3$
$L_d(1):$						
$dq_1 = d_1$	$1/2$	1	$1/2$	$-1/3$	0	$-2/3$
$dq_2 = d_2$	$-1/2$	1	$-1/2$	$-1/3$	$-1/2$	$1/3$
$L_d(1):$						
$dq_2^- = d_1 d_3$	$1/2$	1	$1/2$	$1/3$	$1/2$	$-1/3$
$dq_1^- = d_2 d_3$	$-1/2$	1	$-1/2$	$1/3$	0	$2/3$

$L_d(1)$:						
$p_1 = d_1d_3d_3$	$1/2$	1	$1/2$	1	1	0
$p_2 = d_2d_3d_3$	$-1/2$	1	$-1/2$	1	$1/2$	1
$Markers:$	$1/2 \cdot h$	u	$\mathfrak{h}_1$	$\mathfrak{h}_2$	I_3	Y_W
$L_d(2)$:						
d_1d_1	1	2	1	$-2/3$	0	$-4/3$
$uq^- = d_1d_2$	0	2	0	$-2/3$	$-1/2$	$-1/3$
d_2d_2	-1	2	-1	$-2/3$	-1	$2/3$
$L'_d(2)$:						
$\nu_e = d_1d_1d_3$	1	2	1	0	$1/2$	-1
$n = d_1d_2d_3$	0	2	0	0	0	0
$\nu_e^- = d_2d_2d_3$	-1	2	-1	0	$-1/2$	1
$L_d(3)$:						
$d_1d_1d_1$	$3/2$	3	$3/2$	-1	0	-2
$e_L = d_1d_1d_2$	$1/2$	3	$1/2$	-1	$-1/2$	-1
$e_R = d_1d_2d_2$	$-1/2$	3	$-1/2$	-1	-1	0
$d_2d_2d_2$	$-3/2$	3	$-3/2$	-1	$-3/2$	1

Above, $uq = d_3$ is compared to the up-quark, and $dq_i = d_i, i = 1, 2$ are the two down quarks that we have to play with, and the dq_i^- are the corresponding anti-quarks. Moreover, p_i seems to be related to the protons of SM, n to the neutron, and ν to the corresponding neutrinos, and finally, e_L, e_R should be, respectively, left and right electrons.

To go from left to right handedness, in the physics language, comes out by just exchanging d_1 and d_2. Here, one may see that $e_L + \nu_L \to e_R$, and one easily find reasons for the decay,

$$n \to p + e + \nu_e.$$

Note also that we may, in an obvious way, identify ν_e with the element f and $e+f$ with interchanging d_1, d_2, and note also that particles and anti-particles, which usually are denoted by an "overline" and related to the reversing of time, in our model add up to n, the "neutron" in the model.

However, the decay and the other processes here are defined uniquely with the help of summation of the markers. In our Toy Model, the iterated extensions will take over, and hopefully explain better the results. In fact, the marker $\mathbf{u}$, the Z of the weak force, has been added to the list of markers for this purpose.

The part of the SM that the physicists call *The Electroweak Sector* is concerned with representations of $\tilde{H}(\sigma_g)/\mathfrak{g}$, on the complexified bundles, $B_o^c := \mathbf{C} \otimes_{\mathbf{R}} B_o$, $B_p^c := \mathbf{C} \otimes_{\mathbf{R}} B_p$, and the sum, $B_o^c \oplus B_p^c$, i.e. it is concerned with representations on Weyl Spinors or Dirac Spinors. The possible representations on spinors, i.e. on B_o^c or B_p^c, are then usually written as

$$D_p = \xi_p + \nabla_p, \nabla_p \in U(\mathbf{C} \otimes \mathfrak{sl}(2)).$$

(Note that this is not our Levi-Civita connection, which in physics is usually just denoted $\nabla_{e_i} e_j$.) The typical example is, for a trivial metric g,

$$D_\mu = \imath\delta_\mu - g' 1/2 Y_W B_\mu - g 1/2 \overline{\tau}_L \overline{W}_\mu.$$

Here, B_μ is strange, and it seems that

$$Y_W = 2(\mathfrak{h}_2 - I_3) = (1/2\mathfrak{h}_2 \pm \mathfrak{h}_1),\ \overline{\tau}_L \overline{W}_\mu = \sum_{k=1}^{3} W_\mu^k \sigma^k,$$

where $\pm$ signifies (right/left), and g' and g are coupling constants, not related to any metric. Note that here we have choices, considering left or right-handed systems. From the above, one can easily deduce the:

Electromagnetic Sector, as we have explained above, in Chapters 10 and 11, see also Laudal (2011). The new thing for the Electroweak case is that one considers representations on the Dirac Spinors, of the type,

$$\mathbf{D}_\mu = \gamma^\mu D_\mu.$$

The corresponding energy equation is, therefore,

$$(Q - E^2)\psi = 0,$$

where

$$Q = \sum_{\mu,\nu} g^{\mu,\nu}\mathbf{D}_\mu \mathbf{D}_\nu.$$

Assume we have a trivial metric, and consider the equation

$$\left(\sum_p \gamma_p D_p - E\right)\left(\sum_q \gamma_q D_q + E\right) = \sum_{p,q} \gamma_p D_p \gamma_q D_q - E^2,$$

where we either assume that $[\gamma, D_p] = 0$, so that $\sum_{p,q} \gamma_p D_p \gamma_q D_q = \sum_p D_p D_p$ or accept that the difference is without interest, we find a solution of the energy equation is the same as a solution of the Dirac type equation,

$$\sum_p (\gamma_p D_p \pm E)(\psi) = 0.$$

The consequence is that we should look for $B_\mu, W_\mu^k \in \tilde{H}$, and ψ, such that,

$$\overline{\psi}\left(\sum_p \gamma_p D_p\right)\psi,$$

is constant. This is exactly the Lagrangian in the electroweak sector,

$$\mathbf{L}_{EW} = \sum_\mu \psi^\dagger \gamma^\mu (\imath\delta_\mu - g'1/2Y_W B_\mu - g1/2\overline{\tau}_L \overline{W}_\mu)\psi.$$

The Quantum Chronodynamic Sector, is the theory that takes care of quarks and fermions. Here, we are interested in the representations

ρ of $\tilde{H}(\sigma)(\mathfrak{g}_1)$ on $\tilde{\Delta}_{\mathbf{C}} := \mathbf{C}\otimes_{\mathbf{R}}\tilde{\Delta}$. We know that they are of the form

$$\rho(dt_p) = D_p := \xi_p + \nabla_p, \nabla_p \in U(\mathbf{C}\otimes\mathfrak{su}(3)),$$

treating ∇_p as a potential. In QCD, the corresponding representations are taken as

$$D_\mu = \delta_\mu - \imath g_s G^i_\mu T^i,$$

where $\{T^i\}$ is a basis for $\mathfrak{su}(3)$. The Lagrangian is then guessed to be

$$\mathbf{L}_{QCD} = \imath U^\dagger(\delta_\mu - \imath g_s G^i_\mu T^i)\gamma^\mu U + \imath D^\dagger(\delta_\mu - \imath g_s G^i_\mu T^i)\gamma^\mu D,$$

where U are Dirac spinors associated with up-quarks, and D are Dirac spinors associated with down-quarks, and g_s is the *strong coupling constant.*

The last provisos of the equation seem very peculiar. However, if we understand that the model for this Lagrangian is not the usual energy equation, but the first-order Dirac equation, and moreover, that the operators γ^μ are just defined on $B_o\oplus B_p$, and the up-quark is not an element in $B_o\oplus B_p$, then it becomes reasonable. Recall that all elements in $\mathfrak{g}$ kill the up quark d_3, but do not kill the two down-quarks d_1, d_2. However, there should be a difference between left and right handedness, here as above.

The Higgs Sector, is the last case I shall mention.The Higgs Lagrangian is, before *Symmetry Breaking*,

$$\mathbf{L}_H = \phi^\dagger(\delta^\mu - \imath/2(g'Y_W B^\mu + g\tau\overline{W}^\mu))(\delta_\mu + \imath/2(g'Y_W B_\mu + g\tau\overline{W}_\mu))\phi$$
$$-\lambda^2/4(\phi^\dagger\phi - v^2)^2,$$

where $\phi = (\phi^+, \phi^0)$ (not the coordinate in $\tilde{H}$!) is a Spinor, with the electric charges, Q, as indicated, both components having weak isospin, $Y_W = 1$.

The classical interpretation seems strange, and I propose that this has to do with the restriction to the four-dimensional Minkowski/Einstein space. Working in our six-dimensional moduli-space, we have much more room to explain, as we have done, the difference between left and right handedness, and to treat mass, total, and kinetic energy on the same level, as we have seen above, in Section 12.2, where we introduced the new particle, the *Hebron*, with the hope it might be related to the Higgs particle. The rest, I assume, is physics.

Chapter 14

End Words

This book is an extension of several papers, see Laudal (2011, 2013, 2014), and has been the subject of quite a number of talks, at Paris/ENS, Stockholm/Mittag Leffler, Marseille/Luminy, Oslo, Angers, Trieste, Belfast, Mulhouse, Nice, Praha, Lahore, Tallinn, Casablanca, Rabat, Strasbourg, London, Queen Mary, and Tromsø, during the last 20 years or so. It is based on ideas and a general philosophy that popped up around the time of the Sophus Lie Memorial Week in 1992. It was first presented in Laudal (2005), and then in Laudal (2011), in which the main new idea is TIME, considered as a metric on the moduli space of the phenomena we choose to study, assumed to be expressible within algebraic geometry.

However, in algebraic geometry, the space is modeled as a certain set of representations of algebras, and the notion of moduli space of such a chosen set of representations is, in most cases, dependent upon the notion of a quotient of an action of a Lie group or a Lie algebra, on the space in question. And such quotients do not necessarily exist, in classical algebraic geometry, except in very special cases. Here is where non-commutative algebraic geometry enters. One is led to consider the quotient of the space of representations of an algebra A, on which a gauge Lie algebra operates, as the family of representations with a (generalized) connection defined on the gauge group. Any such family of representations of a general associative algebra should be considered as the set of generalized *points*, parametrized by an algebra of *observables*, computable from the structure of the family considered. This guarantees that the *moduli spaces* we are interested in, exist. The local structure is taken care of by the structure of deformations, and the dynamics of these *moduli spaces* hinges on the existence of the non-commutative phase space functor (NPF) and the construction of *dynamical structures*, see Chapter 4.

There is, in some circles, a belief that the local structure in algebraic geometry, which in the commutative situation is furnished by the local rings of the points, does not have an analog in the non-commutative setting (see e.g. the Wikipedia slot on non-commutative algebraic geometry, March 29, 2017). This is luckily not entirely true. The Generalized Burnside Theorem (GBT), see Chapter 3, is already rather old. And GBT together with NPF form the basis for our kind of non-commutative algebraic geometry NAG. The underlying mathematical tool for this theory is non-commutative deformation theory, see Eriksen *et al.* (2017), and also Laudal (2000, 2002).

This non-commutative algebraic geometry is quite different from the more well-known non-commutative geometry, NCG, proposed by Alain Connes, (2007), see also Schucker (2002) for a nice exposition and a useful bibliography. The application of NAG to modern physics is equally different from that of Connes, see again Connes (2007). In particular, it seems to me that the set-up of NCG, assuming the existence of a C*-algebra of operators on some Hilbert space, and a convenient spectral triple, is lacking the kind of explanatory power that I needed to be able to work with physics. As a growing number of theoretical physicists agree, a unified theory of gravity and quantum theory will need a fundamental new understanding of the notion of time, see e.g. Smolin (2002), Chapter 15. This seems also to be outside the reach of NCG.

The first challenge of my series of papers on the subject was to make up my mind about what one should mean by the notion of a *mathematical model*, capable of explaining something. My choice of method here, see Chapter 1, was natural for me, but may also be due to my very limited intuition in physics (whatever that should mean), and to my respect for mathematical aesthetics. Anyway, the result is a mathematical model for everything, modestly called a "Toy Model."

14.1 Relations to Non-Commutative Geometry (NCG)

It is easy to see a formal relationship between the theory exposed in this note, and the NCG. A spectral triple, see Connes (2007) or Chamseddine and Mukanov (2014), $(\mathfrak{A}, \mathfrak{H}, \mathfrak{D})$, is formally a (unitary) representation, of an associative complex C*-algebra, $\mathfrak{A}$ in a complex Hilbert space, $\mathfrak{H}$,

$$\rho : \mathfrak{A} \to \mathrm{End}_{\mathbf{C}}(\mathfrak{H})$$

together with an operator, called the Dirac operator, $\mathfrak{D} \in \mathrm{End}_{\mathbf{C}}(\mathfrak{H})$. This induces a derivation

$$\mathrm{ad}(\mathfrak{D}) : \mathfrak{A} \to \mathrm{End}_{\mathbb{C}}(\mathfrak{H}), \ \ \mathrm{ad}(\mathfrak{D})(a) = [\mathfrak{D}, \rho(a)],$$

corresponding to a vanishing element of $\mathrm{Ext}^1_{\mathfrak{A}}(\mathfrak{H}, \mathfrak{H})$, such that the extension of ρ with itself, defined by $\mathrm{ad}(\mathfrak{D})$, is trivial. Clearly, it follows from this that there is an extension of ρ from $\mathfrak{A}$ to $\mathrm{Ph}(\mathfrak{A})$, i.e. a representation

$$\rho_1 : \mathrm{Ph}(\mathfrak{A}) \to \mathrm{End}_{\mathbb{C}}(\mathfrak{H}), \ \ \rho_1(da) = [\mathfrak{D}, \rho(a)],$$

thus, the Dirac operator induces a derivative or momentum, ρ_1, of the given representation ρ.

Now, assume given a metric g on an affine space defined by the polynomial k-algebra $C = k[t_1, ..., t_d]$ (so in the commutative case), k being the real numbers, then a Dirac operator could be given by the Dirac derivation $\delta = \mathrm{ad}(g - T)$, as defined in $C(\sigma_g)$. Consider now the canonical representation

$$\rho : C(\sigma_g) \to \mathrm{End}_k(C)$$

given by $\rho(t_i) := t_i, \rho(dt_i) := \sum_k g^{i,k}\delta_{t_k}$. Let as above, $Q = \rho(g - T)$, then the triple $(C(\sigma_g), \rho, Q)$ is a "spectral triple," and as we have seen in Section 4.4, the metric, or distance-function, in $\mathrm{Spec}(C)$ may be calculated, just like in NCG,

$$d(o,p) = \sup_{f \in C_{<g}} |f(p) - f(o)|,$$

where the subset $C_{<g} \subset C$ is defined by

$$C_{<g} = \left\{ f \in C \| \| \sum_{i,j} \frac{\partial f}{\partial t_i} \frac{\partial f}{\partial t_j} g^{i,j} \| \leq 1 \right\}.$$

This may be generalized to define metrics in "spaces" defined by more general "spectral triples," see Connes (2018).

Note the differences between the NCG and our use of NAG. In our set-up, δ is universally given; the Dirac operator of NCG is not. The infinitesimals are the elements of the algebra of observables, $dt_i \in C(\sigma_g)$, or the operators $\rho(dt_i) \in \mathrm{End}_k(V)$, not only the compact operators in a Hilbert space.

Moreover, the set-up of NCG seems to lack a reasonable relationship between gauge groups and invariant theory, and it does not explain the "local gauge groups" of the Standard Model, nor the necessity of putting a "spin structure" on the space to be studied.

And, most importantly, NCG does not, in a credible way, define time and the maximal velocity of light, nor the parity operator P, the γ usually attached to a spectral triple (see Chamseddine and Mukanov, 2014). The nature of the elementary particles of the Standard Model is also somewhat arbitrary. Moreover, the essential fact, that for our Toy Model, $\Theta_{\tilde{H}}$ is a canonical $\mathfrak{g}*$-bundle, where $\mathfrak{g}*$ contains the gauge groups of the Standard Model, seems not to have an analogue within the NCG model.

These comments should, however, not be understood as reproaches against the very beautiful mathematical theory of non-commutative geometry. They simply show that our set-up, of the Toy Model, is very special. Because we insist on working within purely algebraic geometry, it is reasonable to find that the tools introduced in NCG and in our Toy Model, for the same purpose, may come out differently.

14.2 Models for Quantum Gravitation

There is a whole library of models for "Quantum Gravity." These models are assumed to "explain" both Quantum Field Theory and Curved Space–Time, that is, Einstein's General Relativity. Most of them try to introduce some non-commutativity in the space–time, not only in the phase space, as we have seen above. They also find some related symmetry, like quantum groups, Hopf algebras, or other generalizations of the standard gauge groups of classical quantum theory. The result is then referred to as "Quantum Geometry."

It seems that there is a widely accepted view among these researchers that the classical continuum of space–time should "break down at the Planck scale." The argument for this seems to be based on the notions of "probes," on their energies, and on the ineluctable formation of Black Holes, for very high energies. Some of these models also call for possible variable light velocities.

We have seen in Chapter 4 that considering the dynamical structure $C(\sigma_g)$ related to a metric g on an affine space, $\mathrm{Spec}(C)$, with "no furniture," i.e. considering the Universe as empty, classical General Relativity would emerge. However, when the Universe is not empty, i.e. when we consider the furniture given by some representation V of $C(\sigma_g)$, then the time operator, $[\delta] = \mathrm{ad}(Q)$ operating on $\mathrm{End}_k(V)$, may have a discrete spectrum, forming an additive monoid in the reals, with a positive generator, h, the Planck "constant." Clearly, since space is measured with the help of time, and the constant light velocity here is put equal to 1, we find that space also becomes discrete with a minimal measurable distance given by the Planck constant (for this furniture). The variability of the velocity of light is, however, non-sensical in our model.

Here, we have copied the physicists' way of peeling off part of a given model, to obtain a less complicated version, to check out with experiments, and then see if the results can be explained in the full model. Recall, from Chapter 4, the general Force Laws, of $C(\sigma_g)$, expressed in the phase space $\mathrm{Ph}(C)$ of a polynomial algebra, $C = k[t_1, .., t_d]$, dependent upon a non-degenerate metric g. In $C(\sigma_g)$, where the Dirac derivation δ is defined, we have the following Force

Law in $\mathrm{Ph}(C)$:

$$\begin{aligned} d^2t_i := \delta^2(t_i) =& [g - T, dt_i] \\ =& -\sum_{p,q} \Gamma^i_{p,q} dt_p dt_q - \sum_{p,q} g_{p,q} F_{i,p} dt_q \\ &+ 1/2 \sum_{p,q} g_{p,q} [F_{i,q}, dt_p] \\ &+ 1/2 \sum_{l,p,q} g_{p,q} [dt_p, (\Gamma^{i,q}_l - \Gamma^{q,i}_l)] dt_l + [dt_i, T]. \end{aligned}$$

We shall, as in Chapter 4, consider the above formula as a general Force Law, in $\mathrm{Ph}(C)$, induced by the metric g. This means the following.

First, assume given a representation

$$\rho_0 : C \to \mathrm{End}_k(V),$$

and pick any tangent vector (momentum) of the formal moduli of the C-module V, i.e. an extension of ρ_0

$$\rho_1 : \mathrm{Ph}(C) \to \mathrm{End}_k(V),$$

then, if ρ_1 can be extended to a representation

$$\rho_2 : \mathrm{Ph}^2(C) \to \mathrm{End}_k(V)$$

with $\rho(d_2(d_1 t_i)) = \rho_1(d^2 t_i)$, given by the formula of the Force Law, this means that the Force Law has induced a second-order momentum in the formal moduli space of the representation ρ_0. We have already seen that the reduction of this force law to the commutative case gives us General Relativity, see also Laudal (2011). Moreover, the same model reduces to the Dirac equation for an "electron," i.e. for the Dirac Spinors (our electron).

14.3 The General Dynamical Model

The purpose of this book has been to construct a model for the study of most natural phenomena, given a reasonable mathematical model for the primary observation. The Toy Model, $\underline{\tilde{H}}$ studied above, is one simple choice, and it would be a miracle if it turned out to be more than a small intermediate step on the long road towards understanding (some non-trivial part of) nature.

Nevertheless, it has a series of interesting properties. The choice of Minkowski, and of Einstein, of the metric $ds^2 = dt^2 - (dx_1^2 + dx_2^2 + dx_3^2)$, considering time as a fourth dimension, works fine in explaining the maximality of the velocity of light, but it also contains the mathematical model of a tachyon, which seems to have been forgotten, regretted, or denied.

In the Toy Model $\underline{\tilde{H}}$, the mathematical notion of velocity is different, and the set of velocities turn out to be a compact space, so there exists an absolute maximal speed, the speed of light.

Treating Einstein's proper time as one of the zero-velocities in the Toy Model, we see that the Minkowski metric turns into the trivial metric of a four-dimensional Riemannian space, $dt^2 = dx_1^2 + dx_2^2 + dx_3^2 + ds^2$, and the Schwartzchild metric turns into a metric very close to ours. Deforming these metrics into metrics on the corresponding blow-up of the ds-line, makes it part of the restriction of the Toy Model $\underline{\tilde{H}}$, to $M(B)$, but the interpretation of this space as a moduli space of an observer observing an observed, in three-space, gets lost. For this, we need more dimensions, at least six.

This interpretation is basic, as we shall identify states of our simple $\mathfrak{g}*$ bundles (composed of tangent vectors at a point of $\underline{\tilde{H}}$) as (glimpses of) elementary particles at that point. This provides models for three *charged and mass-less*, and three *charged and massy*, elementary states, generating all the other state-spaces of particles in our natural fauna.

The moduli spaces of the mathematical models of the phenomena we are interested in, and their dynamical superstructure, create our algebras of *observables*. Their representations, or their measurements, are the main objects of our study. Therefore, the basic elements of the theory are the representations of the algebras of observables. The theory of non-commutative algebraic geometry furnishes a framework for this theory, by making clear the role of the symmetries, the gauge groups, in the choice of representations to be studied.

The evolution equations defined by the universal Dirac derivation, and a choice of Dynamical Structure, give us an analytic continuation of a properly prepared representation. We have not, in this book, ventured into concrete calculations of solutions to the equations that pop up. This will be the subject of later works. The main theoretical problem that is not yet covered is related to the physical notion of Interaction, see above. We have included a short sketch of a mathematical model for interactions, for the general model, including a bold proposal for a mathematical explanation of the electroweak force. But this is not entirely satisfactory. What happens when two elementary particles "come close" to each other? How should we interpret this in terms of the representations representing the particles? Here, the non-commutative algebraic geometry and the theory of non-commutative deformations of families of representations will, certainly, be essential, but the detailed treatment of this has to wait, as will the many other questions that pop up when one wants to use the model in "real life." In relation to the short Section 5.2, on Cosmology, it is certainly tempting to make comparisons with the very speculative literature treating notions like *wormholes, firewalls, complementarity, entanglement*, and the likes. The exceptional fibers, the black holes $E(\underline{\lambda})$, in our Toy Model, could be used to "make wormholes," and certainly to give examples of hologram-like phenomena, and the solution of the Furniture equation in Section 4.1 might look like a mathematical model of a firewall at the Horizon. The fact that we, in this model, see what "happened before" in cosmological time, where the gravitational mass-density was bigger, could also be used as an argument for the non-existence of both Black Energy and Black Matter, but compare with Section 10.4. This may all be interesting, but my goal for this work has been much more modest, making a reasonable tool-box for studying mathematically well-defined phenomena, and picking one, related to my intuitive feeling of information, interaction, and dynamics, all related to deformation theory in algebraic geometry. The goal was to try to "understand" the basic notions in physics, in particular.

14.4 Time, Lagrangians, Probabilities, Reality

I wanted a "simple" but mathematically sound model in which the notions of "classical" physics made sense, treating the kind of physics I never understood as a student, with the hope of, through this purely mathematical construct, eventually to understand a little more of the "Nature" we seem to live in. As examples of questions I had, and never saw explained, let me mention the following.

What is *Time*? What do we really mean when we use notions like *object, particle, state, energy, mass, and observing something in reality*? And what is the basis of the *dynamics* we want to uncover? Why should we trust the Parsimony Principle, and the ugly looking Lagrangians, to guide us with respect to the future of changing phenomena?

There are several old, and still very interesting, texts treating these problems. One is the book "*La Science Moderne, et sonétat actuel*," published in 1914 by Picard (1914). He is omnipotent, and treats almost all of Science, known at the time, and he ends his treatment of physics with "relativity," where he cites Einstein's new claims that the "light might have a weight." He makes clear his general view, that; "ce n'est qu'en adoptant des points de vue divers, quelque fois opposées, que les sciences progressent," but ends with the hope: "Souhaitons qu'il sorte quelque choses de précis de ces speculations hardies."

The following year Eddington had "proved" the general theory of relativity, but then Einstein had once more a "point de vue divers," and the problem became the new quantum theory.

The classical physics literature, like Weineberg's *The Quantum Theory of Fields*, Weinberg (2005), is singularly void of a discussion of the meaning of it all. An exception is the last letters between Einstein and Born from the years between 1948 until Einstein's death in 1955, see Einstein and Born (1969). They adopt "des points de vue divers, quelque fois opposées," and they have ideas. One might hope that one of their decedents could "sorte quelque choses de précis de ces speculations hardies."

At this point, it may be fitting to recall the starting point of our study, formulated in Chapter 1 as: "If we want to study a natural phenomenon, called π, we must in the present scientific situation, describe π, in some mathematical terms, say as a mathematical object, X, depending upon some parameters, in such a way that the changing aspects of π, would correspond to altered parameter-values for X. This object would be a *model for* π, if, moreover, X with any choice of parameter-values, would correspond to some, possibly occurring, aspect of π," This is very close to the "EPR Criterion of Reality," the basic definition in Einstein–Podolsky and Rosen's fundamental paper: "Can Quantum Mechanical Description of Physical Reality be Considered Complete?," published in *Physical Review* on May 15, 1935.

14.4.1 *Unsolved problems in physics*

Still today, the "List of unsolved problems in physics" is long. Let us look at some of them; see for a complete list the references.

Wikipedia: Time and Arrow of Time, and links to CP violation in Weak force decay, related to second law of thermodynamics, Quantum Reality, Quantum Field Theory, Quantum Gravity Grand Unification Theory, Yang Mills Physical information, or destruction, Fine-tuning the Universe, Cosmic Inflation, Horizon problem, Dark Energy, Matter Generation of Matter, etc. Let us now go through some of these problems from a very down-to-Earth philosophical point of view.

Time does not exist as a coordinate in our Space. Clocks are "mechanical" objects that are used to "measure" change of phenomena. We have accepted the interpretation of an object, as a representation of our algebra of observables. Note that these representations are all built upon the simple representations of our photons and our quarks. Observing something means to identify a state of the representation V representing this something, and concluding that at our point in the cosmos, this state has a positive energy. NASA's David Mazza, flashes the following, under the headline "Observing or seeing a star." "What you see as a star, is actually the result of a quantum interaction between the local field and the retina of your eye." Of course, in our model the retina of my eye is modeled as the local space, home of the local field, at the point representing me and the observed, i.e. (o, p). Lagrangians, ruling our use of, and beliefs in, the Parsimony Principle, are amazing mathematical tools, covering some kind of universal deformational machinery. Probabilities are real, since they are part of our Geometry. The field of real numbers, constructed from our natural integers, cannot parametrize all changes, not even in an "object" just containing a finite number of given "Points." There is no mathematical exact treatment of the whereabouts of any one special of these points, even in the most simple dynamical setting. The changes of the "object," however, can be beautifully described up to isomorphisms. Reality, outside the mathematical model we use to explain what we are doing, is a psychological anchor developed in our brain, which we, probably, should cherish.

And finally, what experimental argument can one find for the proposed theory? The last enthusiastic claims in physics, of having observed gravitation waves, seems to me to indicate that time, the metric of our moduli space, may be wobbling, measured by some clock of our construction. We should expect this to happen as a result of the process described in Chapter 2, and in Section 6.3.

Stephen Hawking has in his last book the following "statement:" It won't be possible to probe down to the Planck scale length in the lab, though we can study the Big Bang to get some ideas. However, to a large extent we shall have to rely on mathematical beauty and consistency to find the ultimate theory of everything.

He was a student of Dirac, one of my heroes, and must have learned this lesson very early. I decided to call the universal derivation δ of the Ph^{∞} functor, the Dirac derivation, not only because Dirac's equation is closely related to δ, but also because of Dirac's strong belief in mathematical beauty, and simplicity, as a guide to understanding the world we live in. I, therefore, agree with Hawking, although I am not so optimistic as he seems to have been, expecting science to uncover the ultimate theory of everything before the end of this century!

14.5 Relations to Classical Cosmologies

To be able to move from some of the classical models of cosmology to our Toy Model, and see the relations, it is fair to start with Special Relativity, and its geometrical formulation via the four-dimensional Minkowxki space, M, parametrized by $\{t, x_1, x_2, x_3\}$, given by the metric

$$ds^2 = -dt^2 + \sum_1^3 dx_i^2$$

The condition for ds being real is of course

$$dt^2 \leq \sum_1^3 dx_i^2,$$

defining the "future" cones. Light moves along the "null-lines," i.e. along the lines with $ds = 0$. So, for every point $m \in M$, we have a double cone, $\tilde{c}(m)$ in the tangent space of that point. "Blowing up the point m," i.e. considering the projective space corresponding to $\tilde{c}(m)$, we find a sphere $E(m)$ with the coordinates $\underline{\omega}$. Now, drawing for every $\underline{\omega} \in E$ the corresponding directed line in the $\underline{x}$-space, we obtain a three-dimensional subspace of M, homeomorphic to our cylinder $c(\underline{\lambda})$, the "Future null-infinities," of Trautman–Bondi in the theory of gravitational waves.

It can now be seen that the disjoint union of all these cylinders $c(\underline{\lambda})$, for $\underline{\lambda} = (t_1, t_2, t_3) \in \underline{\Delta} \subset \underline{H}$, is isomorphic to $\underline{\tilde{H}}$, and the corresponding union for $|\lambda| = t = \sqrt{(\sum_{i=1,2,3} t_i^2)}$ surjects onto M, and is homeomorphic to $M(B)$, see Section 5.2, and the definitions and remark there. The paper of Manin and Marcolli (2014) uses the affine Minkowski space, to introduce the future and past infinities and explain the factor S^2 as blow-ups in algebraic geometry. They use this to comment on Penrose's theory of a cyclic universe, where "eons" follow "eons."

There is a funny theological historical background for this "belief," interestingly found in the apocryphal gospel of Judas, see Flammarion (2006). There, the present eon, in which we live, must die and vanish for the next to come to life. And this notion of "eon" is probably meant to include the whole Universe!

And there are more to find of interesting analogies between ancient Cosmologies and this Toy Model. We have already seen that the Conformally Trivial metric of Section 8.3 pre-dictates a circular time, and a kind of old Hindu cosmology.

But the most amazing analogy of this kind comes from the first five verses of the Bible. The origin of the World is described as an action of God (some undefined Force), on the space consisting of darkness covering a deep sea, that we might identify with a three-dimensional space, $\underline{\Delta}$, but provided by some bundle, $\tilde{E}$, of rank two related to the Force. Then God (the Force) creates light, disjoint from $\underline{\Delta}$. For each point $\underline{\lambda} \in \underline{\Delta}$, the fiber of $\tilde{E}$ is $E(\underline{\lambda})$, and the light-space created is, of course, $c(\underline{\lambda})$. This $\tilde{c}(\underline{\lambda})$ was called day, and the space $\tilde{\Delta}(\underline{\lambda})$ was named night.

The only thing we miss in this story is the creation of The Force itself. In the Toy Model, this is given by a universal point, its tangent structure, and the Magic of Deformation Theory.

14.6 So What?

Did I succeed; do I understand a little more of the physics Hylleraas and his assistants tried to teach me in 1954–1956? Yes, I think so. That is not a falsifiable statement, but I do not care.

Bibliography

Augustin, S. (1861). *Les confessions de saint augustin,* Janet. Charpentier, Libraire-Éditeur, Paris.

Beĭlinson, A. A. (1978). Coherent sheaves on $\mathbf{P}^n$ and problems in linear algebra, *Funktsional. Anal. i Prilozhen.* **12**, 3, pp. 68–69.

Bjar, H. and Laudal, O. A. (1990). Deformation of Lie algebras and Lie algebras of deformations, *Compositio Math.* **75**, 1, pp. 69–111, http://www.numdam.org/item?id=CM_1990__75_1_69_0.

Cerchiai, B. L., Madore, J., Schraml, S., and Wess, J. (2000). Structure of the three-dimensional quantum Euclidean space, *Eur. Phys. J. C Part. Fields* **16**, 1, pp. 169–180, doi: 10.1007/s100520050013, https://doi.org/10.1007/s100520050013.

Chamseddine, A., Ali, H., Connes, A., and Mukanov, V. (2014). *Quanta of geometry,* arXiv:1409.2471v3[hep-th].

Connes, A. (1988). *On the Foundation of Noncommutative Geometry,* www.alainconnes.org/docs/gelfand.pdf.

Connes, A. (2007). Non-commutative geometry and the spectral model of space-time, *Quant. Spaces, Prog. Math. Phys.,* **53** pp. 203–227, doi: 10.1007/978-3-7643-8522-4_5, https://doi.org/10.1007/978-3-7643-8522-4_5.

Connes, A. (2018). *Geometry and the Quantum,* in *Foundations of Mathematics and Physics One Century after Hilbert* (Springer, Cham), pp. 159–196.

Deligne, P. (1999). *Quantum Fields and Strings: A Course for Mathematicians* (AMS institute for Advanced Study), ISBN 0821820125.

Dodelson, M. and Silverstein, E. (2017). String-theoretic breakdown of effective field theory near black hole horizons, *Phys. Rev. D.* **96**, 6, pp. 066010, 19, doi: 10.1103/physrevd.96.066010, https://doi.org/10.1103/physrevd.96.066010.

Earman, J. (ed.) (1989). *World Enough and Space-Time, Absolute versus Relational Theories of Space and Time, A Bradford Book*, Vol. 1 (The MIT Press, Cambridge, Massachusetts).

Einstein, A. and Born, M. (1969). *Briefwechsel 1916-1955* (Nymphenburger Verlagshandlung Gmbh).

Eriksen, E., Laudal, O. A., and Siqveland, A. (2017). *Noncommutative Deformation Theory*, Monographs and Research Notes in Mathematics (CRC Press, Boca Raton, FL), ISBN 978-1-4987-9601-9, doi: 10.1201/9781315156057, https://doi.org/10.1201/9781315156057.

Ferguson, K. (2011). *Pythagoras. His Lives and the Legacy of a Rational Universe* (Icon Books Ltd.), ISBN 1848312318.

Fernflores, F. (2019). The equivalence of mass and energy, *The Stanford Encyclopedia of Philosophy.* https://plato.stanford.edu/archives/fall2019/entries/equivME/.

Flammarion (2006). *L'evangile de judas, Flammarion.*

Gooth, J. (2017). Experimental signatures of the mixed axial–gravitational anomaly in the weyl semimetal nbp, *Nature*, **547**, p. 324–327.

Hadfield, T. and Majid, S. (2007). Bicrossproduct approach to the Connes-Moscovici Hopf algebra, *J. Algebra.* **312**, 1, pp. 228–256, doi:10.1016/j.jalgebra.2006.09.031, https://doi.org/10.1016/j.jalgebra.2006.09.031.

Hawking, S. (1988). *A Brief History of Time* (Bamtham Books, Transworld Publishers, London), ISBN 9780857501004.

Hugget, S. A., Mason, L. J., Tod, K. P., Tsou, S., and Woodhouse N. M. J. (1988). *The Geometric Universe* (Oxford University Press), ISBN 0198500599.

Lasenby, A., Doran, C., and Arcaute, E. (2004). Applications of geometric Algebra in electromagnetism, quantum theory and gravity, in *Clifford Algebras, Prog. Math. Phys.*, **34**, pp. 467–489.

Laudal, O. A. (1979). *Formal Moduli of Algebraic Structures, Lecture Notes in Mathematics*, Vol. 754 (Springer, Berlin), ISBN 3-540-09702-3.

Laudal, O. A. (1986). *Matric Massey products and formal moduli. I,* in *Algebra, Algebraic Topology and Their Interactions (Stockholm, 1983), Lecture Notes in Mathematics*, Vol. 1183 (Springer, Berlin), pp. 218–240, doi: 10.1007/BFb0075462, https://doi.org/10.1007/BFb0075462.

Laudal, O. A. (2000). Noncommutative algebraic geometry, *Max-Planck-Institut für Mathematik, Bonn* **115**.

Laudal, O. A. (2002). *Noncommutative Deformations of Modules,* pp. 357–396, doi: 10.4310/hha.2002.v4.n2.a17, https://doi.org/10.4310/hha.2002.v4.n2.a17, the Roos Festschrift volume 2.

Laudal, O. A. (2003). *Noncommutative Algebraic Geometry,* in *Proceedings of the International Conference on Algebraic Geometry and Singularities (Spanish) (Sevilla, 2001)*, Vol. 19, pp. 509–580, doi: 10.4171/RMI/360, https://doi.org/10.4171/RMI/360.

Laudal, O. A. (2004). The structure of $Simp_n(A)$, Preprint, Institut Mittag-Leffler, in *Proceedings of NATO Advanced Research Workshop, Computational Commutative and Non-Commutative Algebraic Geometry.* Chisinau, Moldova, June 2004.

Laudal, O. A. (2005). Time-space and space-times, in Noncommutative Geometry and Representation Theory in Mathematical Physics, *Contemp. Math.*, **391** (Amer. Math. Soc., Providence, RI), pp. 249–280, doi: 10.1090/conm/391/07334, https://doi.org/10.1090/conm/391/07334.

Laudal, O. A. (2008). Phase spaces and deformation theory, *Acta Appl. Math.* **101**, 1–3, pp. 191–204, doi: 10.1007/s10440-008-9192-8, https://doi.org/10.1007/s10440-008-9192-8.

Laudal, O. A. (2011). *Geometry of Time-Spaces* (World Scientific Publishing Co. Pte. Ltd., Hackensack, NJ), ISBN 978-981-4343-34-3; 981-4343-34-X, doi: 10.1142/9789814343350, https://doi.org/10.1142/9789814343350, noncommutative algebraic geometry, applied to quantum theory.

Laudal, O. A. (2013). Cosmos and its furniture. In *Mathematics in the 21st Century, Springer Proceedings in Mathematics and Statistics*, ed. P. Cartier *et al.*, Springer, Basel 2014.

Laudal, O. A. (2014). The structure of Ph*, generalised de rham, and entropy, in *Proceedings from the QQQ Conference 2012, J. of Phys.: Conference Series* **532** 012013.

Laudal, O. A. and Pfister, G. (1988). *Local Moduli and Singularities, Lecture Notes in Mathematics*, Vol. 1310 (Springer-Verlag, Berlin), ISBN 3-540-19235-2, doi: 10.1007/BFb0078937, https://doi.org/10.1007/BFb0078937, with an appendix by B. Martin and Pfister.

Mandl, F. and Shaw, G. (2010). *Quantum Field Theory* (John Wiley and Sons, Ltd), ISBN 978-0-471-49683-0.

Manin, Y. I. and Marcolli, M. (2014). Big Bang, blowup, and modular curves: algebraic geometry in cosmology, *SIGMA Symmetry Integrability Geom. Methods Appl.* **10**, pp. Paper 073, 20, doi: 10.3842/SIGMA.2014.073, https://doi.org/10.3842/SIGMA.2014.073.

Meschini, D. (2008). *A Metageometric Enquiry Concerning Time, Space, and Quantum Physics*, Ph.D. Thesis, University of Jyvaskyla, Suomi Finland.

Newman, E. T. (2017). *Surprising Structures Hiding at Penrose's Future Null Infinity,* arXiv:1701.08406.

Nima Arkani-Hamed, T. R. S., Douglas P. Finkbeiner, and Weiner, N. (2009). A theory of dark matter, *Phys. Rev. D* **79**.

Øhrstrøm, P. (1985). W. R. Hamilton's view of algebra as the science of pure time and his revision of this view, *Historia Math.* **12**, 1, pp. 45–55, doi: 10.1016/0315-0860(85)90067-9, https://doi.org/10.1016/0315-0860(85)90067-9.

Penrose, R. (2010). *Cycles of time* (Alfred A. Knopf, Inc., New York), ISBN 978-0-307-26590-6, an extraordinary new view of the universe. Collected Works: Oxford University Press, Vol. 2.

Penrose, R. and Hameroff, S. (2011). Consciousness in the Universe: Neuroscience, Quantum Space-Time Geometry and Orch OR Theory, *J. Cosmol.* **14**.

Picard, E. (1914). *La Science Moderne, et son état actuel*, Paris (Ernest Flammarion).

Procesi, C. (1967). Non-commutative affine rings, *Atti Accad. Naz. Lincei Mem. Cl. Sci. Fis. Mat. Natur. Sez. Ia (8)* **8**, pp. 237–255.

Procesi, C. (1973). *Rings with Polynomial Identities* (Marcel Dekker, Inc., New York), p. 17.

Refsdal, S. (1964). The gravitational lens effect, *Royal Astronomical Society*, pp. 295–305.

Reichenbach, H. (1961). *The Rise of Scientific Philosophy* (University of Calefornia Press, Berkeley and Los Angeles), ISBN 9780520010550.

Reiten, I. (1985). An introduction to the representation theory of Artin algebras, *Bull. London Math. Soc.* **17**, 3, pp. 209–233, doi: 10.1112/blms/17.3.209, https://doi.org/10.1112/blms/17.3.209.

Sachs, R. K. and Wu, H. (1977). General relativity and cosmology, *Bull. Amer. Math. Soc.* **83**, 6, pp. 1101–1164, doi: 10.1090/S0002-9904-1977-14394-2, https://doi.org/10.1090/S0002-9904-1977-14394-2.

Schlessinger, M. (1968). Functors of Artin rings, *Trans. Amer. Math. Soc.* **130**, pp. 208–222, doi: 10.2307/1994967, https://doi.org/10.2307/1994967.

Schucker, T. (2002). *Forces from connes' geometry*, arXiv:hep-th/0111236v2.

Smolin, L. (2002). *Rien ne va plus en physique*, Quai des Sciences (Dunod), ISBN 978-0-471-49683-0.

Sormani, P. N. L. B. D. G. N. Y. and Denson Hill, C. (2017). The mathematics of gravitational waves, *Notices of the AMS.*

Volk, S. (2018). Can Quantum Physics Explain Conciousness? *Science*, AAAS, SchienceMag.org, March 1, 2018.

Weinberg, S. (2005). *The Quantum Theory of Fields.* Vol. III (Cambridge University Press, Cambridge), ISBN 0-521-66000-9; 978-0-521-67055-5; 0-521-67055-1, supersymmetry.

Index